SCALING OF STRUCTURAL STRENGTH

SCALING OF STRUCTURAL STRENGTH

Second edition

Zdeněk P Bažant

ELSEVIER
BUTTERWORTH
HEINEMANN

AMSTERDAM • BOSTON • HEIDELBERG • LONDON • NEW YORK • OXFORD
PARIS • SAN DIEGO • SAN FRANCISCO • SINGAPORE • SYDNEY • TOKYO

Elsevier Butterworth-Heinemann
Linacre House, Jordan Hill, Oxford OX2 8DP
30 Corporate Drive, Burlington, MA 01803

First published by Hermes Penton Ltd 2002
Second edition 2005

Copyright © 2005, Elsevier Ltd. All rights reserved

The right of Zdeněk P Bažant to be identified as the author of this Work has been asserted
in accordance with the Copyright, Designs And Patents Act 1988

No part of this publication may be reproduced in any material form (including
photocopying or storing in any medium by electronic means and whether
or not transiently or incidentally to some other use of this publication) without
the written permission of the copyright holder except in accordance with the
provisions of the Copyright, Designs and Patents Act 1988 or under the terms of
a licence issued by the Copyright Licensing Agency Ltd, 90 Tottenham Court Road,
London, England W1T 4LP. Applications for the copyright holder's written
permission to reproduce any part of this publication should be addressed
to the publisher.

Permissions may be sought directly from Elsevier's Science and Technology Rights
Department in Oxford, UK: phone: (+44) (0) 1865 843830; fax: (+44) (0) 1865 853333;
e-mail: permissions@elsevier.co.uk. You may also complete your request on-line via
the Elsevier homepage (http://www.elsevier.com), by selecting 'Customer Support'
and then 'Obtaining Permissions'.

British Library Cataloguing in Publication Data
A catalogue record for this book is available from the British Library

Library of Congress Cataloguing in Publication Data
A catalogue record for this book is available from the Library of Congress

ISBN   0 7506 6849 0

For information on all Elsevier Butterworth-Heinemann
publications visit our website at http://books.elsevier.com

Typeset by Newgen Imaging Systems (P) Ltd., Chennai, India
Printed and bound in Great Britain

**Working together to grow
libraries in developing countries**

www.elsevier.com | www.bookaid.org | www.sabre.org

ELSEVIER   BOOK AID International   Sabre Foundation

Contents

Foreword   ix

Author's Preface   xi

About the Author   xiii

1   Introduction   1
    1.1. Nature of Problem and Approach   1
    1.2. Classical History   3
    1.3. Recent Developments in Quasibrittle Materials   6
    1.4. Basic Theories of Size Effect   10
    1.5. Power Scaling in Absence of Characteristic Length   11
    1.6. Transitional Size Effect Bridging Power Laws for Different Scales   13
    1.7. Deductions from Dimensional Analysis   17
    1.8. Stability of Structures and Size Effect   19

2   Asymptotic Analysis of Size Effect   21
    2.1. Asymptotic Analysis of Size Effect in Structures with Notches
         or Large Cracks   21
    2.2. Energetic Size Effect Law and Its Asymptotic Matching
         Character   25
    2.3. Size Effect Law in Terms of LEFM Energy Release Function . .   27
    2.4. Use of J-Integral for Asymptotic Scaling Analysis   27
    2.5. Identification of Fracture Parameters from Size Effect Tests . .   29
    2.6. Validation by Fracture Test Data and Numerical Simulation . .   33
    2.7. Size Effect for Crack Initiation via Energy Release   38
    2.8. Stress Redistribution Caused by Boundary Layer of Cracking .   42
    2.9. Strain Gradient Effect on Failures at Crack Initiation   45
    2.10. Universal Size Effect Law   46
    2.11. Asymptotic Scaling and Interaction Diagram for the Case of
          Several Loads   48

vi   Scaling of Structural Strength

    2.12. Size Effect on Approach to Zero Size 50

3 Randomness and Disorder   **53**
    3.1. Is Weibull Statistical Theory Applicable to Quasibrittle Structures? 53
    3.2. Nonlocal Probabilistic Theory of Size Effect 57
    3.3. Energetic-Statistical Formula for Size Effect for Failures at Crack Initiation 62
    3.4. Size Effect Ensuing from J-Integral for Randomly Located Cracks 67
    3.5. Could Fracture Fractality Be the Cause of Size Effect? 69
    3.6. Could Lacunar Fractality of Microcracks Be the Cause of Size Effect? 72

4 Energetic Scaling for Sea Ice and Concrete Structures   **77**
    4.1. Scaling of Fracture of Floating Sea Ice Plates 77
    4.2. Size Effect on Softening Inelastic Hinges in Beams and Plates . 89
    4.3. Size Effect in Beams and Frames Failing by Softening Hinges . 93
    4.4. Size Effect in Floating Ice Subjected to Line Load 99
    4.5. Steel-Concrete Composite Beams and Compound Size Effect . 101
    4.6. Size Effect Formulae for Concrete Design Codes 110
    4.7. Size Effect Hidden in Excessive Dead Load Factor in Codes . . 114
    4.8. No-Tension Design of Concrete or Rock from the Size Efffect Viewpoint 117

5 Energetic Scaling of Compression Fracture and Further Applications to Concrete, Rock and Composites   **121**
    5.1. Propagation of Damage Band Under Compression 121
    5.2. Size Effect in Reinforced Concrete Columns 124
    5.3. Fracturing Truss (Strut-and-Tie) Model for Shear Failure of Reinforced Concrete 130
    5.4. Breakout of Boreholes in Rock 132
    5.5. Asymptotic Equivalent LEFM Analysis for Cracks with Residual Bridging Stress 133
    5.6. Application to Compression Kink Bands in Fiber Composites . 134
    5.7. Effect of Material Orthotropy 135

6 Scaling via J-Integral, with Application to Kink Bands in Fiber Composites   **137**
    6.1. J-Integral Analysis of Size Effect on Kink Band Failures 137
    6.2. J-integral Calculations 140
    6.3. Case of Long Kink Band 144

6.4. Failure at the Start of Kink Band from a Notch or Stress-Free Crack 145
6.5. Comparison with Size Effect Tests of Kink Band Failures . . . 147

7 Time Dependence, Repeated Loads and Energy Absorption Capacity 151

7.1. Influence of Loading Rate on Size Effect 151
7.2. Size Effect on Fatigue Crack Growth 154
7.3. Wave Propagation and Effect of Viscosity 156
7.4. Ductility and Energy Absorption Capacity of Structures 157

8 Computational Approaches to Quasibrittle Fracture and Its Scaling 165

8.1. Eigenvalue Analysis of Size Effect via Cohesive (Fictitious) Crack Model 165
8.2. Microplane Model 167
8.3. Spectrum of Distributed Damage Models Capable of Reproducing Size Effect 169
8.4. Simple, Practical Approaches 170
8.5. Nonlocal Concept and Its Physical Justification 171
8.6. Prevention of Spurious Localization of Damage 172
8.7. Discrete Elements, Lattice and Random Particle Models 174

9 New Asymptotic Scaling Analysis of Cohesive Crack Model and Smeared-Tip Method 177

9.1. Limitations of Cohesive Crack Model 179
9.2. K-Version of Smeared-Tip Method for Cohesive Fracture 182
9.3. Nonstandard Cohesive Crack Model Defined by a Fixed K-Profile. 185
9.4. Asymptotic Scaling Analysis 189
   9.4.1. Case 1. Positive Geometry with Notch or Stress-Free Initial Crack, for Fixed K-Density ($g_0 > 0, g'_0 > 0$) . . . 189
   9.4.2. Case 2. Fracture Initiation from Smooth Surface, for Fixed K-Density ($g_0 = 0, g'_0 > 0$) 195
   9.4.3. Cases 1 and 2 for Standard Cohesive Crack Model or First Three Terms of Asymptotic Expansion 198
   9.4.4. Case 3. Negative-Positive Geometry Transition ($g_0 > 0, g'_0 = 0, g''_0 > 0$) 199
9.5. Small-Size Asymptotics of Cohesive Crack Model 204
9.6. Nonlocal LEFM—A Simple Approach to Cohesive Fracture and Its Scaling 206
9.7. Broad-Range Size Effect Law and Its Dirichlet Series Expansion 208

viii   Scaling of Structural Strength

9.8	Size Effect Law Anchored in Both Small and Large Size Asymptotics	213
9.9	Recapitulation	215

10 Size Effect at Continuum Limit on Approach to Atomic Lattice Scale ... 219

10.1.	Scaling of Dislocation Based Strain-Gradient Plasticity	219
10.2.	Scaling of Original Phenomenological Theory of Strain-Gradient Plasticity	228
10.3.	Scaling of Strain-Gradient Generalization of Incremental Plasticity	236
10.4.	Closing Observations	237

11 Future Perspectives ... 239

Addendum ... 241

A1.	Size Effect Derivation by a New Method for Asymptotic Matching	241
A2.	Size Effect of Finite-Angle Notches	249
A3.	Size Effect on Flexural Strength of Fiber-Composite Laminates	252
A4.	Size Effect in Fracture of Closed-Cell Polymeric Foam	254
A5.	Variation of Cohesive Softening Law Tail in Boundary Layer of Concrete	255
A6.	Can Fracture Energy Be Measured on One-Size Specimens with Different Notch Lengths?	258
A7.	Notched–Unnotched Size Effect Method and Standardization of Fracture Testing	260
A8.	Designing Reinforced Concrete Beams against Size Effect	262
A9.	Scaling of Probability Density Distribution (pdf) of Failure Load	264
A10.	Recent Debates on Fractal Theory of Size Effect and Their Impact on Design Codes	264
A11.	Boundary Layer Size Effect in Thin Metallic Films at Micrometer Scale	265
A12.	Size Effect on Triggering of Dry Snow Slab Avalanches	267

Bibliography ... 269

Subject Index ... 313

Reference Citation Index ... 321

Foreword

Up to the 1980s, all the experimentally observed size effects in solid mechanics were generally attributed to material strength randomness. Bažant revisited the scaling theory beginning with his 1984 discovery of the scaling law for the size effect caused by the release of stored energy due to stable growth of large fractures or large damage zones prior to failure. Using asymptotic matching arguments, he derived a deceptively simple law of surprisingly broad applicability, bridging the power scaling laws of classical fracture mechanics and plasticity. With his assistants, he experimentally verified his law for various materials, and showed how to use the scaling law to identify the cohesive fracture characteristics from experiments. Later, using extreme value statistics, he formulated a probabilistic generalization describing the transition to the classical statistical size effect in very large structures failing at fracture initiation. He also extended his size effect law to compression fracture, including kink band propagation in fiber composites. Recently, he used similar asymptotic arguments to show that the currently accepted dislocation-based strain-gradient theory of metal plasticity for micron scale needs a fundamental revision because of unreasonable asymptotic properties on approach to nanoscale.

Scaling of structural strength remains, however, a topic that many researchers in solid mechanics seem to have temporarily set aside. It is indeed striking to see that scaling and dimensional analysis have a tendency to disappear from curricula and from the scientific literature in solid mechanics. Is it because computers are allowing large size calculation today that scale extrapolations have become useless? This topic reflects upon the relationship between the experiments, material characteristics and structural engineeering. As in statistical physics, it sheds new lights on the existing theories and helps in building new, consistent engineering models. Above all, I am sure that when reading the conclusion of the book, the reader will be convinced that scaling ought to play a pivotal role into the understanding new problems such as earth dynamics and nanomechanics (to mention just two extremes).

I am very glad that Zdeněk Bažant agreed to take the time to write this volume. It is an excellent and condensed presentation of the author's pioneering works in this field. Zdeněk decided also to include some new, unpublished results in his manuscript. I am indebted to him for this mark of esteem and trust.

Finally, thanks should also be extended to my graduate student Bruno Zuber who helped in the preparation of the final manuscript.

Nantes, November 2001 $\hspace{4cm}$ Gilles Pijaudier-Cabot

Author's Preface

In 1973, while browsing in the library, one paper in the *Indian Concrete Journal* caught my eye. P.F. Walsh, a young Australian then unknown to me, was reporting remarkable experiments. They revealed, in concrete specimens, a strong size effect. But that size effect did not conform to a power law and thus was in conflict with the Weibull statistical theory, then reigning supreme and sacrosanct.

At about the same time, luckily, the late Stanley Fistedis invited me to consult his group at Argonne National Laboratory in matters of failure analysis of concrete vessels and containments under various hypothetical scenarios of nuclear accidents in a liquid-metal-cooled breeder reactor. The objective was reliable extrapolation from normal-scale laboratory specimens to these very large (and politically very sensitive) structures. In view of dense distributed reinforcement, it was necessary to somehow take realistically into account the distributed cracking, for which it seemed unavoidable to postulate strain-softening. This phenomenon, as we know today, gives rise to a deterministic size effect.

Then, in the early 1970s, there was the luck of my getting to teach advanced topics in structural stability to some very inquisitive students in our solid mechanics program at Northwestern, who argued about stability of softening structures, and of hearing a great seminar by Jim Rice on the triggering of localization instability by geometrically nonlinear plastic deformations, which is analogous to the strain-softening trigger.

Somehow all these stimuli set me at the beginning of the 1970s on an initially controversial but exciting path which has not yet reached its end. It has been struggle and fun—struggle because most solid mechanics sages regarded at that time the strain softening (the cause of deterministic size effect) as a lowly crime of ignorants (fortunately, I was no longer behind the Iron Curtain where the mechanics bosses actually managed to get any funding for strain-softening models proscribed by a ruling of the academy)—and fun because it led at NSF Workshops to all these lively polemics about improperly posed boundary value problems, uniqueness, regularization, mesh sensitivity, material stability, etc.

In my efforts leading to this monograph, I am indebted to many. I wish to thank my friend Jacky Mazars for advising Pijaudier-Cabot to become my doctoral student. Gilles often dropped to my office cheerfully but debated forcefully and gave me hard time. He contributed some key ideas of the nonlocal damage concept and its stability foundations, which finally made the concept of strain softening noncontroversial. Aside from Gilles, I have been blessed in my studies of size effect and localization with the collaboration of a long line of bright and hard-working doctoral students —E. Becq-Giraudon, S. Beissel, M. Brocca, F.C. Caner, G. Cusatis, M. Cyr, R. Desmorat, R. Gettu, Z. Guo, M. Jirásek, M.E. Karr, M.T. Kazemi, J.-J. H. Kim, J.K. Kim, Z. Li, F.-B. Lin, G. de Luzio, B.-H. Oh, P.A. Pfeiffer, P.C. Prat, W.F. Schell, M. Thoma, S. Şener, Y. Xi, K. Xu, Q. Yu, Y. Zhou and G. Zi, as well as postdoctoral associates and visiting scholars—I. Carol, L. Cedolin, J. Červenka, D. Ferretti, Y.-N. Li, P. Kabele, Y.W. Kwon, D. Novák, J. Ožbolt, J. Planas and J. Vítek. Their enormous help to my research leading to this monograph is deeply appreciated.

I cannot thank enough my esteemed colleague Isaac Daniel for his invaluable advise and help in fracture testing of fiber composites, sandwiches and foams. To John Dempsey, aside from provocative discussions, I am indebted for the truly unique experience of taking part of his 'expedition' to the Arctic Ocean in which size effect tests of sea ice specimens, up to the record-breaking size of 80 m, were successfully carried out. The great research environment that we have at Northwestern University has been a big plus, but my escapes to the calm atmospheres of hotels Maria in Sils, Le Calette in Cefalú, Paraiso del Mar in Nerja, Parador Aiguablava on Costa Brava and others were conducive to thinking through some more challenging sections of this monograph.

Thanks are further due to E.-P. Chen for funding, from his applied mechanics program at Sandia National Laboratories, my work on a review of scaling on which much of this book is based. The present monograph would not have happened had Gilles not pressed me gently but persistently. It certainly would not have happened without generous financial support for my research at Northwestern, which was initially granted by the National Science Foundation and Air Force Office of Scientific Research, and during the 1990s came mainly from the solid mechanics program directed at the Office of Naval Research by Yapa D.S. Rajapakse. I am grateful to Yapa for inducing me to take more fundamental viewpoints and pushing me to shift my focus from the scaling problems of concrete and geomaterials to those of sea ice and, more recently, fiber composites, rigid foams and sandwich structures for ships.

Evanston, September 2001   Zdeněk P. Bažant

About the Author

Born and educated in Prague, Dr. Zdeněk P. Bažant joined Northwestern University faculty in 1969, became Professor in 1973, was named to the distinguished W.P. Murphy Chair in 1990, and served during 1981-87 as Director of the Center for Concrete and Geomaterials. In 1996 he was elected to the *National Academy of Engineering*, Washington. He has published over 400 refereed journal articles; authored books on *Stability of Structures* (1991, with L. Cedolin), *Fracture and Size Effect*, (1997, with J. Planas), *Concrete at High Temperatures* (1996, with M. Kaplan), *Inelastic Structural Analysis* (2001, with M. Jirásek) and *Creep of Concrete* (1966, in Czech); and edited 14 books. He served as *Editor* (in chief) of *ASCE J. of Engrg. Mech.* (1988-94) and is Regional Editor of *Int. J. of Fracture*, and member of editorial boards of 16 other journals. He was founding president of IA-FraMCoS, president of Soc. of Engrg. Science (SES), and chairman of IA-SMiRT Div. H. An Illinois Registered Structural Engineer, he has served as consultant to Argonne National Laboratory and many firms. He chaired various committees in ASCE, RILEM and ACI. His honors include: *honorary doctorates* from Czech Techn. Univ., Prague, Universität Karlsruhe, Germany, University of Colorado, Boulder, and Politecnico di Milano. He is a *Foreign Member* of Austrian Academy of Sciences, and of Engrg. Academy of Czech Rep. He received *Prager Medal* from SES; Warner Medal from ASME; *Newmark Medal, Croes Medal, Huber Prize* and *T.Y. Lin Award* from ASCE; *L'Hermite Medal* from RILEM; *Torroja Gold Medal* (Madrid); *D.M. Roy Lecture Award* from the American Ceramic Society; *ICOSSAR Lecture Award* from the International Association for Structural Safety and Reliability; *Humboldt Award; Guggenheim, NATO, Humboldt, JSPS, Kajima and Ford Science Fellowships*; *Meritorious Publ. Award* from SEAOI; *Stodola Gold Medal* from Slovak Acad. of Sci.; *Best Enqrq. Book of the Year* Award from SAP; *Medal of Merit* from Czech Soc. of Mech.; and China-Taiwan Gov. *Lectureship Award*; etc. He was elected *Fellow* of Am. Soc. of Civil Engrs., Am. Soc. of Mechanical Engrs., Am. Concrete Inst. and Am. Academy of Mechanics; and *Honorary Member* of the Czech Soc. of Civil Engrs. and of Building Res. Inst., Spain. He is a *Fellow* of Am. Academy of Mech., SES, ASME, ASCE, ACI and RILEM.

Chapter 1

Introduction

1.1. Nature of Problem and Approach

Scaling is the quintessential aspect of every physical theory. If scaling is not understood, the theory itself is not understood. So it is not surprising that the question of scaling has occupied a central position in many problems of physics and engineering. Following Prandtl's (1904) development of the boundary layer theory, the study of scaling acquired during the last century a particularly prominent role in fluid mechanics.

In solid mechanics, the scaling problem of main interest is the effect of the size of structure on its strength. This problem is very old, in fact older than the mechanics of materials and structures. Its discussion started in the Renaissance. However, after an initial period of keen interest, little progress occured for two and half centuries, until in the first part of the 20-th century the statistical source of size effect became understood.

During the last quarter century, the non-statistical energetic source of size effect emerged as a focus of attention. Rapid progress has been taking place. The purpose of this treatise,[1] which expands an extensive recent review by Bažant and Chen (1997), is to summarize this progress in a concise manner and to describe the current understanding of this rich, multifaceted phenomenon. A few new results are included as well. Emphasis is placed on quasibrittle materials, for which the problem of size effect is most acute and most complex. These are materials incapable of plastic yielding, failing due to fracture that is characterized by a relatively large fracture process zone, in which the material

---

[1] Based on research supported mainly by the Office of Naval Research, Washington, D.C., under Grant N00014-91-J-1109, and monitored by Dr. Yapa D.S. Rajapakse, Manager, Solid Mechanics Program.

undergoes distributed strain-softening damage in the form of microcracking and frictional slip.

The salient feature of quasibrittle structures is a strong and complex size effect on structural strength. The size effect is caused principally by stable growth of a large cohesive fracture or a large fracture process zone with microcracking before the maximum load is attained. The mechanism of the size effect is mainly deterministic, involving stress redistribution and the release of stored energy engendered by a large fracture or a large fracture process zone. In some situations, the randomness of local material strength also intervenes in the size effect.

The quasibrittle material that has been researched most intensely and for the longest period of time, is concrete. Its study provided the stimulus for the development of strain-softening damage theories and their regularization in the form of nonlocal continuum models, and particularly for the study of size effects. The experimental database on damage, fracture and size effect in concrete has become truly enormous.

Aside from normal and high-strength concretes, the quasibrittle materials further include fiber-reinforced concretes, asphalt concretes, fiber-polymer composites, various particulate composites, stiff foams, various polymers, many types of rocks, toughened ceramics, ice (especially sea ice), snow, bone, biological shells, stiff clays or mud, cemented sands, grouted soils, coal, paper, wood, wood particle board, various refractories, filled elastomers, some special tough metal alloys, etc. They also include various materials that are brittle on the normal laboratory scale but quasibrittle on the scale of microelectronic components or micro-electro-mechanical systems (MEMS). Many of these materials are 'high-tech' materials.

This exposition will begin by sketching the long history of size effect studies. Attention will then be focused on three main types of size effects, namely the statistical size effect due to randomness of strength, the energy release size effect, and the possible size effect due to fractality of fracture or microcracks. Definitive conclusions on the applicability of these theories will be drawn.

Subsequently, the applications of the known size effect law for the measurement of material fracture properties will be discussed, and the modeling of the size effect by the cohesive crack model, nonlocal finite element models and discrete element models will be reviewed. Extensions to compression failure and to the rate-dependent material behavior will also be outlined.

Furthermore, the damage constitutive laws used for describing a microcracked material in the fracture process zone will be briefly reviewed. Finally, various applications to quasibrittle materials, including concrete, fiber composites, sea ice, rocks and ceramics, will be presented, and the role of size effect in some famous structural catastrophes will be pointed out.

Some key concepts and results will be explained and derived in this brief treatise in full detail. Many subjects, however, will have to be treated only in a review fashion, without derivations, in order to keep the exposition short. The intent is to provide the reader a broad overview of the field without forcing him to struggle through a bulky volume. Nevertheless, a few very recent results that have not yet been published in periodicals (mainly Chapter 9) will be presented in detail, with full derivation.

The recent textbook by Bažant and Planas (1998) should be considered as complementary to the present compact treatise. Most of the results covered here only in a review style are explained and derived in that book in full detail, in a textbook fashion accessible to anyone with knowledge of mechanics at the level of a B.S. degree in civil or mechanical engineering. That book also gives many additional literature references.

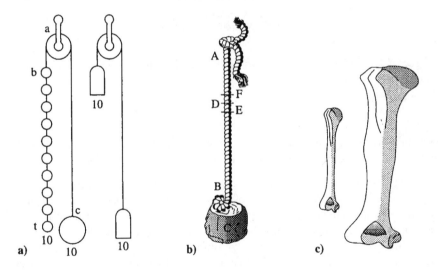

Figure 1.1: Figures illustrating the size effect discussions by (a) Leonardo da Vinci in the early 1500s, and (b, c) Gallileo Galilei in 1638

1.2. Classical History

Let us begin by sketching a bit of the history. Questions of size effect and scaling were discussed already by Leonardo da Vinci (1500s), who stated that "Among cords of equal thickness the longest is the least strong" (Fig. 1.1a). He also wrote that a cord "is so much stronger ... as it is shorter". This rule

Figure 1.2: Title page of the famous book of Galileo (1638), which founded mechanics of materials

implied inverse proportionality of the nominal strength to the length of a cord, which is of course a strong exaggeration of the actual size effect.

More than a century later, the exaggerated rule of Leonardo was rejected by Galileo (1638) in his famous book (Fig. 1.2), in which he founded mechanics of materials. Galileo argued that cutting a long cord at various points (F, D and E in Fig. 1.1b) should not make the remaining part stronger. He pointed out, however, that a size effect is manifested in the fact that large animals have relatively bulkier bones than small ones, which he called the "weakness of giants" (Fig. 1.1c).

Half a century later, a major advance was made by Mariotte (1686). He experimented with ropes, paper and tin and made the observation, from today's viewpoint revolutionary, that "a long rope and a short one always support the same weight unless that in a long rope there may happen to be some faulty place in which it will break sooner than in a shorter". He proposed that this is a consequence of the principle of "the Inequality of the Matter whose absolute Resistance is less in one Place than another". In qualitative terms, he thus initiated the statistical theory of size effect, two and half centuries before Weibull. At that time, however, the theory of probability was at its birth and was not yet ready to handle the problem.

Marriotte's conclusions were later rejected by Thomas Young (1807). He took a strictly deterministic viewpoint and stated that "a wire 2 inches in diameter is exactly 4 times as strong as a wire 1 inch in diameter", and that "the length has no effect either in increasing or diminishing the cohesive strength". This was a setback, but he obviously did not have in mind the random scatter of material strength. Later more extensive experiments clearly demonstrated the presence of size effect for many materials.

The second major advance was the famous work of Griffith (1921). While founding the theory of fracture mechanics, he also introduced fracture mechanics into the study of size effect. He concluded that "the weakness of isotropic solids...is due to the presence of discontinuities or flaws... The effective strength of technical materials could be increased 10 or 20 times at least if these flaws could be eliminated". He demonstrated this conclusion by his experiments showing that the nominal strength of glass fibers was raised from 42,300 psi for the diameter of 0.0042 in. to 491,000 psi for the diameter of 0.00013 in.

In Griffith's view, however, the flaws or cracks triggering failure were only microscopic, which was not characteristic of quasibrittle materials. The sizes and random distribution of these flaws determined the local macroscopic strength of the material but did not affect the global scaling. Thus, Griffith's work represented a physical basis of Mariotte's statistical concept of size effect, rather than a discovery of a new type of size effect.

With the exception of Griffith (and von Mises), the theoreticians in mechanics of materials paid hardly any attention to the question of scaling and size effect. This attitude persisted into the 1980s. The reason doubtless was that all the material failure theories that have existed prior to nonlocal damage mechanics and quasibrittle (nonlinear) fracture mechanics exhibit, as a rule, no size effect because they employ a failure criterion expressed in terms of stresses and strains; this includes the elasticity with allowable stress and plasticity, as well as fracture mechanics of materials with only microscopic cracks or flaws in which the only role of fracture mechanics is to establish or explain the value of the local macroscopic strength (Bažant 1984). Therefore, the mechanicians universally assumed, until mid 1980s, that the size effect, if observed, was inevitably statistical, and should thus be relegated to the statisticians and experimenters. For example, the subject was not even mentioned by Timoshenko in 1953 in his monumental treatise "History of strength of materials".

Significant progress was nevertheless achieved in probabilistic and experimental investigations. Fundamental papers by Fisher and Tippett (1928) and Fréchet (1927) established the extreme value statistics and formulated the weakest-link model for a chain. Their work was early supplemented by the studies of Tippett (1925), and Peirce (1926), and later refined by von Mises (1936) and others (see also Freudenthal, 1968). The capstone on the edifice of statistical size effect initiated by Mariotte was laid by Weibull (1939) in Sweden (see also Weibull 1949, 1956).

Weibull (1939) reached a crucial conclusion: the tail distribution of extremely small strength values with extremely small probabilities cannot be adequately described by any of the known distributions. He proposed for the tail of the extreme value distribution of local strength of a small material element, a power law with a threshold. The distribution of the strength of a chain based on this power law came to be known as the Weibull distribution although in mathematics it was discovered and rigorously proven earlier by Fisher and Tippett (1928). This distribution was also found applicable to various other physical phenomena. Weibull's successors (see, e.g., Freudenthal 1968; Selected Papers 1981) justified this distribution theoretically, by probabilistic modeling of the distribution of microscopic flaws in the material.

With Weibull's work, the basic framework of the statistical theory of size effect became complete. Most subsequent studies until the 1980s dealt basically with refinements, justifications and applications of Weibull's theory (e.g., for concrete, Zaitsev and Wittmann 1974; Mihashi and Zaitsev 1981; Zech and Wittmann 1977; Mihashi 1983; Mihashi and Izumi 1977; see also Carpinteri 1986 1989; Kittl and Diaz 1988, 1989, 1990). It was generally assumed that, if a size effect was observed, it had to be of Weibull type. Today we know this is not the case.

Weibull statistical theory of size effect applies to structures that (1) fail (or must be assumed to fail) right at the initiation of the macroscopic fracture (in detail, see Chapter 3), and (2) have at failure only a small fracture process zone causing only a macroscopically negligible stress redistribution. This is certainly the case for various fine-grained ceramics and for metal structures embrittled by fatigue. But this is not the case for *quasibrittle* materials, the main subject of this short treatise.

Aside from the size effects caused by quasibrittle fracture and by material randomness, complex size effects arise at the continuum limit on approach to the nanoscale. These important size effects, which are caused by surface tension and by dislocations in the lattice structure of metallic crystals, will be discussed at the end, but only very briefly.

1.3. Recent Developments in Quasibrittle Materials

The quasibrittle material of widest use is concrete. The study of its fracture mechanics, initiated by Kaplan (1961), prepared the ground for the discovery of a different type of size effect.

Leicester (1969), Kesler, Naus and Lott (1971) and Walsh (1972) concluded that the classical linear elastic fracture mechanics (LEFM) of sharp cracks does not apply to concrete. Leicester (1969) tested geometrically similar notched beams of different sizes, fitted the results by a power law, $\sigma_N \propto D^{-n}$, and observed that the optimum n was less than $1/2$, the value required by LEFM. The power law with a reduced exponent of course fits the test data in the central part of the transitional size range well but does not provide the bridging of the ductile and LEFM size effects. It was tried to explain the reduced exponent value by notches of a finite angle, which however is objectionable for two reasons: (i) notches of a finite angle cannot propagate (rather, a crack must emanate from the notch tip), (ii) the singular stress field of finite-angle notches gives a zero flux of energy into the notch tip. Same as Weibull theory, Leicester's power law also implied nonexistence of a characteristic length, which cannot be the case for concrete due to the large sizes of its inhomogeneities and its fracture process zone.

The demonstration of inapplicability of LEFM is mainly due to Walsh (1972, 1976), who tested geometrically similar notched beams of different sizes and plotted the results in a double logarithmic diagram of nominal strength versus size (Fig. 1.3). Without attempting a mathematical description, he made the point that this diagram deviates from a straight line of slope $-1/2$, and that this deviation must be regarded as a departure from LEFM.

In 1976 it was analytically demonstrated that localization of strain-softening damage into bands engenders a size effect on post-peak deflections and energy dissipation of structures (Bažant 1976) and the essential idea of the crack band model was proposed as a remedy. In the early 1980s, based on an approximate energy release analysis, a simple formula for the size effect law was derived (Bažant 1984a); it described the size effect on nominal strength of quasibrittle structures containing notches or traction-free (fatigued) large cracks that have formed in a stable manner.

Concerning the evolution of fracture mechanics of concrete, a major step was made by Hillerborg et al. (1976). Inspired by the softening and plastic fracture process zone models of Barenblatt (1959, 1962) and Dugdale (1960) (later extended by Knauss 1973, 1974, Wnuk 1974 and Kfouri and Rice 1977), they formulated for concrete what they called the *fictitious crack model*, now more often called the *cohesive crack model* because of its close similarity to earlier cohesive crack models for other materials. Further they showed by finite element analysis that the failure of unnotched plane concrete beams in bending exhibits a size effect, and that this size effect is deterministic rather than of the statistical Weibull type.

As a refinement of the initial studies of the role of strain softening (Bažant 1976, Bažant and Cedolin 1979, 1980; see also Bažant and Cedolin 1991), the

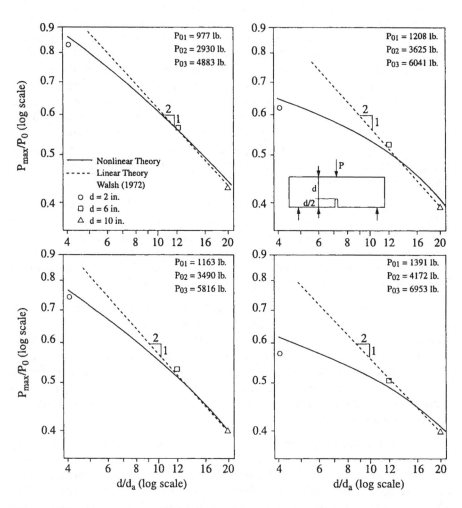

Figure 1.3: Data points obtained by Walsh (1972) in four of his six series of tests of geometrically similar notched three-point bend beams, and the fitting curves obtained by Bažant and Oh (1983) by finite element analysis with the crack band model

crack band model was formulated in detail (Bažant 1982, Bažant and Oh 1983) and was shown to accurately capture by simple finite element analysis the size effect observed by that time on concrete specimens and structures. The crack band model is nowadays almost the only concrete fracture or damage model used in industry and commercial codes (e.g. code DIANA, Rots 1988 ; code SBETA, Červenka and Pukl 1994; or ATENA).

A more general nonlocal approach to strain-softening damage, capable of describing the size effect in quasibrittle materials in a more fundamental and more realistic manner, followed soon (Bažant, Belytschko and Chang 1984; Bažant 1984b; Pijaudier-Cabot and Bažant 1987; Bažant and Pijaudier-Cabot 1988; Bažant and Lin 1988a,b; etc.).

Beginning with the mid 1980s, the interest in the quasibrittle size effect surged enormously and many researchers made noteworthy contributions; to name but a few: Planas and Elices (1988, 1989a, 1989b, 1993); Petersson (1981), and Carpinteri (1986). The size effect has become a major theme at conferences on concrete fracture (Bažant, ed., 1992; Mihashi et al., eds., 1994; Wittmann, ed., 1995; Mihashi and Rokugo 1998; Bažant and Rajapakse 1999).

It was also recognized that measurements of the size effect on the maximum load allow a simple way to determine the fracture characteristics of quasibrittle materials. This line of investigation culminated with the Cardiff workshop (Barr, 1995) at which representatives of American and European societies endorsed a unified recommendation (still pending) for a test standard based on the measurement of maximum loads alone.

An intriguing idea was injected into the current lively debates on size effect by Carpinteri et al. (1993, 1995a,b), Carpinteri (1994a,b) and Carpinteri and Chiaia (1995). Inspired by numerous recent studies of the fractal characteristics of cracks in various materials (Mandelbrot 1984; Brown, 1987; Mecholsky and Mackin 1988; Cahn 1989; Chen and Runt 1989; Hornbogen, 1989; Peng and Tian 1990; Saouma et al 1990; Bouchaud et al. 1990; Chelidze and Gueguen 1990; Issa et al. 1992; Long et al. 1991; Måløy et al. 1992; Mosolov and Borodich 1992; Borodich 1992; Lange et al. 1993; Xie 1987, 1989, 1993; Xie et al. 1994, 1995; Saouma and Barton 1994; Feng et al. 1995; etc.), Carpinteri and Chiaia (1995) proposed, on the basis of strictly geometrical arguments, that the difference in fractal characteristics of cracks or microcracks at different scales of observation is the principal source of size effect in concrete structures. However, recent mechanical analysis by Bažant (1997b) casts serious doubt on this proposition.

1.4. Basic Theories of Size Effect

At present, there exist (aside from the theories for the transition from a continuum to atomic lattice, to be discussed briefly at the end) three basic theories of scaling in solid mechanics:

1. *Weibull statistical theory of random strength* (Weibull 1939)

2. *Theory of stress redistribution and fracture energy release caused by large cracks* (Bažant, 1984a).

3. *Theory of crack fractality,* in which two variants may be distinguished.

    a. *Invasive* fractality of the crack surface (i.e., a fractal nature of surface roughness) (Carpinteri et al. 1993, 1995a,b,c; Carpinteri 1994a,b), and

    b. *Lacunar* fractality (representing a fractal distribution of microcracks) (Carpinteri and Chiaia 1995).

Aside from these basic theories, there exist four indirect size effects:

1. The boundary layer effect, which is due to material heterogeneity (i.e., the fact that the surface layer of heterogeneous material such as concrete has a different composition because the aggregates cannot protrude through the surface), and to the Poisson effect (i.e., the fact that a plane strain state on planes parallel to the surface can exist in the core of the test specimen but not at its surface).

2. The existence of a three-dimensional stress singularity at the intersection of crack edge with a surface, which is actually engendered by the aforementioned Poisson effect (Bažant and Estenssoro 1979). This causes the portion of the fracture process zone near the surface to behave differently from that in the interior.

3. Time-dependent size effect caused by diffusion phenomena such as the transport of heat or the transport of moisture and chemical agents in porous solids (this is manifested, e.g., in the effect of size on shrinkage and drying creep, and is captured by the size dependence of the drying half time (Bažant and Kim 1991) and its effect on shrinkage cracking (Planas and Elices 1993, Bažant and Raftshol 1983).

4. Time-dependence of the material constitutive law, particularly the viscosity characteristics of strain softening, which impose a time-dependent length scale on the material (Sluys 1992).

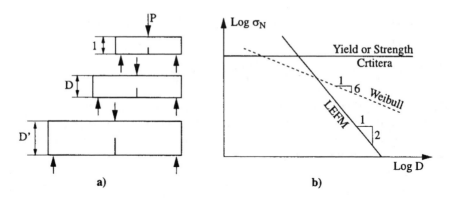

Figure 1.4: (a) Geometrically similar structures of different sizes D and (b) power scaling laws

Today the study of scaling in quasibrittle materials is a lively, rapidly moving field. Despite considerable success in recent research, major questions remain open (Bažant 1999a,b) and polemics abound (e.g. Planas et al. 2001; Bažant and Rajapakse 1999). The brief exposition that follows in this book will focus on the three main theories of size effect and the indirect ones will be left out of consideration.

1.5. Power Scaling in Absence of Characteristic Length

The basic and simplest type of scaling characterizes any physical theory that lacks a characteristic length. Let us consider geometrically similar systems, for example the beams shown in Fig. 1.4a, and seek to deduce the response Y (e.g., the maximum stress or the maximum deflection) as a function of the characteristic size (dimension) D of the structure. Imagine now structures of three sizes D_0, D_1 and D, with the corresponding responses Y_0, Y_1 and Y. Given that there is no characterisitc length, we seek the dimensionless function $f(\lambda)$ of the dimensionless scaling ratio $\lambda = D/D_0$ representing the scaling law, i.e. $Y/Y_0 = f(\lambda)$; D_0 is considered as the reference size. This implies that $Y_1/Y_0 = f(D_1/D_0)$.

The ratio of the responses for sizes D and D_1 is

$$\frac{Y}{Y_1} = \frac{f(D/D_0)}{f(D_1/D_0)}. \tag{1.1}$$

Now, if and only if there is no characteristic length, size D_1 may be taken as the reference size instead of D_0, which means that $Y/Y_1 = f(D/D_1)$. Thus we acquire the condition

$$\frac{f(D/D_0)}{f(D_1/D_0)} = f\left(\frac{D}{D_1}\right) \tag{1.2}$$

(Bažant 1993, Bažant and Chen 1997; for fluid mechanics, see Barenblatt 1979, Sedov 1959). This is a functional equation for the unknown scaling law $f(D)$. To solve it, one may differentiate it with respect to D and then substitute $D_1 = D$. This yields the differential equation

$$\frac{df(\lambda)}{f(\lambda)} = m\frac{d\lambda}{\lambda}, \quad \text{with} \quad m = \frac{df(1)}{d\lambda} = \text{const.} \tag{1.3}$$

which can be easily solved by separation of variables. Because $Y = Y_0$ for $D = D_0$, the initial condition is $f(1) = 1$. One thus concludes that (1.2) has one and only one solution, namely the power law with unknown constant exponent m:

$$f(\lambda) = \lambda^m \tag{1.4}$$

That is why power scaling plays such a prominent role.

When, for example, the scaling law is $f(\lambda) = \ln \lambda$, equation (1.2) is not satisfied; $Y/Y_1 = (Y/Y_0)/(Y_1/Y_0) = \ln\lambda/\ln\lambda_1 \neq \ln(\lambda/\lambda_1)$. This means that a characteristic dimension of the structures, or a characteristic length, does exist in this case. The scaling law $f(\lambda)$ can be valid only for one reference structure size, namely D_0, and the characteristic length is related to D_0.

In conclusion, the power scaling must apply for every physical theory in which there is no characteristic length. In solid mechanics, such failure theories include all the theories of elasto-plasticity, elasticity with a strength limit, as well as LEFM (in which the FPZ is implied to be shrunken into a point). But for LEFM this is true if not only the structures but also the cracks in them are geometrically similar (this excludes metallic structures with small flaws that are a material property and do not scale up with the structure size).

To determine exponent m, the failure criterion of the material must be taken into account. For elasticity with a strength limit (strength theory), or plasticity (or elasto-plasticity) with a yield surface that is expressed in terms of stresses or strains, or both, one finds that $m = 0$ when the response Y represents the stress or strain (for example the maximum stress, or the stress at certain homologous points, or the nominal stress at failure); Bažant (1993). Thus, if there is no characteristic length or dimension, all geometrically similar structures of different sizes must fail at the same nominal stress. By tradition, this came to be known as the case of *no size effect*.

In LEFM, on the other hand, $m = -1/2$, provided that one deals with geometrically similar structures containing geometrically similar cracks or notches.

Introduction   13

This may be generally proven with the help of Rice's J-integral (see Bažant 1993).

If $\log \sigma_N$ is plotted versus $\log D$, the power law is a straight line (Fig. 1.4b). For plasticity, or elasticity with a strength limit, the vanishing of the power law exponent means that the slope of this line is 0, i.e., the line is horizontal. For LEFM, the slope is $-1/2$.

For quasibrittle materials and structures, which are a recently emerged 'hot' subject, the size effect bridges these two power laws, i.e., represents a transition from one power law to another. What is the law that governs this transitional law for scale bridging? This is a complex question, which is still currently the subject of various polemics and lies in the core of the subjects to be discussed next.

We may digress at this point to mention that, for Weibull-type statistical theories (in which the threshold value may usually be taken as 0), the scaling law for the nominal strength of the structure is also a power law. Based on a preliminary study by Zech and Wittmann (1977), the power law exponents for concrete have been thought to be $-1/6$ for 2D or $-1/4$ for 3D similarity (see Fig. 1.4b), however, a detailed study taking into account the simultaneous energetic size effect in concrete now shows that these exponents are $-1/12$ and $-1/8$ (Bažant and Novák 2000b).

By the inverse of the preceding analysis of power scaling (Eqs. 1.2–1.4), the fact that the Weibull-type scaling is a power law implies that the Weibull statistical theories have no characteristic length. This immediately invites a question with regard to the applicability of these theories to quasibrittle materials such as concrete or composites, which obviously possess a certain characteristic length corresponding to the dimension of the inhomogeneities in the microstructure of the material. This is one reason why the Weibull-type statistical theory of size effect is not applicable to quasibrittle materials, except on scales so large that the size of their inhomogeneities becomes negligible. The large-scale material behavior changes from quasibrittle to brittle (see Chapter 3).

1.6. Transitional Size Effect Bridging Power Laws for Different Scales

Because $m = 0$ for plasticity or strength criterion, the size effect in structures is measured by the nominal strength σ_N, which is the value of nominal stress at maximum (ultimate) load. When the nominal strength is independent of D, we say that there is no size effect. The size effect is the dependence of the nominal strength on the structure size. Depending on whether the geomet-

14   Scaling of Structural Strength

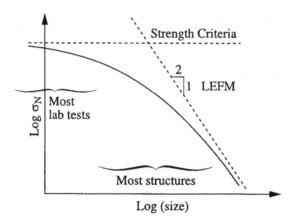

Figure 1.5: Transitional scaling of the nominal strength of quasibrittle structures failing only after large fracture growth

ric similarity is two-dimensional (2D) or three-dimensional (3D), the nominal strength is a parameter of the maximum load P defined as

$$\sigma_N = c_n P/bD \quad \text{for 2D}, \qquad \sigma_N = c_n P/D^2 \quad \text{for 3D} \qquad (1.5)$$

Here b is the structure thickness in the third dimension, for the case of 2D similarity, and c_n is a dimensionless convenience constant which may depend on structure shape but not size and may be exploited to make σ_N coincide for example with the maximum stress or the average stress, or the stress at any particular point.

In quasibrittle materials, the problem of scaling is more complicated because the material possesses a characteristic length that matters. It is nevertheless clear that, for a sufficiently large size, the scale of the material inhomogeneities, and thus the material length, should become unimportant. So the power scaling law should apply asymptotically for sufficiently large sizes. If there is a large crack at failure, the exponent of this asymptotic power law must be $-1/2$, which is represented by the dashed asymptote in Fig. 1.5.

The characteristic length of the material (called also the material length) must also become unimportant for very small structure sizes, for example when the size of concrete specimen is only several times the aggregate size. This means that, for very small sizes, the size effect should again asymptotically approach a power law. Because, for such small sizes, a discrete crack cannot be discerned (as the entire specimen is occupied by the fracture process zone), the exponent of the power law should be 0, corresponding to the strength criterion (see the horizontal dashed asymptote in Fig. 1.5). The difficulty is that most applications of quasibrittle materials fall into the transitional range between

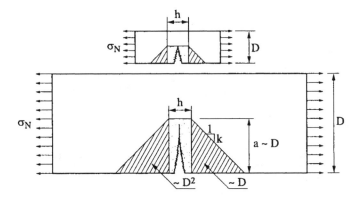

Figure 1.6: Approximate zones of stress relief caused by fracture in small and large specimens

these two asymptotes, for which the scaling law that bridges the two power laws may be expected to follow some transitional curve (Fig. 1.5) whose law must be found by some other kind of arguments.

Let us now offer a simple explanation of the deterministic size effect due to energy release (Bažant 1984). Consider the rectangular panel in Fig. 1.6, which is initially under a uniform stress equal to the nominal stress σ_N. Introduction of a crack of length a with a fracture process zone of a certain length and width h may be approximately imagined to relieve the stress and thus release the strain energy from the areas of the shaded triangles and the crack band shown in Fig. 1.6. The slope of the effective boundary of the stress relief zone, k, is a constant when the size is varied. We may assume that, for the range of interest, the length of the crack at maximum load is approximately proportional to the structure size D while the size h of the fracture process zone is essentially a constant, related to the inhomogeneity size in the material (this assumption is usually, but not always, verified by experiment or nonlocal finite element analysis).

For very large structure sizes, the crack band width h becomes negligible compared to the structure dimensions, and then the energy is getting released only from the shaded triangular zones (Fig. 1.6), whose area is proportional to D^2. This means that the energy release is proportional to $D^2 \sigma_N^2 / E$ ($E =$ Young's modulus). At the same time, the energy consumed is proportional to the area of the band of constant width h, which in turn is proportional to D. So the energy consumed and dissipated by fracture is proportional to $G_f D$ where G_f is the fracture energy—a material property representing the energy dissipated per unit length and unit width (i.e. unit area) of the fracture surface. Thus, $\sigma_N^2 D^2 / E \propto G_f D$. Then it immediately follows that the size effect law

16   Scaling of Structural Strength

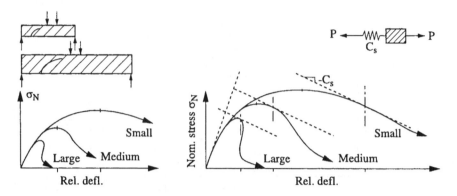

Figure 1.7: Left: Load-deflection curves of quasibrittle structures of different sizes. Right: stability is lost at the tangent points of lines of slope $-C_s$, with $C_s =$ stiffness of loading device

for very large structures is $\sigma_N \propto D^{-1/2}$.

When, on the other hand, the structure is very small, the triangular stress relief zones have a negligible area compared to the area of the crack band, which means that the energy release is proportional to $D\sigma_N^2/E$. Therefore, energy balance requires that $D\sigma_N^2/E \propto G_f D$, from which it follows that $\sigma_N =$ constant. So, asymptotically for very small structures, there is no size effect.

The foregoing analysis (given in more detail in Bažant, 1983, 1984) is predicated on the assumptions that the crack lengths in small and large structures are similar. According to experimental observations and finite element simulations, this is often true for the practically interesting range of sizes. However, there are some cases where the similarity of cracks does not hold true, and then of course the scaling becomes different and more complex.

The curves of nominal strength versus the relative structure deflection (normalized so that the initial slope in Fig. 1.7 be independent of size) have, for small and large structures, the shapes indicated in Fig. 1.7. Aside from the size effect on the maximum load, there is a size effect on the shape of the post-peak descending load-deflection curves. For small structures, the post-peak curves descend slowly; for larger structures, steeper; and for sufficiently large structures they may even exhibit a snapback, that is, a change of slope from negative to positive.

If a structure with post-peak softening is loaded by an elastic device with a spring constant C_s, it loses stability at the point where the load-deflection diagram first attains the slope $-C_s$ (if ever), as seen in Fig. 1.7. These tangent points indicate failure. The ratio of the deflection at these points to the elastic deflection characterizes the ductility of the structure.

Small quasibrittle structures obviously have a high ductility while large quasibrittle structures have a low ductility. The areas under the load-deflection curves characterize the energy absorption. The energy absorption capability of a quasibrittle structure decreases, in relative terms, as the structure size increases. This is important for blast loads and impact.

The progressive steepening of the post-peak curves in Fig. 1.7 with increasing size and the development of a snapback can be most simply explained by the series coupling model, which assumes that the response of a structure may be in essence approximately modeled by the series coupling of the cohesive crack or damage zone with the elastic behavior of the structure (Bažant and Cedolin, 1991, Sec. 13.2).

1.7. Deductions from Dimensional Analysis

The exponents of the asymptotic power laws of the transitional size effect can be easily deduced from dimensional analysis. The number of dimensionless variables governing a physical phenomenon can be determined from Buckingham's Π theorem (Buckingham 1914, 1915; see also Bridgman, 1922; Porter, 1933; Giles, 1962; Streeter and Wylie, 1975; and Barenblatt, 1979, 1987). This theorem states that the number of governing dimensionless variables is equal to the total number of variables minus the number of parameters with independent dimensions (in these cases just two, length and force). It readily follows that the condition of failure governed only by material strength or yield limit σ_0, with no role for the energy release rate, must have the form:

$$f\left(\frac{\sigma_N}{\sigma_0}, \frac{L_1}{D}, \frac{L_2}{D}, \ldots\right) = 0 \qquad (1.6)$$

where f is a certain function and $L_1, L_2, \ldots$ are spatial dimensions whose ratios to D characterize the structure geometry. Since σ_0 is a constant and, for geometrically similar structures $L_1/D, L_2/D, \ldots$ are constants, too, it follows that the nominal stress at failure, σ_N, must be proportional to σ_0, and therefore a constant when the structure size D is varied. Note that the material parameters present in the failure condition, which consist of σ_0 alone, imply no characteristic length.

In linear elastic fracture mechanics (LEFM), the failure is determined by the critical stress intensity factor K_c (fracture toughness), the metric dimension of which is N m$^{-3/2}$. It is straightforward to figure out that the dimensionless

18   Scaling of Structural Strength

failure condition must now have the form

$$\Phi\left(\frac{\sigma_N\sqrt{D}}{K_c}, \frac{L_1}{D}, \frac{L_2}{D}, \dots\right) = 0 \qquad (1.7)$$

where $L_1 = a =$ notch length. Since K_c is a material constant and the ratios $L_1/D, L_2/D, \dots$ are constant as well, for geometrically similar structures, it follows that $\sigma_N\sqrt{D}$ must also be constant. Hence, $\sigma_N \propto D^{-1/2}$ (e.g., Bažant, 1983, 1984; Carpinteri 1984, 1986). Note again that the material parameters present in the failure condition, which consist of K_c alone, imply no characteristic length (together with material strength σ_0, of course, K_c does imply a material length, $l_{ch} = K_c^2/\sigma_0^2$, but σ_0 is not a parameter in LEFM; it is a parameter in ductile-brittle or quasibrittle fracture mechanics).

The dimensional analysis, unfortunately, becomes ambiguous in some more complex problems, for example the bending failure of floating ice plates, to be discussed later.

For systems with no characteristic dimension, the size effect can also be deduced, without recourse to physics, simply by converting the mathematical formulation of the boundary value problem to a dimensionless form. To this end we introduce the dimensionless variables, labeled by an overbar;

$$\bar{x}_i = x_i/D, \quad \bar{u}_i = u_i/D, \quad \bar{\sigma}_{ij} = \sigma_{ij}/\sigma_0 \qquad (1.8)$$

$$\bar{p}_i = p_i/\sigma_N, \quad \bar{f}_i = f_i D/\sigma_N, \quad \bar{E}_{ijkl} = E_{ijkl}/\sigma_0 \qquad (1.9)$$

where $x_i =$ Cartesian coordinates $(i = 1, 2, 3)$, $\sigma_{ij} =$ stress tensor components, $p_i =$ given surface tractions, $f_i =$ body forces, and $E_{ijkl} =$ elastic moduli. The strain components are $\epsilon_{ij} = \frac{1}{2}(u_{i,j} + u_{j,i})$, the field equilibrium equations are $\sigma_{ij,i} + f_j = 0$, and the stress boundary conditions are $n_i\sigma_{ij} = p_i$ on Γ_s, where n_i is the unit surface normal, Γ_s is the surface domain where stresses are prescribed, and the derivatives with respect to x_i are denoted by subscript i preceded by a comma. Denoting the derivatives with respect to dimensionless coordinates as $\partial_i = \partial/\partial \bar{x}_i$ and noting that $\partial/\partial x_i = (1/D)\partial_i$, we can transform the foregoing equations to the following dimensionless form:

$$\bar{\epsilon}_{ij} = \tfrac{1}{2}(\partial_j \bar{u}_i + \partial_i \bar{u}_j), \qquad \partial_j \bar{\sigma}_{ij} + (\sigma_N/\sigma_0)\bar{f}_i = 0 \qquad \text{(in } \mathcal{V}\text{)} \quad (1.10)$$

$$n_j\bar{\sigma}_{ij} = (\sigma_N/\sigma_0)\bar{p}_i \quad \text{(on } \Gamma_s\text{)}, \qquad \bar{u}_i = 0 \quad \text{(on } \Gamma_d\text{)} \qquad (1.11)$$

where $\bar{\epsilon}_{ij} = \epsilon_{ij}$; $\mathcal{V}$ is the domain of structure volume and Γ_d is the surface domain where the displacements are prescribed as zero. These equations must be complemented by the constitutive law and the material failure condition.

In plasticity, or elasticity with a strength limit, the constitutive law and material failure conditions are expressed as equations and inequalities involving functions of the type $F(\sigma, \epsilon)$ or, in dimensionless form,

$$F(\sigma_N, \bar{\sigma}_{ij}, \bar{\epsilon}_{ij}) \qquad (1.12)$$

Because D does not appear in (1.10) and (1.11), $\bar{\sigma}_{ij}$ $\bar{\epsilon}_{ij}$ are proportional to σ_N, and because at least some of the functions F are not homogeneous functions, the conditions in terms of these functions can remain valid for all D only if σ_N is a constant. This demonstrates that there is no size effect in plasticity or strength based theories (or any theory in which the material failure condition is expressed solely in terms of stress and strain).

In LEFM, the constitutive law and the failure condition at crack (or notch) tip may be written as $\sigma_{ij} = E_{ijkl}\epsilon_{ij}$ and $\lim(\sigma_{22}\sqrt{2\pi x_1}) = K_c$ for $x_1 \to 0$; here we assume that the origin of coordinates x_i is placed into the crack tip, x_1 is the direction of propagation and x_2 is normal to the crack plane. Transformation to dimensionless coordinates yields:

$$\bar{\sigma}_{ij} = \bar{E}_{ijkl}\bar{\epsilon}_{ij}, \qquad \lim_{x_1 \to 0}\left(\bar{\sigma}_{22}\sqrt{2\pi\bar{x}_1}\right) = K_c/\sigma_0\sqrt{D} \qquad (1.13)$$

The second of these two equations is valid for all sizes D if and only if $\sigma_{22} \propto 1/\sqrt{D}$. Since, as we observed from (1.10) and (1.11), the stresses $\bar{\sigma}_{ij}$ must be proportional to σ_N, it follows that $\sigma_N \propto 1/\sqrt{D}$. The first equation is then valid for all D if and only if $\epsilon_{ij} = \bar{\epsilon}_{ij} \propto \sigma_N$ or $\epsilon_{ij} \propto 1/\sqrt{D}$. The scaling law of LEFM is thus demonstrated again. Note that instead of the second equation in (1.13), one could use the condition of critical energy release rate, with the same result (Bažant 1983).

1.8. Stability of Structures and Size Effect

Failure of a structure under static loading is synonymous to loss of stability. Generally, stability analysis must take into account the evolution of fracture or distributed damage, which is the subject of much of this treatise. However, for slender beams and frames, and thin plates and shells, fracture or damage begins at (or very near) the smallest elastic critical load, P_{cr}. For slender beams or frames, $P_{cr} = EI\pi^2/L^2$ where E = Young's modulus, I = centroidal moment of inertia of cross section (assumed uniform), and L = effective length = half-wavelength of the deflection curve. For structures geometrically similar in two dimensions, $L = k_0 D$ and $I = k_1 b D^2$ where k_0 and k_1 are constants. It follows that the elastic critical stress, which represents the nominal strength,

$$\sigma_{Ncr} = \frac{P_{cr}}{bD} = \frac{k_1}{k_0^2}\pi^2 = \text{const.} \qquad (1.14)$$

Hence, there is no size effect in elastic buckling of beams and frames. The same can be shown to be true for elastic plates and shells, as well as for elasto-plastic buckling, provided that fracture and damage localization are not involved. This conclusion also follows by dimensional analysis from the fact that the problem of elastic or elasto-plastic stability involves no characteristic length.

Curiously, a different conclusion is obtained for beams or plates on elastic foundation. The best example is an ice plate floating on water, which behaves exactly as the Winkler elastic foundation, its foundation modulus being equal to the specific (unit) weight ρ_w of water. Considering periodic buckling in one direction only, we have the critical normal force $N_{cr} = 2\sqrt{\rho EI}$ where $I = ED^3/12(1-\nu^2)$, $D =$ ice thickness, $E =$ Young's modulus and $\nu =$ Poisson ratio of ice (e.g. Bažant and Cedolin 1991, Eq. 5.2.7). So the critical nominal stress of the ice plate, $\sigma_{Ncr} = N_{cr}/D$ or

$$\sigma_{Ncr} = C_0\sqrt{D}, \qquad C_0 = \sqrt{\frac{\rho E}{3(1-\nu^2)}} \qquad (1.15)$$

where C_0 is a material constant.

The point to note is that the nominal strength increases, rather than decreases, with the thickness of ice. So there is a reverse size effect. Consequently, only sufficiently thin ice plates can fail by buckling. Thicker ones fail by fracture.

Even though the axisymmetric buckling of an axially compressed elastic cylindrical shell is mathematically equivalent to the buckling of an elastic beam on Winkler elastic foundation, the foregoing conclusion does not apply to shells. The reason is that, for geometrically similar shells, the equivalent foundation modulus of the shell is not constant but inversely proportional to D. As a result, the nominal critical stress of the shell is independent of its size.

The foregoing conclusion could have also been easily deduced by dimensional analysis (Sedov 1959, Barenblatt 1987). Compared to the buckling problem of normal beams, plates and shells, already discussed, the floating ice problem involves the specific weight of water, ρ, which is a parameter of a different dimension, force/(length)^3. In contrast to these normal buckling problems, a combination of this parameter with Young's modulus implies the existence of a characteristic length, which in turn implies the existence of size effect.

Chapter 2

Asymptotic Analysis of Size Effect

2.1. Asymptotic Analysis of Size Effect in Structures with Notches or Large Cracks

The scaling properties for the nominal strength σ_N of a structure containing a notch or a stably grown large crack can be most generally deduced by an asymptotic analysis of the energy release (Bažant 1997a). We will now outline it briefly, restricting attention to two-dimensional similarity although the case of three-dimensional similarity could be analyzed similarly. The fracture may be characterized by the dimensionless variables $\alpha_0 = a_0/D$, $\alpha = a/D$, $\theta = c_f/D$, in which a = the total crack length which gives (according to LEFM) the same specimen compliance as the actual crack with its fracture process zone; a_0 = length of the traction-free crack or the notch; $c_f = a - a_0$ = effective size of the fracture process zone (or the effective length of the R-curve); and D is the characteristic size (dimension) of the structure, for example taken as the depth of the notched three-point bend beam shown in Fig. 2.1.

Note that in our analysis the interpretation in the sense of a cohesive crack or an R-curve model is not essential. We can equally well assume that c_f is

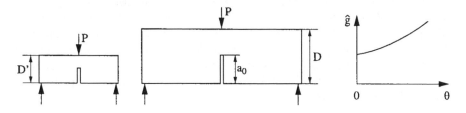

Figure 2.1: Similar structures with large cracks and function $\hat{g}$

in general any kind of material length, for example $c_f = G_f/W_d$ where G_f = fracture energy of the material (dimension J/m^2), and W_d = energy dissipated by distributed cracking in the fracture process zone per unit volume (dimension J/m^3) which is represented by the area under the total stress-strain curve with strain softening in the sense of continuum damage mechanics. Or we can assume that $c_f = EG_f/f_t'^2$, where f_t' is the tensile strength of the material. The last expression is the characteristic size of the fracture process zone of the material according to Irwin (1958).

The energy release from the structure can be analyzed either on the basis of the change of the potential energy of the structure Π at constant load-point displacement, or the change of the complementary energy of the structure Π^* at constant load. We choose the latter, and express Π^* in the following, dimensionally correct, form:

$$\Pi^* = \frac{\sigma_N^2}{E} bD^2 f(\alpha_0, \alpha, \theta) \tag{2.1}$$

Here E = Young's elastic modulus of the material, and f is a dimensionless function characterizing the geometry of the structure. Two further conditions for the maximum load must now be introduced.

First, the fracture at maximum load is propagating. This means that the energy release rate $\mathcal{G}$ must be equal to the energy consumption rate R, which may be interpreted in the sense of the R-curve (resistance curve) giving the dependence of the critical energy release rate required for fracture growth of the crack of length a. Most generally, the resistance to fracture can be characterized as $R = G_f r(\alpha_0, \alpha, \theta)$ in which r is a dimensionless function of the relative crack length α, the relative notch length α_0, and the relative size of the fracture process zone θ, having the property that $r \to 1$ when $\theta \to 0$ and $\alpha \to \alpha_0$.

Obtaining the energy release rate $\mathcal{G} = (\partial \Pi^*/\partial a)/b$ from Eq. (2.1) by differentiation at constant nominal stress, we thus obtain the following first condition for the maximum load

$$b^{-1}[\partial \Pi^*/\partial \alpha]_{\sigma_N} = G_f r(\alpha_0, \alpha, \theta) \tag{2.2}$$

The second condition is that, under load control conditions, the maximum load represents the limit of stability. If the rate of growth of the energy release rate is smaller than the rate of growth of the R-curve, the fracture propagation is stable because the energy release change does not suffice to compensate for the rate of the energy consumed and dissipated by fracture. In the limit, both are equal, and so the second condition of the maximum load, corresponding to the stability limit, reads:

$$\left[\frac{\partial \mathcal{G}}{\partial \alpha}\right]_{\sigma_N} = \frac{\partial R}{\partial \alpha} \tag{2.3}$$

Geometrically, this represents the condition that the curve of the energy release rate must be tangent to the R-curve.

Substituting now the expression for the complementary energy into Eq. (2.1), one can show from the foregoing two conditions of maximum load that the nominal strength of the structure is given in the form:

$$\sigma_N = \sqrt{\frac{EG_f}{D\hat{g}(\alpha_0,\theta)}} \qquad (2.4)$$

in which $\hat{g}$ is a dimensionless function expressed in terms of functions f and r and their derivatives (see Bažant 1996a).

For fracture situations of positive geometry (an increasing $\hat{g}$), which is the usual case, the plot of function $\hat{g}$ at constant relative notch length α_0 looks roughly as shown in Fig. 2.1. This function has the meaning of the dimensionless energy release rate modified according to the R-curve.

Obviously, function $\hat{g}$ must be smooth, and so it can be expanded into Taylor series with respect to the relative material length θ about the point $(\alpha_0, 0)$. In this way the following series expansion of the nominal strength of the structure is obtained:

$$\sigma_N = \sqrt{\frac{EG_f}{D}} \left[\hat{g}(\alpha_0,0) + \hat{g}_1(\alpha_0,0)\frac{c_f}{D} + \frac{1}{2!}\hat{g}_2(\alpha_0,0)\left(\frac{c_f}{D}\right)^2 + \cdots\right]^{-1/2}$$

$$= \frac{Bf'_t}{\sqrt{D}} \left(D_0^{-1} + D^{-1} + \kappa_2 D^{-2} + \kappa_3 D^{-3} + \cdots\right)^{-1/2} \qquad (2.5)$$

Here $\hat{g}_1$ and $\hat{g}_2$ are the first, second, etc., derivatives of function $\hat{g}$ with respect to θ, and $D_0, \kappa_2, \kappa_3, \ldots$ represent certain constants expressed in terms of function $\hat{g}$ and its derivatives at $(\alpha_0, 0)$.

The series expansion is obviously an asymptotic expansion because the powers of size D are negative. So the expansion may be expected to be very accurate for very large sizes, but must be expected to diverge for $D \to 0$. In fact, for $D \to 0$, the truncated large size expansions with more than two terms generally approach ∞ or $-\infty$, which is incorrect.

In this regard, it is helpful to deduce a small-size asymptotic expansion. To this end, one needs to use, instead of θ, the parameter $\eta = \theta^{-1} = D/c_f$. By a similar procedure as before, one can show that the nominal strength of the structure may be written (for the R-curve model) in the form:

$$\sigma_N = \sqrt{\frac{EG_f}{c_f}} \left[\tilde{g}(\alpha_0,\eta)\right]^{-1/2} \qquad (2.6)$$

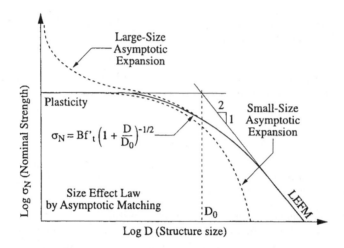

Figure 2.2: Large-size and small-size asymptotic expansions of size effect (dashed curves) and the size effect law as their asymptotic matching (solid curve)

This function again has the meaning of the dimensionless energy release rate (modified by the R-curve). But this function is now expressed as a function of the inverse relative size of the process zone, η. Function $\tilde{g}$ must also be sufficiently smooth to permit expansion into Taylor series with respect to parameter θ about the point $(\alpha_0, 0)$. This yields an asymptotic expansion of the following form:

$$\sigma_N = \sigma_P \left[1 + \left(\frac{D}{D_0}\right) + b_2 \left(\frac{D}{D_0}\right)^2 + b_3 \left(\frac{D}{D_0}\right)^3 + \cdots \right]^{-1/2} \quad (2.7)$$

in which $\sigma_P, D_0, b_2, b_3, \ldots$ are certain constants depending on the shape of the structure.

The small size expansion (2.7) shows that σ_N approaches a horizontal asymptote as $D \to 0$. This fact has been demonstrated numerically with high accuracy by Planas using the cohesive crack model (Bažant and Planas 1998) and also proven for that model analytically. Finite element solutions with the cohesive crack model support this fact, too.

The results obtained may be illustrated by Fig. 2.2 showing the logarithmic size effect plot (for the case of geometrically similar structures with similar and large cracks). The large-size and small-size expansions in Eqs. (2.5) and (2.7) are shown by the dashed curves. The large-size expansion asymptotically approaches the straight line of slope $-1/2$, corresponding to the scaling according to LEFM for the case of large and similar cracks. The small-size expansion approaches on the left a horizontal line, which corresponds to scaling according to the theory of plasticity or any strength theory.

2.2. Energetic Size Effect Law and Its Asymptotic Matching Character

It must be emphasized that the large size and small size asymptotics are mere theoretical extrapolations. Obviously, at some very large size, the mechanism of failure will change, and so an inifinite size is a mathematical abstraction. So is a zero size because a size smaller than the inhomogeneities of the material is meaningless. Thus the purpose of the asymptotic expansions is not to describe the behavior for the limits of infinite and zero sizes. Rather, it is to anchor at opposite inifinities the size effect curve for the size range of practical interest.

So the problem is how to 'interpolate' between the foregoing two expansions in order to deduce an approximate size effect law applicable over the full size range. This is the subject of the well-known theory of matched asymptotics.

We face here a situation in which the asymptotic behaviors (in our case those for the large and small sizes) are relatively easy to obtain while the intermediate behavior (in our case, for the intermediate sizes) is very difficult to obtain. This is a typical situation in which the technique of asymptotic matching is effective (Bender and Orszag 1978, Barenblatt 1979, Hinch 1991). This technique was introduced at the beginning of the century in fluid mechanics by Prandtl in his famous development of the boundary layer theory.

As it turns out, the asymptotic matching is in our case very simple because the first two terms of both asymptotic series expansions (Eqs. 2.5 and 2.7) lead to a formula of the same general form, namely

$$\sigma_N = \frac{Bf'_t}{\sqrt{1+\beta}}, \qquad \beta = \frac{D}{D_0} \qquad (2.8)$$

where D_0 is a constant called the *transitional size*, B is a dimensionless constant, and the tensile strength f'_t is introduced for reasons of dimensionality (it should however be pointed out that this is not asymptotic matching in a pure sense because the coefficients of both asymptotic expansion are not fixed numbers, known and separately determined a priori, but are adjusted so as to match one and the same formula).

The last formula is the energetic size effect law, which was derived initially by Bažant (1983, 1984) on the basis of simplified energy release arguments. The ratio β in this formula is called the *brittleness number* (Bažant 1987, Bažant and

Pfeiffer, 1987) because the case $\beta \to \infty$ represents a perfectly brittle behavior, and the case $\beta \to 0$ represents a perfectly nonbrittle (plastic, ductile) behavior. Because the constant D_0, representing the point of the intersection of the two asymptotes in Fig. 2.2, depends on structure geometry, this definition of brittleness number is not only size independent but also shape independent (which is not true of other brittleness definitions in the literature). The brittleness is understood as the proximity to LEFM scaling.

The asymptotic analysis can be made more general by considering function $\hat{g}$ or $\tilde{g}$ to be a smooth function of θ^r or η^r, rather than θ or η, where r is some constant. Furthermore, it is also possible that, for very large sizes, there is a transition to a ductile failure mechanism which endows the structure with an additional residual nominal strength, σ_r (this may, for example, happen in the Brazilian split-cylinder test, due to friction on sliding wedges under the platens). These modifications can be shown to lead to the following generalized formula:

$$\sigma_N = \sigma_P \left(1 + \beta^r\right)^{-1/2r} + \sigma_r \qquad (2.9)$$

in which σ_P = constant = small-size nominal strength. Exponent r is often more effective in approximating broad-range experimental results than adding higher-order terms of the series expansion. Eq. (2.9) allows close approximation of numerical results obtained by nonlocal finite element analysis of the cohesive crack model for a very broad size range, at least 1:1000. The optimum values of exponent r depend on geometry (e.g., $r = 0.44$ for standard three-point bend beams and 1.5 for a large center-cracked panel loaded on the crack).

It must be emphasized that the simple size effect formula (2.5) is not intended to cover all the possibilities. In fact, more general laws have been developed (for a review, see Bažant and Planas 1998, Ch. 9). The important point, however, is that an equation for the size effect that would be valid over the full size range $(0, \infty)$ is difficult, if not impossible, to derive merely by refining the asymptotic expansion (i.e., adding more terms) at only one end. The reason, of course, is that the radius of convergence may be limited and, even if it were infinite (which is hard to prove), the number of terms required for a good approximation far from the asymptote would become too large to be practical (for example, in the expansion $\exp(x^{-1}) = 1 + x^{-1} + x^{-2}/2! + x^{-3}/3! + ...$, the number of terms for small x that need to be taken to achieve good accuracy is $n \gg x^{-1}$; for $x = 0.01$, this means $n \gg 100$).

To sum up, the size effect formula (2.5) must be understood as an *asymptotic matching* formula. The crucial point to realize is that, in the plot of $\log \sigma_{Nu}$ versus $\log D$, all the physically admissible size effect formulae must approach for $D \to 0$ a horizontal asymptote.

2.3. Size Effect Law in Terms of LEFM Energy Release Function

The coefficients of the size effect law can also be expressed in terms of LEFM functions and material parameters, in the sense of an equivalent LEFM approximation. To this end, one may introduce the approximation $\hat{g}(\alpha_0, \theta) = g(\alpha_0 + \theta)$. With this approximation, which is asymptotically exact for large D, the size effect law corresponding to the asymptotic matching formula in Eq. (2.8) acquires the form:

$$\sigma_N = \sqrt{\frac{EG_f}{g'(\alpha_0)c_f + g(\alpha_0)D}} = Bf'_t \left(1 + \frac{D}{D_0}\right)^{-1/2} \qquad (2.10)$$

The parameters in the last size effect expression are:

$$D_0 = c_f \frac{g'(\alpha_0)}{g(\alpha_0)}, \qquad Bf'_t = \sqrt{\frac{EG_f}{c_f g'(\alpha_0)}} \qquad (2.11)$$

Note that the transitional size D_0, delineating the brittle behavior from nonbrittle behavior, is proportional to the effective size of the fracture process zone.

Further note that D_0 is proportional to the ratio g'/g, and that Bf'_t depends on g'. These functions depend only on the geometry of the structure, not on its size. Thus, the size effect law in Eq. (2.10) expresses not only the effect of size but also the effect of structure geometry (shape). It might better be called the *size-shape* effect law. This law can be applied to structures or specimens that are not geometrically similar.

2.4. Use of J-Integral for Asymptotic Scaling Analysis

Rice's J-integral allows the most fundamental derivation of the scaling law and lends itself naturally to a generalization for compressive fracture in which normal stresses are transmitted across the cracking band. However, there is a disadvantage. Unlike the LEFM energy release function $q(\alpha)$, the J-integral does not capture the shape (geometry) effect and thus does not provide a method to determine the coefficients in the size effect formulae.

We consider geometrically similar structures scaled in two dimensions (the treatment for three dimensions, however, would be analogous). We introduce dimensionless cartesian coordinates (Fig. 2.3) $\xi_i = x_i/D$ and dimensionless displacements $\zeta_i = u_i$ where $i = 1, 2$. For two-dimensional similarity, the elastic material compliances scale as $C_{ijkl} = c_{ijkl}/E$ where c_{ijkl} are constant.

28 Scaling of Structural Strength

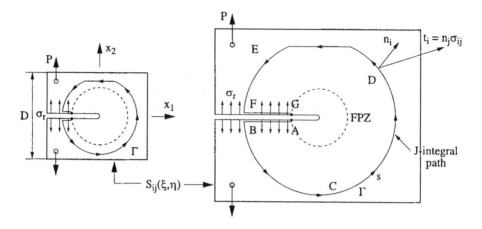

Figure 2.3: Path of J-integral used in size effect analysis

If c_f were zero, the stresses would scale as $\sigma_{ij} = \sigma_N S_{ij}(\boldsymbol{\xi})$ where $\boldsymbol{\xi} =$ coordinate vector of ξ_i, and S_{ij} are size-independent functions. However, the presence of nonzero material length c_f will influence the stress distributions. Based on the principles of dimensional analysis, this influence and the influence on the displacement field must have the form:

$$\sigma_{ij} = \sigma_N S_{ij}(\boldsymbol{\xi},\theta), \qquad u_i = (\sigma_N/E)D\zeta_i(\boldsymbol{\xi},\theta) \qquad (2.12)$$

where $\theta = c_f/D$ and ζ_i are dimensionless functions. The flux of energy into a fracture process zone advancing in the direction of x_1 (Fig. 2.3) can be calculated by Rice's J-integral:

$$\begin{aligned}
J &= \int_\Gamma (W n_1 - n_j \sigma_{ij} u_{i,1}) ds & (2.13) \\
&= \int_\Gamma \left(\frac{1}{2} C_{ijkl} \sigma_{ij} \sigma_{kl} n_1 - n_j \sigma_{ij} \frac{\partial u_i}{\partial x_1} \right) ds & (2.14) \\
&= \int_{\bar\Gamma} \left(\frac{\sigma_N^2}{2E} c_{ijkl} S_{ij} S_{kl} n_1 - n_j \sigma_N S_{ij} \frac{\sigma_N D}{E} \frac{\partial \zeta_i}{\partial \xi_1} \right) D d\bar s & (2.15) \\
&= \frac{\sigma_N^2 D}{E} \mathcal{J}(\theta), & (2.16) \\
\mathcal{J}(\theta) &= \int_{\bar\Gamma} \left(\frac{1}{2} c_{ijkl} S_{ij}(\boldsymbol{\xi},\theta) S_{kl}(\boldsymbol{\xi},\theta) n_1 - n_j S_{ij}(\boldsymbol{\xi},\theta) \frac{\partial \zeta_i(\boldsymbol{\xi},\theta)}{\partial \xi_1} \right) d\bar s & (2.17)
\end{aligned}$$

Here Γ are geometrically scaled closed integration contours BCDE (Fig. 2.3) with length coordinate s, starting and ending on the crack and passing outside the fracture process zone, $\bar\Gamma$ = chosen fixed contour in dimensionless coordinates, with length coordinate $\bar s$ ($ds = Dd\bar s$), n_i = unit normal to the contour (which does not change with scaling), W = strain energy density; $\mathcal{J}(\theta)$ is the dimensionless J-integral. This integral may be expanded in Taylor series, providing

$$\mathcal{J}(\theta) = \mathcal{J}_0 + \mathcal{J}_1\theta + \mathcal{J}_2\theta^2 + ... \tag{2.18}$$

$$\mathcal{J}_0 = \int_{\tilde{\Gamma}} [c_{ijkl}S^0_{ij}S^0_{kl}/2 - n_j S^0_{ij}\zeta^0_{i,1}]d\bar{s}, \tag{2.19}$$

$$\mathcal{J}_1 = \int_{\tilde{\Gamma}} [c_{ijkl}(S^0_{ij}S^0_{kl,\theta} + S^0_{kl}S^0_{ij,\theta})/2 - n_j(S^0_{ij}\zeta^0_{i,1\theta} + S^0_{ij,\theta}\zeta^0_{i,1})]d\bar{s} \tag{2.20}$$

Superscript 0 labels the values or fields evaluated for $\theta = 0$ (which is the case of LEFM). Substituting this into (2.16), truncating the series after the second term, and noting that, at failure, J must be equal to the fracture energy G_f of the material, one gets again the same size effect law as (2.10):

$$\sigma_N = \sqrt{\frac{2EG_f}{[\mathcal{J}_0 + \mathcal{J}_1(c_f/D) + \mathcal{J}_2(c_f/D)^2 + ...]D}} \approx \frac{Bf'_t}{\sqrt{1 + (D/D_0)}} \tag{2.21}$$

where $Bf'_t = \sqrt{2EG_f/\mathcal{J}_1 c_f}$ and $D_0 = c_f \mathcal{J}_1/\mathcal{J}_0$. The truncation leading to this formula is of course justified only if $\mathcal{J}$ is non-zero and non-negligible, which means that a notch of stress-free crack is assumed to exist at the outset.

The foregoing derivation has been simplified in the sense that the length parameter influencing J has been considered as a known constant. Although this seems a good approximation, one could more generally consider J to depend on c/D instead of c_f/D, where c is a variable crack extension. One could then also introduce a variable fracture resistance in the form of an R-curve, and impose the maximum load condition as the condition of the tangency of the R-curve to the energy release curve, in the same manner as used in equivalent LEFM analysis (Bažant 1986, Bažant and Planas 1998). The size effect law ensuing from such refined analysis is found to be the same.

For the case of LEFM, corresponding to the limit $\theta = c_f/D \to 0$, the foregoing J-integral analysis (which can be simplified) proves in general that the power law for stress scaling has the exponent $m = -1/2$.

2.5. Identification of Fracture Parameters from Size Effect Tests

One useful application of the size effect in Eq. (2.10) has proven to be the determination of the nonlinear fracture parameters of the material. To this end one must test a set of specimens with a sufficiently large range of the brittleness number β. The range depends on the degree of statistical scatter of the results. If the scatter is very small, a small range of β is sufficient, and if the scatter

30   Scaling of Structural Strength

Figure 2.4: Similar three-point bend specimens tested by Bažant and Pfeiffer (1987)

is very large, a large range of β is needed. For the typical scatter observed in concrete and many other materials, the minimum range of the brittleness number is 1:4, and preferably, for more accurate results, 1:8. The broader the range, the more accurate the results.

To achieve a sufficient range of brittleness numbers, one may test geometrically similar notched fracture specimens of sufficiently different sizes, as illustrated in Fig. 2.4. However, geometric similarity is not necessary, although the results for geometrically similar specimen are somewhat more accurate because the effect of the changes of geometry is described by Eq. (2.10) only approximately.

To determine the material fracture characteristics from the measured maximum loads of specimens of different brittleness numbers, one may rearrange Eq. (2.10) into a linear regression plot (Fig. 2.5):

$$Y = AX + C \quad \text{with} \quad Y = \frac{1}{g'(\alpha_0)\sigma_N^2}, \quad X = \frac{Dg(\alpha_0)}{g'(\alpha_0)} \tag{2.22}$$

evaluated at α_0. After identifying A and C by regression, the fracture characteristics are then obtained as

$$G_f = 1/AE, \quad c_f = C/A \tag{2.23}$$

In carrying out the regression, proper weighting of the data is helpful; see Bažant and Planas (1998). The optimum weighting is achieved by fitting the test data directly in the plot of $\log \sigma_N$ versus $\log D$ with the help of a nonlinear optimization algorithm such as Levenberg-Marquardt algorithm.

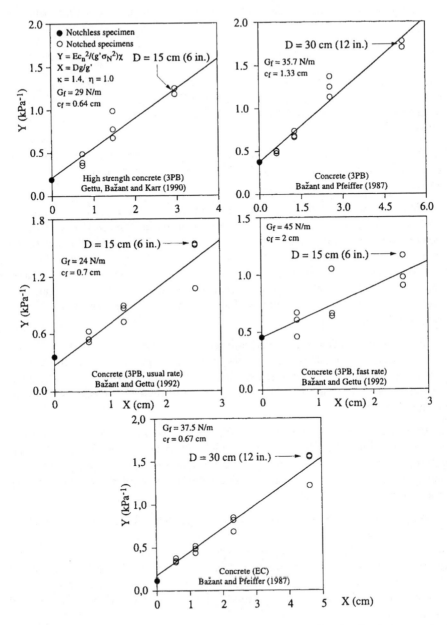

Figure 2.5: Linear regressions (according to the size effect law) of the nominal strength values of notched concrete specimens measured by Bažant and Pfeiffer (1987), Bažant and Gettu (1992) and Gettu et al. (1990)

From G_f and c_f, one can also obtain the critical crack-tip opening displacement

$$\delta_{CTOD} = (1/\pi)\sqrt{8G_f c_f/E} \qquad (2.24)$$

(Bažant and Gettu 1990; Bažant 1995d). The fracture parameter δ_{CTOD} was introduced in the early 1960s in the models of Wells (1961) and Cottrell (1963) for metals, and was co-opted for a similar model for concrete by Jenq and Shah (1985).

The size effect method has been adopted as a standard recommendation for concrete fracture testing by RILEM (1990). A proposal to incorporate it in a new ASTM standard is pending.

There is another definition of fracture energy, introduced for concrete by Hillerborg. It defines the fracture energy, denoted as G_F, as the area under the complete softening stress-displacement curve of the cohesive (or fictitious) crack model. G_F is deduced from the area under the measured complete load-deflection curve of notched fracture specimens of one size, sufficiently large. This is called the *work-of-fracture method* (Nakayama 1965; Tattersall and Tappin 1966), which was pioneered for concrete by Hillerborg et al. (1976) (see also Hillerborg 1985a,b).

Measurement of G_F is, however, quite uncertain. One reason is the difficulty in measuring the far-off tail of the load-deflection diagram. This difficulty is doubtless the explanation why the statistical data on G_F exhibit a much higher coefficient of variations that those on G_f (see the review in Bažant and Becq-Giraudon 2001).

From test results, Planas et al. (1992), Elices et al. (1992) and Guinea et al. (1992) deduced that $G_F \approx 2.5 G_f$. This is explained by the finding that G_f, determined from the size effect on the maximum load, corresponds to the area under the initial tangent of the stress-displacement diagram of the cohesive crack model, and not to the complete area under the softening curve.

When the values of material fracture parameters are determined by a method that is not based on the size effect, one faces the question of spurious size dependence of these values. The fracture energy G_F determined from the area under the measured load-deflection diagram has been found to depend on the size of the specimen (Bažant 1996a, 1995d; Bažant and Kazemi 1991b). Methods to eliminate this dependence were pointed out by Planas and Elices (Bažant and Planas 1998).

As mentioned, the specimens may differ in both size and shape. Then it is convenient to write the size (or size-shape) effect law as $\sigma_i = [E'G_f/(g_i'c_f + g_i D_i)]^{1/2}$ where integer subscript $_i$ refers to specimen number i and the following abbreviated notations are introduced: $\sigma_i = \sigma_{Ni}$, $g_i = g_i(\alpha_{0i})$ and

$g'_i = g'_i(\alpha_{0i}) = dg_i(\alpha)/d\alpha$ at $\alpha = \alpha_{0i}$. D_i is the size of specimen i; and $g_i(\alpha) = k_i^2(\alpha)$ = dimensionless energy release function of specimen i, $k_i(\alpha)$ being the dimensionless stress intensity factor. Least-square fitting of the foregoing formula to the measured σ_i values provides the values of G_f and c_f. This is a nonlinear problem which can be easily handled, for instance, by applying the standard library subroutine for the Levenberg-Marquardt optimization algorithm. However, conversion to a linear regression problem can be achieved by rearranging the formula as follows:

$$F_i(G_f, c_f) = g_i D_i + g'_i c_f - E' G_f \sigma_i^{-2} = 0 \qquad (2.25)$$

where $F_i(G_f, c_f)$ is a function of G_f and c_f. The right-hand side is zero only for theoretically perfect data. In practice it is nonzero and its square should be minimized; $\Phi = \sum_{i=1}^{N} F_i^2(G_f, c_f) = \min$. The minimizing conditions are $\partial \Phi / \partial G_f = 0$ and $\partial \Phi / \partial c_f = 0$. They provide the following linear regression equations for the unknowns G_f and c_f:

$$A_{11} G_f + A_{12} c_f = C_1,$$
$$A_{21} G_f + A_{22} c_f = C_2 \qquad (2.26)$$

in which

$$A_{11} = E'^2 \sum_i \sigma_i^{-4}, \quad A_{12} = A_{21} = -E' \sum_i g'_i \sigma_i^{-2}, \quad A_{22} = \sum_i g_i'^2 \qquad (2.27)$$
$$C_1 = E' \sum_i g_i D_i \sigma_i^{-2}, \quad C_2 = -E' \sum_i g_i g'_i D_i \qquad (2.28)$$

Multiplying the expression for F_i by σ_i or σ_i^2, or dividing it by g_i or g'_i, one can obtain different linear regression equations. For perfect data, they would of course give the same results but, because of the scatter in the real data, the results are slightly different. The reason is that these different regressions imply different weights for different i. Compared to the case with the expression for F_i multiplied by σ_i^2 (or divided by g_i), the regression defined by Eq. (2.27) and Eq. (2.28) gives higher weights for the domain of larger specimen sizes and not too short notches. This is desirable because the experimental scatter in that domain is generally lower.

2.6. Validation by Fracture Test Data and Numerical Simulation

Fig. 2.6 shows the comparison of the size effect law with the data points obtained in the testing of Indiana limestone, carbon-epoxy fiber composites, silicone oxide ceramic and sea ice. The data for sea ice, obtained by Dempsey et al. (1995, 1999) cover an unprecedented, large size range (also Mulmule et al. 1995). In Dempsey's tests, floating notched square specimens of sea ice of

34   Scaling of Structural Strength

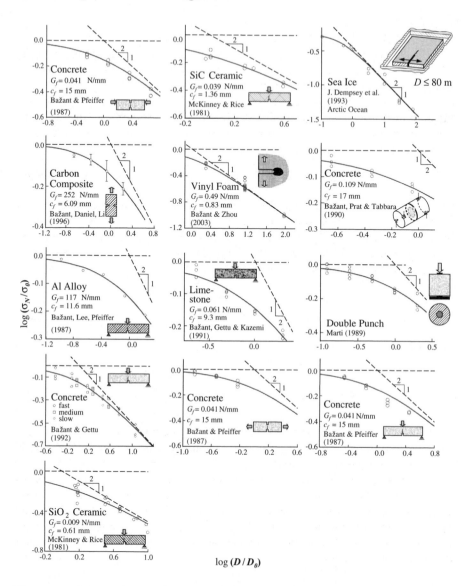

Figure 2.6: Nominal strength data and their fits by size effect law for notched specimen tests of carbon-epoxy composites (Bažant, Daniel and Li 1996), sea ice (Dempsey et al. 1999; record size range, up to 80 × 80 × 1.8 m), SiC and Si)O$_2$ ceramics (McKinney and Rice 1981); Indiana limestone (Bažant, Gettu and Kazemi 1991), vinyl foam Divinycell 100 (Bažant and Zhou 2003), normal concrete (Bažant and Pfeiffer 1987, Bažant, Prat and Tabbara 1990) and tough aluminum alloy (Bažant, Lee and Pfeiffer 1987); all tests are in mode I except the cylinder test, which is in mode III

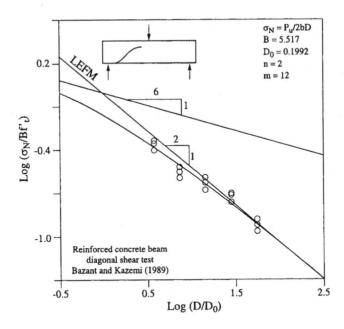

Figure 2.7: Nominal strength data from Bažant and Kazemi's (1991a) tests of diagonal shear failure of reduced-scale concrete beams with longitudinal reinforcement (of size range 1:16), their fit by the size effect law, and comparison with prediction of statistical Weibull-type theory (with m=12)

sizes from 0.5m to the record size of 80m (and thickness 1.8m) were tested *in situ* in the Arctic Ocean (West of Resolute, Cornwallis Island).

Dempsey et al.'s results (1999) revealed a very strong size effect, closely approaching the LEFM asymptote for sizes in excess of $3m$. These results indicated a high brittleness of sea ice at large scales. They led to a complete revision of the previous widespread conviction, based on the testing of small laboratory tests, that sea ice was non-brittle, notch-insensitive, and free of size effect. These results apparently explain why the forces measured exerted on an oil platform by an ice floe, typically several kilometers in size, are usually two orders of magnitude less than those predicted by means of plastic analysis or strength theory. They also confirm that fracture mechanics, which inevitably involves size effect, ought to be used in predicting the load capacity of the floating ice plate in the Arctic.

Fig. 2.7 illustrates the comparison with the size effect law for data obtained on specimens without notches (tests of diagonal shear failure of geometrically similar reinforced concrete beams by Bažant and Kazemi (1991a), with size range 1:16). Fig. 2.8 shows a comparison of the size effect law with data

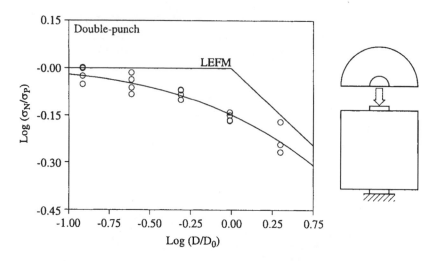

Figure 2.8: Nominal strength data from Marti's (1989) tests of double punch failure of concrete cylinders (of size range 1:16), and their fit by size effect law

obtained on unnotched and unreinforced specimens (cylinders in double-punch loading, size range 1:16; Marti 1989).

The size effect law also closely agrees with the results of finite element analysis using the nonlocal damage concept (e.g., Fig. 2.9, Ožbolt and Bažant 1996), the crack band model (see the curves in Fig. 1.3, Bažant and Oh 1983, Bažant and Lin 1988a , or the cohesive crack model (Bažant and Li 1995b). Furthermore, the size effect law was shown to approximately agree with the mean trend of maximum load values calculated by the discrete element method (random particle simulation of concrete, Bažant, Tabbara et al. 1990, and of sea ice Jirásek and Bažant 1995a, 1995b, Fig. 2.10).

There are nevertheless some instances in which the simple size effect law in Eq. (2.8) or (2.10) is insufficient because the logarithmic size effect plot of the data exhibits a positive curvature, as illustrated in Fig. 2.11. This is for example observed for the Brazilian split cylinder test. The cause seems to be that, for a very large structure, the load to produce the diagonal cracks in a cylinder becomes negligible but failure cannot occur because the wedge regions under the load must slide frictionally, which imposes a certain residual strength σ_r. Another reason may be that the crack length at failure ceases to increase in proportion to the specimen size. Such data can be well described by the generalized size effect law in Eq. (2.9) in which D_0 is very small, smaller than the smallest D in the data set (see Fig. 2.11).

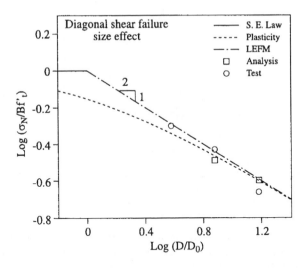

Figure 2.9: Nominal strength values obtained by finite element analysis using the nonlocal model with crack interactions (Ožbolt and Bažant 1996) compared to test data of Bažant and Kazemi (1991a) for diagonal shear failure of longitudinally reinforced concrete beams and to the size effect law (dashed curve)

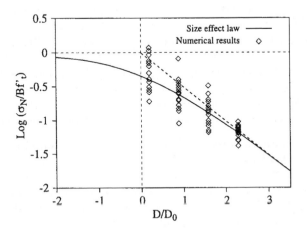

Figure 2.10: Nominal strength values obtained by discrete element method (random particle simulation of the specimens shown) and their comparison with size effect law, exploited for determining the fracture characteristics of the random particle system (Jirásek and Bažant 1995a,b)

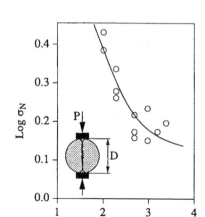

Figure 2.11: Nominal strength data from Brazilian split-cylinder tests of Hasegawa, Shioya and Okada (1985) and their fit by the size effect law with residual strength in Eq. (12)

2.7. Size Effect for Crack Initiation via Energy Release

Aside from notched fracture test specimens, the foregoing analysis applies only to structures that contain at maximum load a large stably grown crack that is traction-free (a state that may be produced in cohesive materials by previous fatigue due to repeated loads). Formation of a large crack before the maximum load is reached is typical for quasibrittle materials. It is in fact the objective of a good design because the large stable crack growth endows the structure with a large energy dissipation capability and a certain measure of ductility. For example, the objective of reinforcing concrete structures, of toughening ceramics, of putting fibers in composites, etc., may be seen in the attainment of the capacity of a structure to grow large cracks prior to failure.

In some situations, however, quasibrittle structures fail at crack initiation. For example, this happens for a plain concrete beam. This nevertheless does not mean that the fracture process zone size would be negligible. Because of heterogeneity of the material, the process zone size is still quite large, as illustrated in Fig. 2.12 (top left). The maximum load is obtained typically when this large cracking zone coalesces into a continuous crack capable of further growth. Because a large cracking zone forms in the boundary layer prior to the maximum load, one cannot expect the Weibull theory to be applicable, as will be explained in Chapter 3.

As described in detail in Bažant (1995d), the failure at crack initiation from a smooth surface can also be analyzed on the basis of the expansions in Eq.

Asymptotic Analysis of Size Effect 39

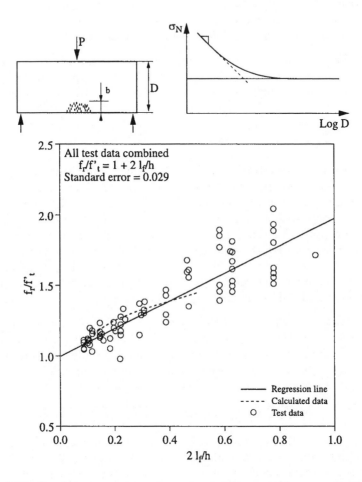

Figure 2.12: Cracking zone at maximum load P in a notchless quasibrittle specimen (top left); law of the size effect for quasibrittle failures at crack initiation (Bažant and Li 1995b, 1996) (top right); and use of this law in linear regression of test data for concrete obtained in eight different laboratories (Bažant and Li 1994b) (bottom)

(2.5) or (2.7), however, with one modification. Since the expansions are made with respect to a zero value of the relative process zone size $\theta = c_f/D$, the argument of the energy release function $g(\alpha)$ is $\alpha = 0$. This means that the energy release rate $g(\alpha) = g(0) = 0$, and so the first term of the large-size expansion in Eq. (2.5) vanishes. Thus, if the series were truncated after the second term, as before, no size effect would be obtained.

Therefore, one must in this case also include the third term of the large-size asymptotic expansion. This leads to the following approximation for the nominal strength of structures failing at crack initiation from a smooth surface:

$$\sigma_N = \sqrt{\frac{EG_f}{g'(0)c_f + \frac{1}{2}g''(0)c_f^2 D^{-1}}} \tag{2.29}$$

A rather general derivation of Eq. (2.29) can alternatively be given on the basis of energetic aspects of fracture mechanics. Using the approach of equivalent linear elastic fracture mechanics (LEFM), one can approximate a cracked structure with a large fracture process zone by a structure with a line crack whose tip is placed approximately in the middle of the fracture process zone (the exact location could be determined from a condition of compliance equivalence). The fracture process zone at maximum load, having depth $2c_f$, is attached to the tensile face of beam, and so the actual crack length is $a_0 = 0$. The equivalent LEFM crack length is $a \approx a_0 + c_f = c_f$.

As stated before, equivalent LEFM in general yields for the nominal strength σ_N of the structure the general expression $\sigma_N = \sqrt{EG_f/Dg(\alpha_0 + c_f/D)}$ where $\alpha_0 = a_0/D$. The function g is sufficiently smooth to allow expansion into a Taylor series in terms of c_f/D, which represents an asymptotic expansion:

$$\sigma_N = \sqrt{\frac{EG_f}{D[g(\alpha_0) + g'(\alpha_0)(c_f/D) + \frac{1}{2!}g''(\alpha_0)(c_f/D)^2 + ...]}} \tag{2.30}$$

Note that Eq. (2.30) describes not only the size effect but also the shape effect, which is embedded in the LEFM function $g(\alpha)$. Because the energy release rate for a zero crack length is zero, i.e. $g(0) = 0$, the first term of the series expansion in Eq. (2.30) vanishes, and so the series must be truncated no earlier than after the third, quadratic term. This yields the asymptotic expansion:

$$\sigma_N = f_r = \lim_{\alpha_0 \to 0} \sqrt{\frac{EG_f}{Dg(\alpha_0 + \frac{c_f}{D})}}$$

$$= \sqrt{\frac{EG_f}{g'(0)c_f + \frac{1}{2!}g''(0)c_f^2 D^{-1} + \frac{1}{3!}g'''(0)c_f^3 D^{-2} + ...}} \tag{2.31}$$

$$= \frac{f_{r\infty}}{\sqrt{1 - (q_1/D) + (q_2/D)^2 - (q_3/D)^3 + ...}} \tag{2.32}$$

in which the nominal strength σ_N is now represented by the modulus of rupture f_r, and

$$f_{r\infty} = \sqrt{\frac{EG_f}{c_f g'(0)}}, \quad q_1 = c_f \frac{-g''(0)}{2! g'(0)}, \quad q_2^2 = c_f^2 \frac{g'''(0)}{3! g'(0)}, \dots \quad (2.33)$$

The objective is not merely a large-size asymptotic approximation but also, and mainly, a generally applicable approximate formula of the asymptotic matching type that has admissible behavior also at the opposite infinity (ln $D \to -\infty$ or $D \to 0$) and provides a smooth interpolation between the opposite infinities. The asymptotic behavior of Eq. (2.32) for $D \to 0$ is not acceptable because it yields an imaginary value. To get a proper asymptotic matching formula, Eq. (2.32) must be modified in such a manner that at least the first two terms of the asymptotic expansion of σ_N in terms of $1/D$ remain unchanged. This modification can be accomplished as follows. Eq. (2.32) may be rewritten as

$$f_r = f_{r\infty} \left[(1-x)^{-r/2} \right]^{1/r} \quad (2.34)$$

where r is an arbitrary positive constant (which is related to the third term in the expansion of function $g(c_f/D)$), and

$$x = (q_1/D) - (q_2/D)^2 + (q_3/D)^3 - \dots \quad (2.35)$$

Then, according to the binomial series expansion

$$f_r = f_{r\infty} \left[1 + \binom{-r/2}{1}(-x) + \binom{-r/2}{2}(-x)^2 + \dots \right]^{1/r} \quad (2.36)$$

$$= f_{r\infty} \left[1 + \frac{r}{2}x + \frac{r(r+2)}{8}x^2 + \dots \right]^{1/r} \quad (2.37)$$

$$= f_{r\infty} \left[1 + \frac{r}{2}\frac{q_1}{D} + r\left(\frac{r+2}{8}q_1^2 - \frac{1}{2}q_2^2\right)\frac{1}{D^2} + \dots \right]^{1/r} \quad (2.38)$$

In contrast to Eq. (2.32), this formula is admissible for $D \to 0$; it gives for f_r a real rather than imaginary limit value. The feature that $f_r \to \infty$ is shared by the famous, widely used Petch-Hall formula for the yield strength of polycrystalline metals. One might prefer a finite limit for f_r but this does not matter in practice because D cannot be less than about three maximum aggregate sizes (as the material could no longer be treated as a continuum). The limit $D \to 0$ is an abstract extrapolation.

Keeping just the first two terms, we obtain from Eq. (2.38) the final deterministic energetic size effect formula:

$$\sigma_N = f_r = f_{r\infty} q(D), \quad q(D) = \left(1 + \frac{rD_b}{D}\right)^{1/r} \quad (2.39)$$

where $q(D)$ is a positive dimensionless decreasing function of size D having a finite limit for $D \to \infty$, and

$$D_b = \left\langle \frac{-c_f g''(0)}{4g'(0)} \right\rangle \qquad (D_b > 0) \qquad (2.40)$$

As will be shown later, D_b roughly represents twice the thickness of the boundary layer of cracking. In the last expression, the signs $\langle .. \rangle$, denoting the positive part of the argument, have been inserted [$\langle X \rangle = \text{Max}(X, 0)$]. The reason is that $g''(0)/g'(0)$ can sometimes be positive, in which case there is no size effect, and this is automatically achieved by setting $D_b = 0$. In the modulus of rupture test, $g''(0)/g'(0) < 0$ and $D_b > 0$. Note that for uniform tension (zero stress gradient, as in the direct tensile test) there is no deterministic size effect according to Eq. (2.40) because $g''(0) = 0$ or $D_b = 0$.

The special case of Eq. (2.39) for $r = 1$ (derived in Bažant, 1995) reads:

$$\sigma_N = f_r = f_{r\infty}\left(1 + \frac{D_b}{D}\right) \qquad (2.41)$$

The case $r = 2$, derived in Bažant (1998), yields the formula:

$$f_r = \sigma_N = \sqrt{A_1 + \frac{A_2}{D}} \qquad (2.42)$$

in which

$$A_1 = f_{r\infty}^2 = \frac{EG_f}{[c_f g'(0)]^2}, \qquad A_2 = f_{r\infty}^2 q_1 = 2 f_{r\infty} D_b = -\frac{EG_f g''(0)}{2c_f [g'(0)]^3} \qquad (2.43)$$

Formula (2.42) was proposed and used to describe some size effect data by Carpinteri et al. (1994, 1995). These authors named this formula the 'multifractal' scaling law (MFSL) and tried to justify it by fracture fractality using, however, strictly geometric (non-mechanical) arguments. The term 'multifractal', though, seems questionable because, as shown in Bažant (1997b and 1997c), the mechanical analysis of fractality leads to a formula different from Eq. (2.42) (this is the case whether one considers the invasive fractality of the crack surface or the lacunar fractality of microcrack distribution in the fracture process zone). No logical mechanical argument for the size effect on σ_N to be a consequence of the fractality of fracture has yet been offered.

2.8. Stress Redistribution Caused by Boundary Layer of Cracking

The size effect formulae (2.41) and (2.39) can also be derived by considering the stress redistribution caused by a finite size zone of distributed cracking, characterized by strain softening (Fig. 2.13 left). The simplest way is illustrated by the bending stress diagram of a beam in Fig. 2.13 (middle). We may assume that failure is not decided by the maximum elastically calculated stress occurring at the bottom face of the beam, but by the average elastically calculated stress $\bar{\sigma}$ in a boundary layer having the thickness D_b, which is

$$\bar{\sigma} = \frac{M}{I}\left(\frac{D}{2} - \frac{D_b}{2}\right) \quad (2.44)$$

where D = beam depth, M = bending moment, and I = centroidal moment of inertia of cross section. Noting that the modulus of rupture f_r, representing the nominal strength, is defined as $f_r = \sigma_N = MD/2I$, we have

$$\sigma_N = f_{r\infty}\left(1 - \frac{D_b}{D}\right)^{-1} \quad (D \gg D_b) \quad (2.45)$$

According to the asymptotic expansion in terms of powers of $1/D$, we may make the replacement $(1 - D_b/D)^{-1} \approx (1 + rD_b/D)^r$ (r being any positive constant), and we see that this leads immediately to the size effect expression in Eq. (2.39). The fact that the result is identical, validates retroactively our simple hypothesis that the average elastic stress in the boundary layer of a fixed thickness D_b is what matters.

Clearly, this hypothesis becomes meaningless if $D < D_b$, and unrealistic if D is close to D_b. So, our hypothesis can have only asymptotic validity for $D_b/D \to 0$. However, the size effect formula we obtained is not unreasonable even for small sizes D, such that $D \geq D_b$ (it makes hardly any sense to consider beams shallower than D_b).

A more realistic hypothesis is to consider that, up to the maximum load, the cracking remains distributed (the discrete crack being formed only at, or after, the maximum load), and that the cracking is described by a bilinear stress-strain diagram with post-peak strain-softening characterized by tangent modulus E_t. The distributed cracking at maximum load is assumed to occupy a boundary layer of a certain fixed thickness denoted as l_f. The corresponding stress distribution is sketched in Fig. 2.13 on the left. The result of such a calculation (Bažant and Li 1995, Bažant and Planas 1998) is a formula that coincides with (2.41) or (2.39) up the second term of the asymptotic expansion of σ_N as a power series in $1/D$, provided that one sets $l_f = D_b/2$. Differences occur only in the third and higher terms. This coincidence lends further support to the hypothesis underlying equation (2.44), namely that the average elastic stress in layer D_b can be considered to be $f_{r\infty}$. It also shows that the thickness of the boundary layer of cracking is about $D_b/2$.

44   Scaling of Structural Strength

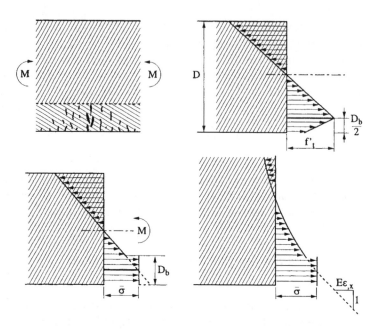

Figure 2.13: Stress redistribution in the boundary layer of distributed cracking and its approximations.

To achieve greater flexibility in the modeling of test data for small sizes, equation (2.39) may further be generalized as

$$\sigma_N = f_{r\infty} q(D), \qquad q(D) = \left(\frac{D + r(s+1)D_b}{D + rsD_b}\right)^{1/r} \qquad (2.46)$$

where s is a non-negative constant and the dimensionless decreasing positive function $q(D)$ generalizes the function originally introduced in Eq. (2.39). This formula is, for large sizes, again asymptotically equivalent to the original formula (2.45) as well as (2.41) up to the second term of expansion in $1/D$. One can verify it by the following approximations, accurate up to the second term of the asymptotic power series in ξ, with $\xi = D_b/D$;

$$\frac{f_{r\infty}}{\sigma_N} = \left(\frac{1 + rs\xi}{1 + r(s+1)\xi}\right)^{1/r} \approx \frac{1 + s\xi}{1 + (s+1)\xi} \qquad (2.47)$$

or

$$\frac{f_{r\infty}}{\sigma_N} \approx (1 + s\xi)\left[1 - (s+1)\xi\right] \approx 1 - \frac{D_b}{D} \qquad (2.48)$$

For a certain sufficiently small size D_{pl}, the modulus of rupture f_r should agree with the prediction of plastic analysis, for which one may assume the stress distribution to be uniform (rectangular) throughout the whole cross section and the bending moment M to be balanced by this stress distribution together with a compressive force at the extreme fiber. For a rectangular cross section, one can deduce from this assumption that $M = f_{r\infty}bD^2/2$ while, by definition of the modulus of rupture, $M = f_r bD^2/6$. Equating these two expressions, one gets $f_r/f_{r\infty} = 3$ for a fully plastic state of a rectangular cross section (with unbounded compression strength). This relation can be used to calibrate the value of s in (2.46), but the question is the proper value of D_{pl}.

According to computational experience with the crack band model as well as the studies of the existing test data on the modulus of rupture of concrete, it appears that the thickness of the boundary layer of cracking is about $2d_a$, d_a being the maximum aggregate size. This means that $D_b \approx 4d_a$ and that plastic behavior is reached for beam depth $D_{pl} \approx 2d_a$, which is just about the shallowest beam that can be cast. Therefore $D_b/D = D_b/D_{pl} \approx 2$. Knowing that $f_r/f_{r\infty} = 3$, we may solve s from Eq. (2.46). It so happens that the solution is $s = 0$.

This suggests that the simpler formula (2.39) ought to be adequate for matching both the brittle large-size asymptotics and the small-size plastic limit, the two basic situations that are easy to analyze. Formula (2.46) nevertheless offers greater freedom, which would allow closer adjustment to test data of a very broad size range once they become available.

2.9. Strain Gradient Effect on Failures at Crack Initiation

The foregoing formulae for size effect at crack initiation from a smooth surface should apply for not only to beams but any solids. For such general situations, they need to be rewritten in terms of gradient $\epsilon_{,n}$ at surface where ϵ = normal strain parallel to the surface and subscript n refers to a derivative in the direction normal to the surface (Fig. 2.13 right).

Since only $\epsilon_{,n}$ matters for stress redistributions near the surface, according to our hypothesis, we need to determine the values of M and D that give in a beam the given $\epsilon_{,n}$. In beam bending, $\epsilon_{,n} = M/EI$, and for failure at the large size limit we have $MD/2I = f_{r\infty}$. Elimination of M from these two expressions gives the relation

$$D = \frac{2f_{r\infty}}{E\epsilon_{,n}} \qquad (2.49)$$

which is second-order accurate asymptotically (here D must be considered as the effective structure size corresponding to the strain gradient); Eq. (2.49) may then be substituted into Eqs. (2.41) or (2.39), or any other among the foregoing formulae for crack initiation.

The most general formula (2.46) thus leads to

$$\sigma_N = f_{r\infty}\left(\frac{2f_{r\infty} + r(s+1)D_b E\langle\epsilon_{,n}\rangle}{2f_{r\infty} + rsD_b E\langle\epsilon_{,n}\rangle}\right)^{1/r} \quad \text{(for } \epsilon > 0\text{)} \quad (2.50)$$

where the value $s = 0$ can probably be used; $\epsilon_{,n}$ is defined as positive if the strain decreases away from the surface, and $\langle.\rangle$ has been introduced to denote the positive part of the argument because tensile failure does not start at the surface when $\epsilon_{,n} < 0$.

The simplest formula (2.41), corresponding to r=1 and s=0, leads to

$$\sigma_N = f_{r\infty}\left(1 + \frac{D_b}{2f_{r\infty}}E\langle\epsilon_{,n}\rangle\right) \quad (2.51)$$

The last two formulae can be used for approximate strength estimation in finite element codes that do not take strain-softening damage into account.

2.10. Universal Size Effect Law

The analysis we have outlined so far yields:

1. the large size expansion of the size effect for long cracks,
2. the small size expansion for long cracks, and
3. the large size expansion for short cracks, while
4. the small size expansion for short cracks could also be obtained.

The question now is whether these expansions could be interpolated, or matched, so as to yield a single formula approximating the intermediate situations and matching all the asymptotic cases. Considering short cracks (crack initiation) according to Eq. (2.39), Bažant (1996a,d) derived the size effect law:

$$\sigma_N = \sigma_0\left(1 + \frac{D}{D_0}\right)^{-1/2}\left\{1 + \left[\left(\bar{\eta} + \frac{D}{D_0}\right)\left(1 + \frac{D}{D_0}\right)\right]^{-1}\right\} \quad (2.52)$$

in which $\bar{\eta}$ and σ_0 are empirical constants. The plot of this formula, which could be called the *universal size effect law*, is shown in Fig. 2.14 (note that the discontinuity of slope on top left of the surface is due to expressing D_b, for the sake of simplicity, in terms of the positive part of the derivative of

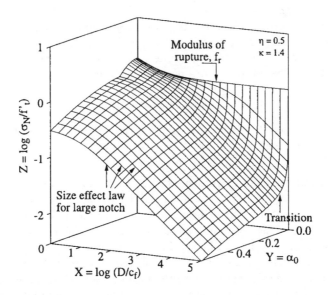

Figure 2.14: Universal size effect law for failure both at crack initiation and after large crack growth (Bažant 1996a,d)

function g; this slope discontinuity could be avoided, but at the expense of a more complicated formula).

The foregoing universal size effect law can probably be exploited for the testing of material fracture parameters (Bažant and Li, 1996; Bažant and Planas 1998). It should allow extending the size effect method to specimens of only one size, provided that both notched and notchless are used in the test series. In that case, it is possible to obtain, for the typical random scatter in concrete testing, a sufficient range of brittleness number β (more than 1:4) without varying the specimen size. On the other hand, if unnotched specimens are not included in the test series, it appears impossible achieve a range broader than about 1:2.7, just by varying the notch length, and this prevents obtaining accurate results.

For the purpose of data fitting, Eq. (2.52) may be reduced to a series of nonlinear regressions (Bažant and Li, 1996). The linear regression plots for some previously reported test data are shown in Fig. 2.5, for which we have already discussed the empty data points that correspond to notched specimens of different sizes. The solid data points correspond to unnotched specimens.

The fact that the solid points are approximately aligned with the trend of the empty data points confirms the approximate applicability of the universal size effect law in Fig. 2.14. Obviously, it is possible to delete the empty data points for specimens of all sizes except the largest and obtain about the same results using only the data points for the notched specimen of the largest size and the unnotched specimen of the same size. This approach may simplify the determination of material fracture parameters from test data.

2.11. Asymptotic Scaling and Interaction Diagram for the Case of Several Loads

The asymptotic analysis presented in Sec. 2.1 may be easily extended to the case of several loads P_i, characterized by nominal stresses $\sigma_i = P_i/bD$. The energy release rates of the individual loads are not additive, but the stress intensity factors of the individual loads, K_{Ii}, are. Therefore, by superposition, $\sum_i \sigma_{Ni}\sqrt{Dg_i(\alpha)} = \sqrt{EG_f}$ where $g_i(\alpha)$ are the dimensionless energy release rates corresponding to loads P_i. The condition of stability limit (tangency of the total energy release rate curve to the R-curve) gives for the maximum load the relative crack length $\alpha = \alpha_m(\alpha_0, \theta)$ (this argument is similar to that which led to Eq. (2.4) but we now consider functions g_i right away as functions of one variable, rather than introducing such a simplification at the end). Inserting this value into the last relation and expanding functions $g_i(\alpha) = g_i(\alpha_0 + \theta)$ into a Taylor series in terms of θ about point $\theta = 0$, we have $g_i(\alpha_0 + \theta) = g_i(\alpha_0) + g'(\alpha_i)\theta + \frac{1}{2}g_i''(\alpha_0)\theta^2 + \ldots$.

For the case of a large crack, we may truncate this series after the second (linear) term. We also consider a positive geometry (i.e. $g_i'(\alpha_0) > 0$ for all i). Furthermore, we may set $\sigma_{Ni} = \mu\sigma_{Di}$ where σ_{Di} are the given (fixed) design loads and μ = safety factor. After rearrangements:

$$\mu = \sqrt{EG_f}\,(\rho_1\sigma_{D1} + \rho_2\sigma_{D2} + \ldots + \rho_n\sigma_{Dn})^{-1/2} \qquad (2.53)$$

$$\rho_i = \sqrt{g_i(\alpha_0)D + g_i'(\alpha_0)c_f}, \quad (i = 1, 2, \ldots n) \qquad (2.54)$$

This equation gives the size effect, as well as the geometry effect, for the case of a large crack. At the same time, it may be regarded as the interaction diagram (failure envelope) for many loads. If α_0 is constant, these interaction diagrams are linear for any given size D (Fig. 2.15 right). But in other than notched fracture specimens of positive geometry, α_0 is not fixed and it is of course possible for α_0 (the traction-free crack length at P_{max}) to depend on the ratios $P_2/P_1, P_3/P_1, \ldots$; then the interaction diagrams are not linear.

For the case of macroscopic crack initiation from a smooth surface, we have $g_i(0) = 0$. Therefore, similar to the case of one load, the series expansions cannot be truncated after the linear term. We may truncate them after the quadratic terms. A similar procedure as in Sec. 2.5 then yields for μ the same

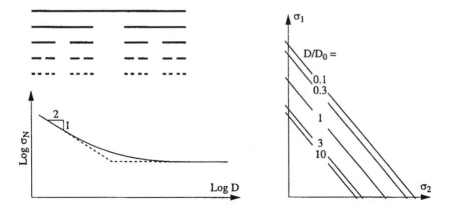

Figure 2.15: Top left: Lines of microcracks as lacunar fractals, at progressive refinements; bottom left: 'MFSL' law proposed by Carpinteri et al. (1995a,b); right: interaction diagrams for different size structures with two loads and constant α_0

expression as (2.53), but with

$$\rho_i = \sqrt{g_i'(0)c_f + \frac{1}{2}g_i''(0)\frac{c_f^2}{D}} \qquad (2.55)$$

Equations (2.53) and (2.55) represent the large-size asymptotic approximations of size effect. Small-size asymptotic approximations for the case of many loads can be derived similarly, replacing the variable θ with $\eta = 1/\theta$.

Similar to the case of one load, it is further possible to find, for the case of many loads, a universal size effect law that has the correct asymptotic properties for large and small sizes and large cracks or crack initiation. It is analogous to (2.52) and may again be written in the form of (2.53) but with

$$\rho_i = r_i \left[1 + \left(\frac{D_{0i}}{D}\right)^r\right]^{1/2r} \left\{1 + s\left[\left(\bar{\eta} + \frac{D}{D_{hi}}\right)\left(1 + \frac{D}{D_{0i}}\right)\right]^{-1}\right\}^{-1/s} \qquad (2.56)$$

$$r_i = [c_f g_i'(\alpha_0)]^{-1/2} \qquad (2.57)$$
$$D_{0i} = c_f g_i'(\alpha_0)/g_i(\alpha_0) \qquad (2.58)$$
$$D_{bi} = c_f \langle -g_i''(\alpha_0)\rangle/4g_i'(\alpha_0) \qquad (2.59)$$

Here r, s and $\bar{\eta}$ are empirical constants whose values may probably be taken as 1 for many practical purposes.

2.12. Size Effect on Approach to Zero Size

For structure sizes smaller than the size of a fully developed fracture process zone, the use of the cohesive crack model is simpler and doubtless more realistic than our previous use of the LEFM functions. Let us now apply this popular model to determine the first two terms of the small-size asymptotic expansion of size effect. Let the coordinates be positioned so that the origin is at the front of the cohesive crack and the crack lies in the plane (x_1, x_3), and let the dimensionless coordinates and variables be defined as in (1.8), Section 1.7. The boundary value problem in the dimensionless form is defined by (1.10) and (1.11). This must be supplemented by two conditions for the cohesive crack: (1) The dimensionless total stress intensity factor $\bar{K}_t = K_t \sqrt{D}/\sigma_N$ produced jointly by the applied load and the stresses $\bar{\sigma} = \bar{\sigma}_{22}$ acting on the crack faces must vanish, $\bar{K}_t = 0$ (which is necessary to ensure the finiteness of the crack-tip stress); and (2) the cohesive (crack-bridging) stresses must satisfy the softening law of the cohesive crack, i.e., the curve relating σ to the opening displacement $w = 2u_2$ on the crack plane. We will consider the law $\sigma = \sigma_0[1-(w/w_f)^p]$ where p, w_f = positive constants, and σ_0 = tensile strength (also denoted as f'_t). In terms of the dimensionless variables corresponding to (1.8), the dimensionless form of the assumed softening law is

$$\bar{\sigma} = 1 - (\bar{D}\bar{w})^p \quad \text{with} \quad \bar{\sigma} = \sigma/\sigma_0, \quad \bar{w} = w/D, \quad \bar{D} = D/w_f \quad (2.60)$$

Let coordinates x_i be positioned so that the crack would lie in the plane (x_1, x_3) and that the tip of the cohesive crack would be at $x_1 = 0$. For a small enough D, the crack-bridging stress $\sigma > 0$ along the whole crack length L, and if D is small enough and the compression strength is unlimited, the cohesive crack at maximum load will occupy the entire cross section or, in the case of a notch, the entire ligament; then $\bar{L} = L/D =$ constant (if the compression strength is limited and the cross section is for instance subjected to bending, L/D will not necessarily be size independent but we may assume it to be such, as an approximation for small D).

To study the dependence of σ_N on D, we will now assume that, for small enough $\bar{D}$,

$$\sigma_N = \sigma_N^0 + \sigma'_N \bar{D}^p, \quad \bar{\sigma}_{ij} = \bar{\sigma}_{ij}^0 + \bar{\sigma}'_{ij}\bar{D}^p, \quad \bar{\sigma} = \bar{\sigma}^0 + \bar{\sigma}'\bar{D}^p \quad (2.61)$$
$$\bar{u}_i = \bar{u}_i^0 + \bar{u}'_i \bar{D}^p, \quad \bar{w} = \bar{w}^0 + \bar{w}'\bar{D}^p, \quad (2.62)$$
$$\bar{K}_t = \bar{K}_t^0 + \bar{K}'_t \bar{D}^p + \bar{\epsilon}''_{ij}\bar{D}^2 + ... \quad (2.63)$$

where $\sigma_N^0, \sigma'_N, \sigma^0, \sigma', \sigma_{ij}^0, ..., \bar{K}'_t$ are size independent. These expressions may now be substituted into (2.60), (1.10), (1.11) and the condition $\bar{K}_t = 0$. The resulting equations must be satisfied for various small sizes $\bar{D}$. For $\bar{D} \to 0$, the dominant terms in these equations are those of the lowest powers of D, which are those with $\bar{D}^0$ and $\bar{D}^p$. By collecting the terms without $\bar{D}$ and those with $\bar{D}^p$, we obtain two independent sets of equations. It so happens that each of these two sets defines a physically meaningful boundary value problem of elasticity for a body with a crack.

Elasticity Problem I: By isolating the terms that do not contain $\bar{D}$ (i.e., contain $\bar{D}^0$), we get:

$$\bar{K}_t^0 = 0, \quad \bar{\sigma}^0 = 1 \quad (\text{for } -\bar{L} \leq \bar{x}_1 < 0, \bar{x}_2 = 0) \tag{2.64}$$

$$\bar{\sigma}_{ij}^0 = \bar{E}_{ijkl} \tfrac{1}{2}(\partial_j \bar{u}_i^0 + \partial_i \bar{u}_j^0), \quad \partial_j \bar{\sigma}_{ij}^0 + \bar{f}_i \, \sigma_N^0/\sigma_0 = 0, \quad (\text{in } \bar{V}) \tag{2.65}$$

$$n_j \bar{\sigma}_{ij}^0 = \bar{p}_i \, \sigma_N^0/\sigma_0 \quad (\text{on } \bar{\Gamma}_s), \quad \bar{u}_i^0 = 0 \quad (\text{on } \bar{\Gamma}_d) \tag{2.66}$$

Elasticity Problem II: By isolating the terms that contain $\bar{D}^p$, we get:

$$\bar{K}_t' = 0 \quad \bar{\sigma}' = -(\bar{w}^0)^p \quad (\text{for } -\bar{L} \leq \bar{x}_1 < 0, \bar{x}_2 = 0) \tag{2.67}$$

$$\bar{\sigma}_{ij}' = \bar{E}_{ijkl} \tfrac{1}{2}(\partial_j \bar{u}_i' + \partial_i \bar{u}_j'), \quad \partial_j \bar{\sigma}_{ij}' + \bar{f}_i \, \sigma_N'/\sigma_0 = 0, \quad (\text{in } \bar{V}) \tag{2.68}$$

$$n_j \bar{\sigma}_{ij}' = \bar{p}_i \, \sigma_N'/\sigma_0 \quad (\text{on } \bar{\Gamma}_s), \quad \bar{u}_i' = 0 \quad (\text{on } \bar{\Gamma}_d) \tag{2.69}$$

The role of stresses and displacements is played by $\bar{\sigma}_{ij}^0$ and $\bar{u}_i^0$ in problem I, and by $\bar{\sigma}_{ij}'$ and $\bar{u}_i'$ in problem II. In problem I, the crack faces are subjected to fixed uniform tractions equal to 1. In problem II, in which σ' plays the role of the cohesive stress, the crack faces are subjected to nonuniform tractions $-(\bar{w}^0)^p$ which can be determined in advance from the $\bar{w}^0$-values obtained in solving problem I.

The magnitude of the loads (surface tractions and body forces) is proportional to σ_N^0 in problem I, and to σ_N' in problem II. These elasticity problems are known to have a unique solution. If σ_N^0 were zero, i.e., if the applied load in problem I vanished, the crack face tractions equal to 1 would cause K_t^0 to be nonzero, in violation of (2.64). Likewise, if σ_N' were zero, i.e., if the applied load in problem II vanished, the nonuniform crack face tractions $-(\bar{w}^0)^p$ in problem II would cause K_t' to be nonzero, in violation of (2.67). If the loads for problems I and II were infinite, then K_t^0 or K_t' would be infinite as well, which would again violate (2.64) or (2.67). Therefore, the only possibility left is that both σ_N^0 and σ_N' are finite (Bažant 2001a).

The initial descending slope of the softening cohesive laws used for quasi-brittle materials such as concrete is finite, i.e., $p = 1$. Consequently, according to (2.61), the size effect law must begin near $D = 0$ as a linear function of D, and as an exponential in the logarithmic plot [the latter ensuing from the approximation $\ln \sigma_N - \ln \sigma_N^0 = \ln(1 + \sigma_N' \bar{D}/\sigma_N^0) \approx (\sigma_N'/\sigma_N^0) \, e^{\ln \bar{D}}$]. The case $p > 1$ means that the softening law begins its descent from a horizontal initial tangent, which is reasonable for ductile fracture. The case $p < 1$ means that the cohesive law begins its descent with a vertical tangent, which seems to be an unrealistic super-brittle behavior.

The condition that $p = 1$, according to the cohesive crack model, is satisfied by the classical size effect law proposed by Bažant (1984). Indeed, $\sigma_N \propto (1+D/D_0)^{-1/2} \approx 1 - D/2D_0$ for small D ($D_0 =$ constant). But this condition is satisfied for neither $\sigma_N = Bf'_t/(1+\sqrt{D/D_0})$ nor $\sigma_N = Bf'_t[1+(D/D_0)^r]^{-1/2r}$ with $r < 1$. As for the case $r > 1$ ($p = r$), the softening law begins its descent from a horizontal asymptote, which means that this case is suitable for ductile fracture of plastically yielding materials.

It must of course be admitted that imposing the small-size asymptotic properties of the cohesive crack model is debatable since, for cross section thicknesses less than several aggregate sizes, the material is not a continuum. Yet imposition of these properties appears advantageous from the viewpoint of asymptotic matching—an approximation for the middle size range will be better if it satisfies the asymptotic properties of the theory applicable for that range.

In the foregoing arguments, we actually did not need to distinguish between the cases of a large crack and of crack initiation from a smooth surface. Therefore, the asymptotic form $\sigma_N = \sigma_N^0 + \sigma'_N D$ is appropriate for the latter case as well. Formula (9.66), which will be derived in Section 9.4, has such an asymptotic property.

Chapter 3

Randomness and Disorder

3.1. Is Weibull Statistical Theory Applicable to Quasibrittle Structures?

The statistical theory of size effect based on the concept of random strength was, in principle, completed by Weibull (1939) (also 1949, 1951, 1956). The Weibull theory has been enormously successful in applications to fine-grained ceramics and metal structures embrittled by fatigue. However, it took until the 1980s to realize that this theory does not really explain the size effect in quasibrittle structures failing after a large stable crack growth.

The Weibull theory rests on two basic hypotheses:

1. The structure fails as soon as one small element of the material representative volume attains the strength limit.

2. The strength limit is random and the probability P_1 that the small element of material does not fail at a stress less than σ has a cumulative distribution with a power law tail:

$$\varphi(\sigma) = \left\langle \frac{\sigma - \sigma_u}{\sigma_0} \right\rangle^m \qquad (\sigma \geq \sigma_u \approx 0) \qquad (3.1)$$

(Weibull 1939) where m, σ_0, σ_u = material constants (m = Weibull modulus, usually between 5 and 50; σ_0 = scale parameter; σ_u = strength threshold, which may usually be taken as 0).

Based on Eq. (3.1), a three-dimensional continuous generalization of the weakest link model for the failure of a chain of links of random strength leads to the Weibull distribution:

$$P_f(\sigma_N) = 1 - \exp\left[-\int_V c[\sigma(x), \sigma_N)]dV(x)\right] \quad (3.2)$$

which represents the probability that a structure that fails at nominal stress σ_N as soon as macroscopic fracture initiates from a microcrack (or a some flaw) anywhere in the structure; σ = stress tensor field just before failure, x = coordinate vectors, V = volume of structure, and $c(\sigma)$ = function giving the spatial concentration of failure probability of the material, which is $V_r^{-1} \times$ failure probability of material representative volume V_r) (Freudenthal 1968); $c(\sigma) \approx \sum_i P_1(\sigma_i)/V_0$ where σ_i = principal stresses ($i = 1, 2, 3$), $P_1(\sigma) = \varphi(\sigma)$ failure probability (cumulative) of the smallest possible test specimen of volume V_0 (or representative volume of the material) subjected to uniaxial tensile stress σ, as expressed by Weibull's expression (3.1); and V_0 = reference volume understood as the volume of specimens on which $c(\sigma)$ was measured.

For specimens under uniform uniaxial stress (and $\sigma_u = 0$), (3.2) and (3.1) lead to the following simple expressions for the mean and coefficient of variation of the nominal strength:

$$\bar{\sigma}_N = s_0 \Gamma(1 + m^{-1})(V_0/V)^{1/m}, \qquad \omega = [\Gamma(1 + 2m^{-1})/\Gamma^2(1 + m^{-1}) - 1]^{1/2} \quad (3.3)$$

where Γ is the gamma function. Since ω depends only on m, it is often used for determining m from the observed statistical scatter of strength of identical test specimens. The expression for $\bar{\sigma}_N$ includes the effect of volume V which depends on size D.

In view of Eq. (3.3), the value $\sigma_W = \sigma_N(V/V_0)^{1/m}$ for a uniformly stressed specimen can be adopted as a size-independent stress measure called the *Weibull stress*. Taking this viewpoint, Beremin (1983) proposed taking into account the nonuniform stress in a large crack-tip plastic zone by the so-called Weibull stress:

$$\sigma_W = \left(\sum_i \sigma_{Ii}^m \frac{V_i}{V_0}\right)^{1/m} \quad (3.4)$$

where V_i ($i = 1, 2, ... N_W$) are elements of the plastic zone having maximum principal stress σ_{Ii}. Ruggieri and Dodds (1996) replaced the sum in (3.4) by an integral. Eq. (3.4), however, considers only the crack-tip plastic zone whose size which is almost independent of D. Consequently, Eq. (3.4) is applicable only if the crack at the moment of failure is not yet macroscopic, still being negligible compared to structural dimensions.

It should be emphasized that distribution (3.1) is only the tail distribution of the extreme values. Of course, far beyond the threshold σ_u there may be a transition to some distribution such as normal, log-normal, or gamma, but

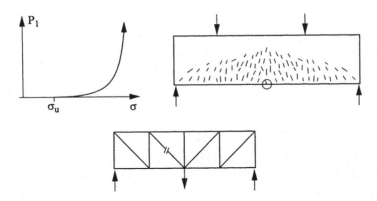

Figure 3.1: Weibull (cumulative) distribution of local material strength (top left), a critical flaw (encircled) in a field of many flaws (top right), and example of a multidimensional statically determinate structure that behaves as a chain and follows Weibull theory (bottom)

on the scale of the drawing in Fig. 3.1 (top left) this occurs in the sky (for a discussion of the implications of extreme value statistics, see Bažant 2001c).

Weibull applied his distribution to the classical problem of a long chain (Fig. 3.1 top right) or cable, for which the weakest-link model obviously applies. This model also applies to any statically determinate structure consisting of many elements (for example bars), because such a structure fails as soon as one element fails. But this is not the case for statically indeterminate structures and multidimensional bodies.

Weibull's theory has been applied to such structures by many researchers. This is correct, however, only if the multidimensional structure (Fig. 3.1 bottom) fails as soon as one small element of the material fails. Such sudden failure occurs in fatigue-embrittled metal structures, in which the critical flaw at the moment at which the sudden failure is triggered is still of microscopic dimensions compared to the cross-section size. But this is not the case for concrete structures and other quasibrittle structures which are designed to fail only after a large stable crack growth. For example, in the diagonal shear failure of reinforced concrete beams, the critical crack grows over 80% to 90% of the cross-section before the beam becomes unstable and fails. During such large stable crack growth, enormous stress redistributions occur and cause a large release of stored energy which, as we already discussed, produces a large deterministic size effect.

The size effect in Weibull theory arises from the fact that, in a larger structure, the probability of encountering a small material element of a certain small strength increases with the structure size. The probability of survival of the structure, $1 - P_f$, is the probability that all the elements survive, which (according to the joint probability theorem) is $[1 - P_1(\sigma)]^N$ where N = number of elements in the structure. Noting that $P_1(\sigma) \ll 1$ and taking logarithms, $\ln(1 - P_f) = N \ln[1 - P_1(\sigma)] \approx -N P_1(\sigma)$ (Fréchet 1927, Fisher and Tippett 1928, von Mises 1936). This is then generalised for nonuniform stress as follows:

$$\ln(1 - P_f) = \int_V \varphi[\sigma(\boldsymbol{x})] \, dV(\boldsymbol{x})/V_r \qquad (3.5)$$

in which P_f = failure probability of the structure, V = volume of the structure, V_r = small representative volume of the material whose strength distribution tail is given by $\varphi(\sigma)$, and $\boldsymbol{x}$ = spatial coordinate vector. By virtue of the fact that the Weibull distribution tail is a power law, the aforementioned probability integral always yields for the size effect a power law. It is of the form

$$\sigma_N = k_v V^{-1/m} = k_0 D^{-n/m} \qquad (3.6)$$

where k_0 = constant characterizing the structure shape, and n = number of dimensions of the structure (1, 2 or 3). For two-dimensional similarity ($n = 2$) and typical properties of concrete, the exponent is approximately $n/m = 1/12$ (Bažant and Novák 2000b), rather than 1/6 as believed on the basis of the classical study by Zech and Wittmann (1977).

As already mentioned, the fact that the scaling law of Weibull theory is a power law implies that there is no characteristic size of the structure, and thus no material characteristic length (this is also obvious from the fact that no material length appears anywhere in the formulation). This observation makes the Weibull-type scaling suspect when one deals with quasibrittle structures whose material is highly heterogeneous, with a heterogeneity characterized by a non-negligible material length.

Applications of the classical Weibull theory to quasibrittle structures face a number of serious objections:

1. The fact that in (3.6) the size effect is a power law implies the absence of any characteristic length. But this cannot be true if the material contains sizable inhomogeneities.

2. The energy release due to stress redistributions caused by macroscopic FPZ or stable crack growth before P_{max} gives rise to a deterministic size effect which is ignored. Thus the Weibull theory is valid only if the structure fails as soon as a microscopic crack becomes macroscopic.

3. Every structure is mathematically equivalent to a uniaxially stressed bar, which means that no information on the structural geometry and failure mechanism is taken into account.

4. The size effect differences between two- and three-dimensional similarities ($n_d = 2$ or 3) are predicted much too large.

5. Many tests of quasibrittle materials (e.g., diagonal shear failure of reinforced concrete beams) show a much stronger size effect than predicted by Weibull theory (see Bažant and Planas, 1998, or the review in Bažant 1997a).

6. The classical theory neglects the spatial correlations of material failure probabilities of neighboring elements caused by nonlocal properties of damage evolution (while generalizations based on some phenomenological load-sharing hypotheses have been divorced from mechanics).

7. When (3.3) is fit to the test data on statistical scatter for specimens of one size (V = const.), and when the Weibull size effect equation is fit to the mean test data on the effect of size or V (of unnotched plain concrete specimens), the optimum values of Weibull exponent m are very different, namely $m = 12$ and $m = 24$, respectively (Bažant and Novák, 2000a). If the theory were applicable, these values would have to coincide.

In view of these limitations, among concrete structures the Weibull theory appears applicable to some extremely thick plain (unreinforced) structures, e.g., the flexure of an arch dam acting as a horizontal beam (but not for vertical bending of arch dams nor gravity dams because large vertical compressive stresses cause long cracks to grow stably before the maximum load). Most other plain concrete structures are not thick enough to prevent the deterministic size effect from dominating. Steel or fiber reinforcement prevents it as well.

To take into account stress redistributions and the inherent energy release, various phenomenological theories of load sharing and redistribution in a system of parallel elements have been proposed (e.g., L. Phoenix). Although they are useful if the redistributions and load-sharing are relatively mild, they appear insufficient to describe the large stress redistributions caused by large stable crack growth and the effects of structure geometry. They lack the fracture mechanics aspects of a large fracture process zone.

3.2. Nonlocal Probabilistic Theory of Size Effect

To take into account the stress redistribution due to a large fracture, one might wish to substitute the LEFM near-tip stress field into the probability integral in (3.5). However, for normal values of the Weibull modulus ($m > 4$), the integral diverges. So this is not a remedy. However, Weibull theory can be

58   Scaling of Structural Strength

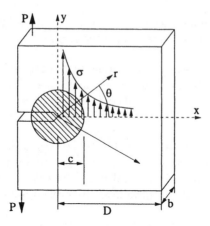

Figure 3.2: Neighborhood, simulating the fracture process zone, over which the strain field is averaged in the nonlocal generalization of Weibull theory (Bažant and Xi, 1991)

extended to capture large stress redistributions approximately—by introducing a nonlocal generalization (Bažant and Xi, 1991), in which the probability integral (3.5) is replaced by the following integral:

$$\ln(1 - P_f) = k_1 \int_V \varphi\left[E\,\bar{\epsilon}(x)\right] dV(x)/V_r \qquad (3.7)$$

Here the stress at a given point in the structure is replaced by the average (over a certain neighborhood, Fig. 3.2) of the strain field, $\bar{\epsilon}$ times the elastic modulus E, to get a quantity of the stress dimension. In other words, the failure probability at a certain point x of the structure is assumed to depend not on the stress (stress according to the continuum theory) at that point but on the average strain in a certain neighborhood of the point, as in nonlocal theories for strain localization in strain-softening materials.

With this nonlocal generalization, the analytical evaluation of the integral (3.7) seems prohibitively difficult, however it is easy to obtain the asymptotic behavior of notched or cracked structures for $D \to \infty$ and $D \to 0$. Also, for $m \to \infty$, the solution should approach the size effect law based on energy release, Eq. (2.8).

It was shown that a simple formula that interpolates between these three asymptotic cases, i.e., achieves asymptotic matching, is as follows (Bažant and Xi, 1991):

$$\sigma_N = \frac{\sigma_P}{\sqrt{\beta^{2n/m} + \beta}} \qquad \beta = \frac{D}{D_0} \qquad (3.8)$$

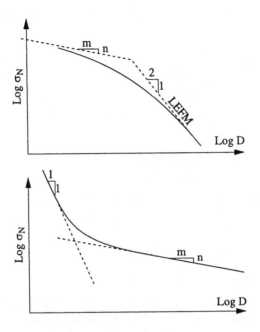

Figure 3.3: Scaling law according the nonlocal generalization of Weibull theory for failures after large crack growth (left) and at crack initiation (right)

This formula, applicable only to structures with notches or large cracks at maximum load, is sketched in Fig. 3.3 (top). This figure also shows the aforementioned asymptotic scaling laws (which turn out to be the same as the Weibull type scaling law for small sizes—a line of slope $-m/n$), and the LEFM scaling law for large similar cracks and large sizes (a line of slope $-1/2$).

According to the foregoing results, the scaling law of the classical Weibull theory is, in the case of notches or pre-existing large cracks, applicable only in the limit of sufficiently small structures. However, comparisons with test data for concrete show that the deterministic size effect law which begins by a horizontal asymptote, and the size effect law in (3.8) which begins by an asymptote of slope $-m/n$, both fit the test data about equally well, relative to the scatter of measurements. So it seems that, for the case of notches or pre-existing cracks, material randomness of Weibull type cannot play a significant role in sized effect. But this is not quite true for structure failing at crack initiation from the surface, as will be discussed in the next section.

It is interesting that the effect of material randomness in notched or pre-cracked structures completely disappears for large structure sizes. This is revealed by the fact that the large size asymptote has the LEFM slope of $-1/2$.

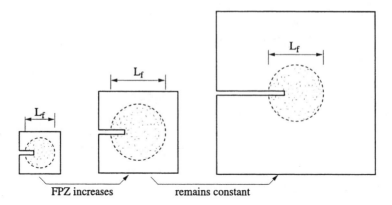

Figure 3.4: Changes of fracture process zone size with increasing structure size

How can it be explained physically?

The explanation is that, when the structure is sufficiently large, a further increase of the structure size is not accompanied by any increase in the size of the fracture process zone (Fig. 3.4). The Weibull-type probability integral in Eq. (3.7) is taken over the entire structure, however, the only significant contribution to the integral comes from the fracture process zone. Since the fracture process zone does not increase with an increase of the structure size, it is obvious that the failure probability should not be affected by a further increase of the structure size if it is already large.

In the nonlocal formulation of Weibull theory for quasibrittle materials such as concrete, originally developed for notched specimens or structures with a large notch or traction-free crack at the moment of failure (Bažant and Xi 1991), the failure probability of a small material element is a function of nonlocal (spatially averaged) continuum strain. Recently, however, it was found more appropriate to use the average of the inelastic part of strain (Bažant and Novák 2000a). This refined formulation was found to work well in general—not only for notched specimens but also for unnotched specimens or structures failing at the initiation of macroscopic fracture, and in particular to the test of flexural strength, called the *modulus of rupture* (Fig. 3.5). The purpose of the nonlocality is not only to prevent spurious localization of cracking but also to introduce spatial correlation of random material strength, governed by a certain finite characteristic length of the material.

As in Bažant and Xi's (1991) analysis of structures containing notches or large traction-free cracks, the size effect on the mean or median of the modulus of rupture is found, for normal size beams, to be essentially deterministic. However, the size range in which the statistical size effect dominates is different—it

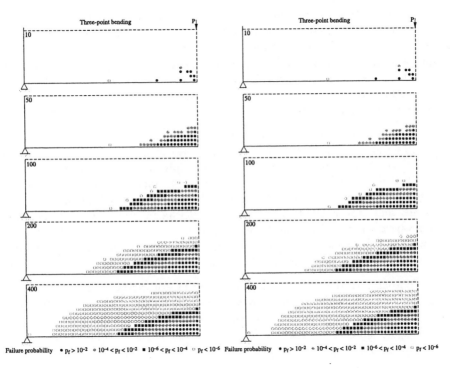

Figure 3.5: Monte Carlo simulation showing how various points in a three-point bent unnotched beam contribute to the failure probability at subsequent loading stages (after Bažant and Novák 2000a)

is the asymptotic range of very large (rather than very small) sizes. While in the case of notches or large cracks this range lies below the range of practical interest, in the case of crack initiation it lies, by contrast, above the normal size range and is important for extrapolations to bending fracture of extremely large structures such as arch dams or foundation plinths (Bažant and Novák 2000a,b).

The main benefit of the nonlocal Weibull theory is the possibility to predict for various structure sizes (and shapes) the full probability distribution of structural strength, and in particular the modulus of rupture. Examples of calculating the 5% and 95% probabilities of structural failure have been given (Bažant and Novák 2000a,b) and a good agreement with the extensive test data available in the literature has been achieved; see Figs. 3.6, 3.7 and 3.8. Calculations show that the coefficient of variation characterizing the scatter of the modulus of rupture decreases with an increasing beam size, in agreement with experimental results.

As a fundamental check of soundness of any probabilistic theory of failure, the classical Weibull theory based on the weakest link model (extreme value distribution) must be recovered as the asymptotic limit when the size of a quasibrittle or strain-softening structure tends to infinity. This requirement is obviously satisfied by the probabilistic Weibull theory (Bažant and Xi 1991, Bažant and Novák 2000a,b). The usual stochastic finite element methods, however, do not satisfy this basic requirement. This casts doubts on their applicability, for calculating loads of an extremely small failure probability such as 10^{-7}, requiring the use of extreme value distribution.

Compared to the usual stochastic finite element methods, a useful simplification achieved by the nonlocal Weibull theory is that the nonlocal structural analysis with strain softening can be conducted deterministicaly because the probability analysis is separated from the stress analysis, similar to the classical Weibull theory.

3.3. Energetic-Statistical Formula for Size Effect for Failures at Crack Initiation

The large size asymptote of the deterministic energetic size effect formula (2.29) is horizontal, $f_r/f_{r,\infty} = 1$. The same is true of all the existing formulae for the modulus of rupture, reviewed in Bažant and Planas (1998). But this is not in agreement with the numerical results of the aforementioned Bažant and Novák's (2000a) nonlocal Weibull theory as applied to modulus of rupture, in which the large-size asymptote in the logarithmic plot has the slope $-n/m$ corresponding to the power law of the classical Weibull statistical theory (Weibull

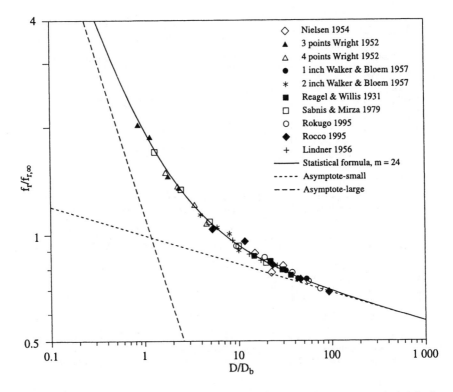

Figure 3.6: Energetic-statistical formula (3.9) for failures at crack initiation, compared with test data from the literature on the modulus of rupture of concrete (with 5 and 95 probability percentiles (after Bažant and Novák 2000b)

64 Scaling of Structural Strength

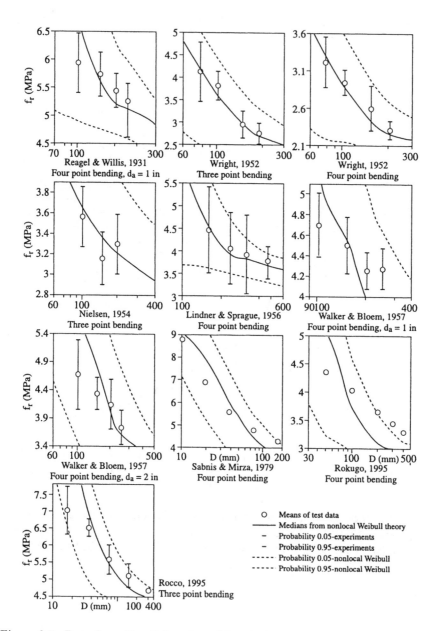

Figure 3.7: Data from Fig. 3.6 replotted in relative coordinates, compared with the same formula (after Bažant and Novák 2000b)

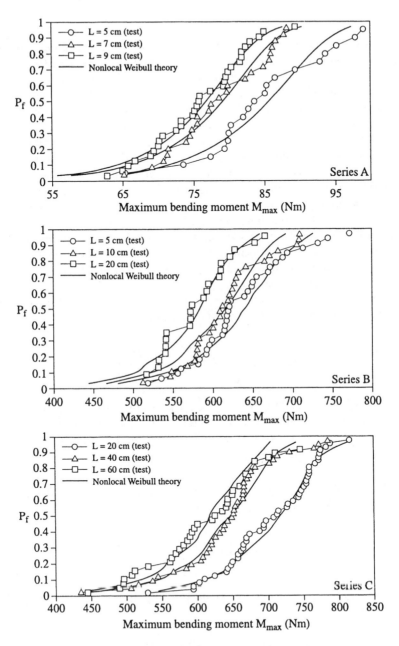

Figure 3.8: Distributions of maximum bending moment M_{max} measured by Koide et al. (1998, 1999) compared with simulations of modulus of rupture tests with nonlocal Weibul theory (after Bažant and Novák 2000a)

1939).

In view of this theoretical evidence, there is a need to amalgamate the energetic and statistical theories, despite the fact that the existing test data can be matched very well by the deterministic theory for failures at crack initiation, already described (Eq. 2.11). Such amalgamation is important, for example, for analyzing the size effect in vertical bending fracture of arch dams, foundation plinths or retaining walls—a phenomenon that must have contributed to the catastrophic failures of Malpasset Dam in the French Maritime Alps in 1959, the Saint Francis Dam near Los Angeles in 1928, or the Schoharie Creek Bridge on New York Thruway in 1987. Briefly, the formula may be deduced as follows. According to the deterministic energetic model, $\Delta^r = (f_r/f_{r,\infty})^r - rD_b/D = 1$, which is the value of the large-size horizontal asymptote. From the statistical viewpoint, this difference, characterizing the deviation of the nominal strength from the asymptotic energetic size effect for a relatively small fracture process zone (large D), should conform to the size effect of Weibull theory, $D^{-n/m}$, where m = Weibull modulus and n = number of spatial dimensions ($n = 1$, 2 or 3, in the present calculations 2). Therefore, instead of $\Delta = 1$, one needs to set $\Delta = (D/D_b)^{-n/m}$. This leads to the following Weibull-type statistical generalization of the energetic size effect formula (2.11):

$$f_r = f_{r,\infty} \left[\left(\frac{D_b}{D}\right)^{rn/m} + \frac{rD_b}{D} \right]^{1/r} \tag{3.9}$$

(Bažant and Novák 2000b) where $f_{r,\infty}$, D_b, r are positive constants, representing the unknown empirical parameters to be determined by experiments. Because in all practical cases $rn/m < 1$ (in fact, $\ll 1$), formula (3.9) satisfies three asymptotic conditions:

1. For small sizes, $D \to 0$, it asymptotically approaches the deterministic energetic formula (2.29);

$$f_r = f_{r,\infty} r^{1/r} \left(\frac{D_b}{D}\right)^{1/r} \propto D^{-1/r} \tag{3.10}$$

2. For large sizes, $D \to \infty$, it asymptotically approaches the Weibull size effect;

$$f_r = f_{r,\infty} \left(\frac{D_b}{D}\right)^{n/m} \propto D^{-n/m} \tag{3.11}$$

3. For $m \to \infty$, the limit of (3.9) is the deterministic energetic formula (2.29).

Equation (3.9) is in fact the simplest formula with these three asymptotic properties. It may be regarded as the asymptotic matching of the small-size deterministic and the large-size statistical size effects.

Based on the preliminary results of Zech and Wittmann (1977), the value of Weibull modulus was long ago fixed as $m = 12$, which implies the final asymptote to have the slope $-n/m = -1/6$ (since $n = 2$ for most of the data). Based on a limited data set available in 1977, these investigators obtained the value $m = 12$. They deduced it in the standard way, which was from the coefficient of variation of strength values measured on specimens of one size and one shape. However, optimization of the much larger set of test data that exist in the literature today and include good size effect data (Rocco 1995, Rokugo et al. 1995), showed the value $m = 24$ to be optimal.

3.4. Size Effect Ensuing from J-Integral for Randomly Located Cracks

Formula (3.9) can also be derived in a more fundamental manner by generalizing the deterministic derivation from the J-integral. To this end, note that, in the case of failure at crack initiation, the point at which the crack may originate is randomly located according to the weakest link theory, and, in the case of large cracks at failure, the location of the crack tip at failure is random. The randomly located fracture process zone (FPZ) at the tip of such a crack is enclosed by the contour of the J-integral, and so the contour is randomly located, too.

For $D \to \infty$, the FPZ, and thus also the smallest possible J-integral contour that surrounds the FPZ, occupy an infinitely small portion of the structure volume V. Therefore, there must exist a certain domain U that is so remote from the crack and simultaneously so much smaller than the structural dimension D that the stress field σ_U in U may be considered nearly uniform. Obviously $\sigma_U \propto \sigma_N$.

It is now helpful to exploit Beremin's (1983) concept of Weibull stress (Eq. 3.4). The Weibull stress σ_W^U for domain U is

$$\sigma_W^U \propto \sigma_U D^{-n/m} \propto \sigma_N D^{-n/m} \qquad (3.12)$$

The weakest link theory is equivalent to comparing the Weibull stress with a fixed strength limit, but since the magnitude of stress σ_U is proportional to the J-integral, the weakest link theory is actually equivalent to comparing the Weibull J-integral (or Weibull energy release rate), defined as

$$J_W \propto J D^{-n/m} \qquad (3.13)$$

to a fixed critical value of the J-integral, that is, to the fracture energy, G_f. According to equation (3.6), an equivalent approach is to replace the nominal strength σ_N with the Weibull nominal strength:

$$\sigma_N^W = \sigma_N (D_a/D)^{n/m} \qquad (3.14)$$

where D_a is some unspecified constant. Making this replacement in (2.16) with (2.18), we can retain (2.16) if we redefine

$$\mathcal{J}(\theta) = (D_a/D)^{n/m}(\mathcal{J}_0 + \mathcal{J}_1\theta + \mathcal{J}_2\theta^2 + ...) \qquad (3.15)$$

Consider now the failures at crack initiation $\mathcal{J}_0 = 0$. If we introduce again the arbitrary random parameter r (which has no effect on the first two asymptotic terms), choose a suitable value for the arbitrary constant D_a, and proceed in the same way as from Eq. (2.18) to Eq. (2.21), we acquire formula (3.9). Thus we have gained a more fundamental derivation of that formula.

To derive the energetic-statistical formula (3.8) for notches or large cracks, expansion (2.21) must be modified. If we consider geometrically similar notches in specimens of various sizes, then the first term $\mathcal{J}_0$ should not be randomized with the Weibull factor $(D_a/D)^{n/m}$, i.e.

$$\mathcal{J}(\theta) = \mathcal{J}_0 + (D_a/D)^{n/m}(\mathcal{J}_1\theta + \mathcal{J}_2\theta^2 ...) \qquad (3.16)$$

where $\mathcal{J}_0$ is now large. Truncating this after the second term of expansion and proceeding in the same way as from (2.18) to (2.21), we readily obtain the energetic-statistical size effect law (3.8) for large similar cracks.

If, however, there is no notch and the material randomness causes the location of the tip of a large crack in specimens of different sizes to deviate randomly from a geometrically similar location, then the fully randomized expansion (3.15) applies. Proceeding similarly as from (2.18) to (2.21), one gets a new formula

$$\sigma_N = \sigma_P \beta^{-n/m} (1+\beta)^{-1/2}, \qquad \beta = D/D_0 \qquad (3.17)$$

The corresponding large-size asymptotic size effect is

$$\sigma_N \propto D^{-(\frac{1}{2}+\frac{n}{m})} \qquad (3.18)$$

This is a stronger size effect than LEFM gives for similar cracks, which is explained by the fact that the randomness of crack tip location intensifies the size effect—in a larger structure, the chance for the crack tip to find a location of a given low strength is greater. It must be recognized, though, that formula (3.18) might be an oversimplification. In reality, the tip of a large crack is not necessarily located at the place of the lowest strength; rather, the tip location is influenced by the previous crack path, which is ignored by equation (3.18).

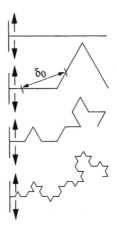

Figure 3.9: Von Koch fractal curve at progressive refinements and measurement of its length by a ruler of length δ_0

The derivation of the deterministic size effect laws from equivalent LEFM can be randomized in a similar manner, and the same formulae result. In that approach, the shape dependence of the formula ensues directly. The shape dependence, however, is introduced into the foregoing formulae by setting $J(\theta) = g(\alpha_0 + \theta)$, which shows that $\mathcal{J}_0 = g(\alpha_0)$, $\mathcal{J}_1 = \frac{1}{2}g'(\alpha_0)$, $\mathcal{J}_2 = \frac{1}{6}g''(\alpha_0)$, etc.

3.5. Could Fracture Fractality Be the Cause of Size Effect?

This intriguing question was recently raised by Carpinteri (1994a,b) (see also Carpinteri et al. 1993, 1995a,b,c; Carpinteri and Ferro 1994; and Carpinteri and Chiaia 1995). The arguments he offered were original and ingenious, however, they were not based on mechanical analysis and energy considerations. Rather they were strictly geometrical and partly intuitive.

Subsequently, Bažant (1995d) attempted a mechanical analysis of the problem, which will now be briefly outlined. The answer has been negative. However, the fact that the surface roughness of cracks in many materials can be described, at least over a certain limited range, by fractal concepts, is not in doubt (e.g., Mandelbrot et al. 1984; Brown 1987; Mecholsky and Mackin 1988; Cahn 1989; Chen and Runt 1989; Hornbogen 1989; Peng and Tian 1990; Saouma et al. 1990; Bouchaud et al. 1990; Chelidze and Gueguen 1990; Issa et al. 1992; Long et al. 1991; Måløy et al. 1992; Mosolov and Borodich 1992; Borodich 1992; Lange et al. 1993; Xie 1987, 1989, 1993; Xie et al. 1994,

70   Scaling of Structural Strength

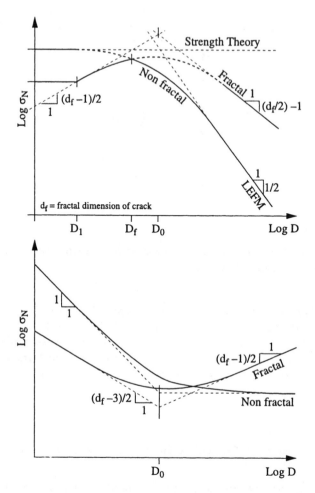

Figure 3.10: Size effect curves predicted by nonfractal and fractal energy-based analyses, for failures after large crack growth (right) or at crack initiation (bottom)

1995; Saouma and Barton 1994; Feng et al. 1995). A fractal description of the crack surfaces might offer one viable way to characterize the dependence of the fracture energy of quasibrittle materials on the crack surface roughness.

In two dimensions, a fractal curve, which can be imagined to represent a crack, can be illustrated, for example, by the von Koch curves shown in Fig. 3.9. Progressive refinements are obtained by inserting self-similar bumps into each straight segment. If the length of this curve is measured by a ruler of a certain resolution δ_0, imagined as the ruler length, the length measured will obviously depend on the length of the ruler and if the length of the ruler approaches zero, the measured length will approach infinity. This is described by the equation

$$a_\delta = \delta_0 (a/\delta_0)^{d_f} \tag{3.19}$$

where a_δ is the measured length along the curve, a is the projected (smooth, Euclidean) crack length. Exponent d_f is called the *fractal dimension*, which is greater than 1 if the curve is fractal, and equal to 1 if it is not.

Obviously the total energy dissipation W_f for the crack length a_δ would be infinite if we assume that a finite amount of energy G_f is dissipated per unit crack length. This is a conceptual difficulty for fracture mechanics of fractal cracks. In a sequel to the study of Mosolov and Borodich (1992), Borodich (1992) proposed to resolve this difficulty by setting

$$W_f/b = G_{fl} a^{d_f} \tag{3.20}$$

in which W_f = total energy dissipation; G_{fl} represents what may be called the *fractal fracture energy* whose dimension is not J/m^2 but J/m$^{d_f+1}$.

Based on this fractal concept of fracture energy, one may carry out a similar asymptotic analysis as we have outlined for non-fractal cracks (Bažant, 1997b). For the case of failure after a large stable crack growth, the matching of the large size and small size asymptotic expansions for the fractal fracture yields, instead of Eq. (2.8), the result:

$$\sigma_N = \sigma_N^0 D^{(d_f-1)/2} \left(1 + \frac{D}{D_0}\right)^{-1/2} \tag{3.21}$$

For failure at crack initiation, the asymptotic analysis yields instead of Eq. (2.29) the result:

$$\sigma_N = \sigma_N^\infty D^{(d_f-1)/2} \left(1 + \frac{D_b}{D}\right) \tag{3.22}$$

These expressions reduce to the nonfractal case when $d_f = 1$. The plots of these equations are shown in Fig. 3.10 in comparison with the size effect formulas for the nonfractal case.

The hypothesis that the fracture propagation is fractal has been made and the consequences have been deduced (Bažant, 1997b). Now, by judging the consequences we may decide whether the hypothesis was correct. Looking at the plots in Fig. 3.10 it is immediately apparent that the fractal case disagrees with the available experimental evidence. For failures after large crack growth, the rising portion of the plot has never been seen in experiments, and there are many data showing that the asymptotic slope is very close to $-1/2$, rather than to the much smaller value predicted from the fractal hypothesis. This is clear by looking at Figs. 2.6–2.8. For failures at crack initiation, the kind of

72   Scaling of Structural Strength

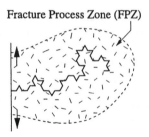

Figure 3.11: Fractal crack curve and its fracture process zone with distributed cracking

plots seen in Fig. 3.10 (bottom), with a rising size effect curve for large sizes, is also never observed. Thus it is inevitable to conclude that the hypothesis of a fractal source of size effect is contradicted by test data and thus untenable. (The existence of fractal characteristics of fracture surfaces in various materials is of course not questioned, and neither is the possibility that these fractal characteristics may influence the value of the fracture energy of the material and may have to be considered in micromechanical models which predict the fracture energy value.)

What is the physical reason that the fractal hypothesis fails? Probably it is the fact that the front of the crack is surrounded by a large fracture process zone consisting of microcracks and frictional slips, as shown in Fig. 3.11. Because the fracture energy G_f of quasibrittle materials is usually several orders of magnitude larger than the surface energy, the whole volume of the fracture process zone of microcracking dissipates far more energy than the formation of the crack surface. Therefore, from the energy viewpoint, the crack curve, which might of course be fractal, cannot matter.

3.6. Could Lacunar Fractality of Microcracks Be the Cause of Size Effect?

Having already discussed Weibull theory, we are ready to tackle another type of fractality—the *lacunar fractality of microcracks*, which is illustrated in Fig. 2.15 (top). From distance we see one crack, but looking closer we see it consists of several shorter cracks with gaps between them, and looking still closer we see that each of these cracks consists of several still shorter cracks with shorter gaps between them, and so forth. Refinement *ad infinitum* generates a Cantor set or a fractal set whose fractal dimension d_f is less than the Euclidean dimension of the space (which is 1 in this example; Fig. 2.15 top left).

It seems that the microcrack systems in concrete do exhibit this type of fractality, but only to a limited extent. Quasibrittle materials are materials with large heterogeneities and a large characteristic length. So obviously the refinement to smaller and smaller cracks must have a cutoff.

The argument that lacunar (or rarefying) fractality is the cause of size effect in quasibrittle structures (Carpinteri and Chiaia, 1995) went as follows. The fractal dimensions of the arrays of microcracks are different at small and large scales of observation. For a small scale, the fractal dimension D_f is distinctly less than 1, and for a large scale it is nearly 1. For the failure of a small structure the small scale matters, and for the failure of a large structure the large scale matters.

Therefore, as it was argued, there should be a transition from a power scaling law corresponding to small scale fractality to another power scaling law corresponding to the large scale fractality, the latter having exponent 0 for the strength, i.e., no size effect. Thus, as it was claimed, the size effect should be given by a transitional curve between the two asymptotes of slope $-1/2$ and 0 shown in Fig. 2.15 (bottom left). The slope of the initial asymptote was assumed to be $-1/2$.

This size effect was described by the formula $\sigma_N = \sqrt{A_1 + A_2/D}$ which coincides with (2.42) and was called the 'multifractal' scaling law (MFSL) (A_1, A_2 = constants).

There exist, however, test data that clearly disagree with the MFSL Law, Eq. (2.42). Many test data exhibit in the logarithmic size effect plot an initial slope much less than $-1/2$, particularly for specimen sizes that are as small as possible for the given size of aggregate. The MFSL law was also proposed for structures with notches and large cracks at the moment of failure, but many test data approach an asymptote of slope $-1/2$ at very large sizes, and there are many others that exhibit a negative rather than positive curvature in the plot of log σ_N and log D, which disagrees with the MFSL law. A comprehensive analysis of test data on the modulus of rupture (Bažant and Novák 2000b) showed that the MFSL law may be improved by adding exponent r, with optimal value about $r = 1.44$, as indicated in Eq. (2.29).

On closer scrutiny, there are also mathematical and physical reasons why the lacunar fractality cannot be the source of the observed size effect. If the failure is assumed to be controlled by lacunar fractality, that is by microcracks, it obviously implies that the failure occurs at crack initiation, in which case the mathematical formulation must be akin to Weibull theory.

Labeling the aforementioned small and large scales of observations by superscripts A and B, the Weibull distributions of the strength of a small material element in the fracture process zone with lacunar microcracks may be written as

$$\varphi\left[\sigma(x); d_f^A\right] = \left\langle \frac{\sigma_N S(\xi) c_f^{1-d_f^A} - \hat{\sigma}_u^A}{\hat{\sigma}_0^A} \right\rangle^m \quad (3.23)$$

$$\varphi\left[\sigma(x); d_f^B\right] = \left\langle \frac{\sigma_N S(\xi) c_f^{1-d_f^B} - \hat{\sigma}_u^B}{\hat{\sigma}_0^B} \right\rangle^m \quad (3.24)$$

Here the stress in the small material element of random strength has been written as $\sigma = \sigma_N S(\xi)$, in which S is the same function for all sizes of geometrically similar structures, and $\xi = x/D$, for the nonfractal (non-lacunar) case.

For the fractal (lacunar) case, this is generalized as $\sigma = \sigma_N S(\xi) c_f^{1-d_f}$ because the stress of the material element, in the case of lacunar microcracks, must be considered to have a non-standard, fractal dimension. Obviously, the Weibull constants $\hat{\sigma}_0$ and $\hat{\sigma}_u$ must now be considered to have fractal dimensions as well, but Weibull modulus m must not. An equation of the type of Eq. (3.23) or (3.24) was written by Carpinteri et al.; however, further analysis consisted of geometric and intuitive arguments. We will now sketch a mechanical analysis (Bažant, 1997b).

In Weibull theory (failure at initiation of macroscopic fracture), every structure is equivalent to a long bar of variable cross section (Bažant, Xi and Reid 1991; Fig. 3.12). Carpinteri et al's argument means that a small structure is subdivided into small material elements (Fig. 3.12a) and a large structure is subdivided into proportionately larger material elements (Fig. 3.12c). However, this is not an objective view of the failure mechanism of two structures made of the same material.

The large elements of the larger structure shown in Fig. 3.12c must be divisible into the small elements considered for the structure in Fig. 3.12a, which are the representative volumes of the material for which the material properties are defined. If the large elements were not divisible into the small ones, it would imply that the material of the small structure is not the same.

Having in mind the subdivision of the large elements into the small elements as shown in Fig. 3.12b, we may now calculate the failure probability of the large structure on the basis of the refined subdivision into the small elements, as shown in Fig. 3.12b, or else it would imply that the small and large structures are not made of the same material. We note that the failure probability P_f of the large structure subdivided into large elements ΔV_{Bj} ($j = 1, 2, \cdots N$), and the failure probability P_{fj}^B of the large element ΔV_{Bj} of the large structure subdivided into small elements ΔV_{Aij}, must satisfy the following relations of Weibull theory:

$$-\ln(1 - P_f) = \sum_j \varphi(\sigma_N S_j^B;\ d_f^B)\Delta V_{Bj}/V_r \quad (3.25)$$

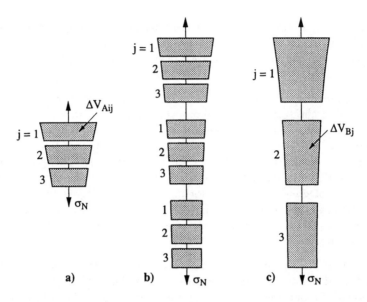

Figure 3.12: Subdivision of: (a) a small structure into small elements, (b) a large structure into small elements, and (c) a large structure into large elements

$$-\ln(1 - P_{fj}^B) = \sum_j \varphi(\sigma_N S_{Aij};\ d_f^A)\Delta V_{Aij}/V_r \qquad (3.26)$$

Now, since we may subdivide each element B of the large structure into the small elements A if the material is the same, we have

$$-\ln(1 - P_f) = -\sum_j \ln(1 - P_{fj}^B) = \sum_j \sum_i \varphi(\sigma_N S_{ij}^A;\ d_f^A)\Delta V_{Aij}/V_r \qquad (3.27)$$

Equating this to Eq. (3.6), we see that, in order to meet the requirement of the objective existence of the same material, the Weibull characteristics on scales A and B must be different and such that

$$\varphi(\sigma_N S_j^B;\ d_f^B) = (\Delta V_j^B)^{-1} \sum_i \varphi(\sigma_N S_{ij}^A;\ d_f^A)\Delta V_{ij}^A \qquad (3.28)$$

Eqs. (3.27) and (3.28) imply that consideration of different scales cannot yield different scaling laws. The same power law must result from the hypothesis of lacunar fractality of microcrack distribution, regardless of the scale considered.

If the argument for the MFSL law were accepted it would imply that a large material volume with no fractality may not be subdivided into smaller representative material volumes exhibiting lacunar fractality. As long as both the small and large volumes are sufficiently larger than the size of material inhomogeneities (the maximum aggregate size in concrete), this implication is selfcontradicting.

The argument would be acceptable if the size of the smaller material volume were less than the size of these inhomogeneities, for instance on the scale of the matrix between the inhomogeneities (e.g., the mortar between the large aggregates in concrete). The fractal dimensions of the systems of the tiny cracks in the matrix and of the cracks in concrete as a composite must of course be different. By definition, the failure of even the smallest concrete structure is not governed by the mortar alone. The aggregates are essential, and the failure is governed by the properties of the composite. So Carpinteri's argument does not appear acceptable as a basis of the scaling law for one and the same material.

To sum up, the scaling law of a structure failing at the initiation of fracture from a fractal field of lacunar microcracks must be identical to the scaling of the classical Weibull theory. The only difference is that the values of Weibull parameters depend on the lacunar fractality. This difference could be taken into account if the values of these parameters could be predicted by micromechanics. But as long as the Weibull parameters are determined by experiments, the concept of lacunar fractality of microcracks contributes nothing. The lacunar fractality can have no effect on the scaling law. However, the lacunar fractality may offer a valid way to describe the influence of microcrack distribution on the fracture energy.

Chapter 4

Energetic Scaling for Sea Ice and Concrete Structures

4.1. Scaling of Fracture of Floating Sea Ice Plates

Thermal Bending Fracture

Different types of size effect are exhibited by sea ice failures. The scaling of failure of floating sea ice plates in the Arctic presents some intricate difficulties. One practical need is to understand and predict the formation of very long fractures (of the order of 1 km to 100 km) which cause the formation of open water leads or serve as precursors initiating the build-up of pressure ridges and rafting zones.

Large fractures can be produced in sea ice as a result of the thermal bending moment caused by cooling of the surface of the ice plate (Fig. 4.1). Due to buoyancy, the floating plate behaves exactly as a plate on elastic Winkler foundation, with the foundation modulus equal to the unit weight of sea water. Under the assumption that the ice plate is infinite and elastic, of constant thickness h, that the temperature profiles for various thicknesses h are similar, and that the thermal fracture is semi-infinite and propagates statically (i.e., with insignificant inertia forces), it was found (Bažant 1992b) that the critical temperature difference

$$\Delta T_{cr} \propto h^{-3/8} \qquad (4.1)$$

This means that the critical nominal thermal stress $\sigma_N \propto h^{-3/8}$. The analysis was done according to LEFM. Despite the existence of a large fracture process zone, LEFM is justified because a steady-state propagation must develop. The fracture process zone does not change as it travels with the fracture front, and thus it dissipates energy at a constant rate, as in LEFM.

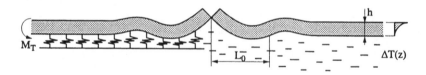

Figure 4.1: Bending fracture of floating sea ice plate caused by temperature difference

It has been shown that the scaling law in (4.1) must apply to failures caused by any type of bending cracks, provided that they are full-through cracks propagating along the plate (created by any type of loading, e.g. by vertical load; Slepyan 1990, Bažant 1993).

It may be surprising that the exponent of this large size asymptotic scaling law is not $-1/2$. However, this apparent contradiction may be explained if one realizes that the plate thickness is merely a parameter but not actually a dimension in the plane of the boundary value problem, that is, the horizontal plane (x, y). In that plane, the problem has only one characteristic length—namely the well-known flexural wavelength of a plate on elastic foundation, L_0. As it happens, L_0 is not proportional to h but to $h^{3/4}$. Thus it follows that the exponent of L_0 in the scaling law is $(-3/8)(4/3) = -1/2$. So the scaling of thermal bending fracture does in fact obey the previously mentioned LEFM scaling law:

$$\Delta T_{cr} \propto L_0^{-1/2}, \qquad (4.2)$$

Simplified calculations (Bažant, 1992b) have shown that, in order to propagate such a long thermal bending fracture through a plate 1m thick, the temperature difference across the plate must be about 25°C, while for a plate 6m thick the temperature difference needs to be only 12°C. This is a large size effect. It may explain why very long fractures in the Arctic Ocean are often seen to run through the thickest floes rather than through the thinly refrozen water leads between and around the floes (as observed by Assur in 1963).

Numerical Simulation of Vertical Penetration

An important practical problem, in which the scaling is different, is the failure caused by vertical (downward or upward) penetration through the floating ice plate (Fig. 4.2). In that case, the fractures are known to form a star pattern of radial cracks (Fig. 4.2, top left) which propagate outward from the loaded area. The failure occurs when the circumferential cracks begin to form, as indicated by the load-deflection diagram in Fig. 4.2 (bottom).

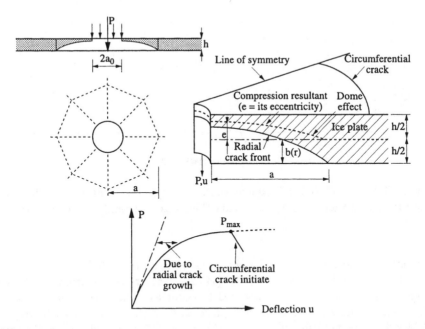

Figure 4.2: Top left: radial and circumferential cracks caused by vertical penetration of an object through floating sea ice plate. Top right: part-through radial crack and shift of compression resultant causing dome effect. Bottom: typical load-deflection diagram

This problem was initially analyzed under the assumption of full-through bending cracks, in which case the asymptotic scaling law for large cracks again appears to be of the type $h^{-3/8}$ (Slepyan, 1990, Bažant 1992a). However, experiments as well as finite element analyses show that the radial cracks before failure do not reach through the full thickness of the ice plate, as shown in Fig. 4.2 (top). This enormously complicates the analysis.

To solve this problem (Bažant and Kim 1997), the elasticity of one half of the sector of the floating plate limited by two adjacent radial cracks is characterized by a compliance matrix obtained numerically. The radial cross section with the crack is subdivided into narrow vertical strips. In each strip, the crack is assumed to initiate through a plastic stage (representing an approximation of the cohesive zone). This is done according to a strength criterion (in the sense of Dugdale model), with constant in-plane normal stress assumed within the cross section part where the strain corresponding to the strength limit is exceeded.

For the subsequent fracture stage, the relationship of the bending moment M and normal force N in each cracked strip to the additional rotation and

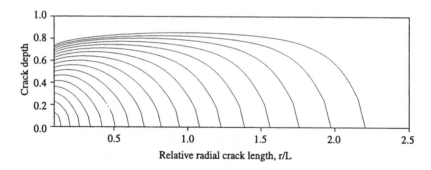

Figure 4.3: Calculated subsequent profiles of the radial part-through crack, after Bažnt and Kim (1998) (the plate thickness is strongly exaggerated)

in-plane displacement caused by the crack is assumed to follow the nonlinear line spring model of Rice and Levy (1972). The transition from the plastic stage to the fracture stage is assumed to occur as soon as the fracture values of M and N become less than their plastic values (to do this consistently, the plastic flow rule is assumed such that the ratio M/N would always be the same as for fracture).

This analysis (Bažant and Kim 1997) has provided the profiles of crack depth shown in Fig. 4.3, where the last profile corresponds to the maximum load (the plate depth is greatly exaggerated in the figure). The figure also shows the radial distribution of the nominal stresses due to bending moment and to normal force. The normal forces transmitted across the radial cross section with the crack are found to be quite significant. They cause a dome effect which helps to carry the vertical load.

An important question in this problem is the number of radial cracks that form. The solution (Bažant and Li, 1995) shows that the number of cracks depends on the thickness of the plate and has a significant effect on the scaling law.

The numerical solution of the integral equation along the radial cracked section, expressing the compatibility of the rotations and displacements due to crack with the elastic deformation of the plate sector between two cracks, yields the size effect plot shown in Fig. 4.4. The numerical results shown by data points can be relatively well described by the generalized size effect law of Bažant, shown in the figure. The top of the figure indicates the number of radial cracks for each range of crack thicknesses. The deviation of the numerical results from the smooth curve, seen in the middle of the range in the figure, is probably caused by insufficient density of nodal points near the fracture front. As confirmed by Fig. 4.4, the asymptotic size effect does not have the

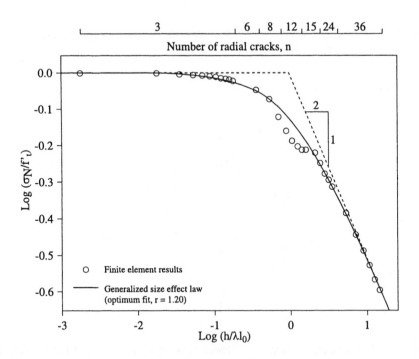

Figure 4.4: Size effect curve calculated by analysis of growth of part-through cracks, with varying number of radial cracks for different thickness ranges

slope $-3/8$ but the slope $-1/2$. Obviously, the reason is that, at the moment of failure, the cracks are not full-through bending cracks but grow vertically through the plate thickness.

Approximate Analytical Solution of Vertical Penetration

Consider the ice plate floating on water shown in Fig. 4.5a,b. Failure under a vertical load involves formation of radial bending cracks in a star pattern (Fig. 4.5c). The radial cracks penetrate at maximum load to an average depth of about $0.8h$ and maximum depth $0.85h$ where h is the ice thickness (Fig. 4.6a). The maximum load is reached when polygonal (circumferential) cracks, needed to complete a failure mechanism, begin to form (dashed lines in Fig. 4.5c). The nominal strength is defined as $\sigma_N = P/h^2$.

Superposing the expressions for the stress intensity factor K_I of the part-through radial bending crack of depth a (Fig. 4.6b,d) produced by bending

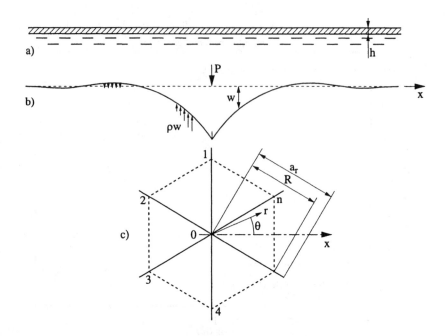

Figure 4.5: Floating ice plate, its deflection under concentrated load and crack pattern

moment M and normal force N (per unit length), one gets

$$K_I = \frac{\sqrt{\pi a}}{h}\left[\frac{6M}{h}F_M(\alpha) + NF_N(\alpha)\right] \quad (4.3)$$

where functions $F_M(\alpha)$ and $F_N(\alpha)$ are found in handbooks. According to Irwin's relation, the energy release rate is

$$\mathcal{G} = \frac{K_I^2}{E'} = \frac{N^2}{E'h}g(\alpha) \quad (4.4)$$

where g is a dimensionless function, $g(\alpha) = \pi\alpha\left[(6e/h)F_M(\alpha) + F_N(\alpha)\right]^2$; ($\alpha = a/h$); $e = -M/N$ = eccentricity of the normal force resultant in the cross section (positive when N is above the mid-plane), and $E' = E/(1 - \nu^2)$.

f4

To relate M and N to vertical load P, consider element 12341 of the plate (Fig. 4.5c and 4.6e,f,g), limited by a pair of opposite radial cracks and the initiating polygonal cracks. The depth to the polygonal cracks at maximum load is zero, as they just initiate, and since the cracks must form at the location of the maximum radial bending moment, the vertical shear force on the planes

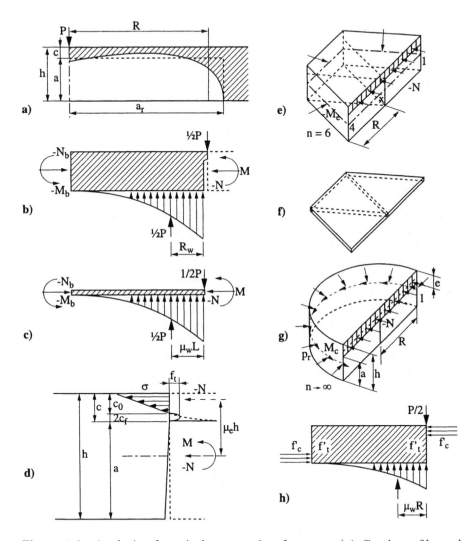

Figure 4.6: Analysis of vertical penetration fracture: (a) Crack profile and (b-h) forces acting on element 123401

of these cracks is zero. The distance R of the polygonal cracks from the vertical load P must be proportional to the characteristic length L since this is the only constant in the differential equation governing the problem, and so we may set $R = \mu_R L$ where dimensionless μ_R is assumed to be a constant.

In each narrow radial sector, the resultant of the water pressure due to deflection w (Fig. 4.6b,c) is located at a certain distance r_w from load P. Since r_w can be solved from the differential equation for w, and since the solution depends only on one parameter, the characteristic length L, it is necessary that r_w be proportional to L. Integration over the area of a semi-circle of radius r_w yields the resultant of water pressure acting on the whole element 12341. Again the distance of this resultant, whose magnitude is $P/2$, from load P must be proportional to L, i.e., may be written as $R_w = \mu_w L$ where μ_w is a constant that can be solved from the differential equation of plate deflections. Of course, μ_w is a constant only as long as the behavior is elastic; this is true only if the crack depth a is constant, which is what is assumed.

The normal force N and bending moment M on the planes of the radial cracks and the polygonal cracks may be assumed to be approximately uniform. The condition of equilibrium of horizontal forces acting on element 12341 in the direction normal to the radial cracks requires the normal forces on the planes of the polygonal cracks to be equal to the normal force N acting in the radial crack planes. The axial vectors of the moments M_c acting on the polygonal sides are shown in Fig. 4.6e,g by double arrows. Summing the projections of these axial vectors from all the polygonal sides of the element, one finds that their moment resultant with axis vector in the direction 14 is $2RM_c$, regardless of the number n of radial cracks. So, upon setting $R = \mu_R L$, the condition of equilibrium of the radial cracks with respect to the moments about axis 14 (Fig. 4.6b,c,e,g) located at mid-thickness of the cross section may be written as:

$$2(\mu_R L)M + 2(\mu_R L)M_c - \tfrac{1}{2}P(\mu_w L) = 0 \tag{4.5}$$

Furthermore, condition (4.4) of vertical propagation of the radial bending cracks must be taken into account; it may be written as $\mathcal{G} = G_f$ where G_f is the fracture energy of ice. Thus, the critical value of normal force (compressive, with eccentricity e) may be written as

$$N = -\sqrt{\frac{E'G_f h}{g(\alpha)}} \tag{4.6}$$

The polygonal cracks are initiated when the normal stress σ reaches the tensile strength f'_t of the ice. As in all heterogeneous brittle materials, one must expect that a layer of distributed microcracking, of some effective constant thickness D_b that is a material property, will form at the top face of ice plate before the polygonal crack will start propagating. According to the nonlocal concept for distributed damage, we may assume that the polygonal cracks initiate when the average stress in this layer reaches the strength f'_t, and since the average stress is roughly the elastically calculated stress for the middle of layer D_b, the criterion of initiation of the polygonal cracks simply is

$$\frac{M_c}{h^3/12}\left(\frac{h}{2} - \frac{D_b}{2}\right) + \frac{N}{h} = f'_t \qquad (4.7)$$

This criterion, however, can be valid only when h is sufficiently larger than D_b, i.e., asymptotically for $h/D_b \to \infty$. The case $h < D_b$ is physically meaningless. For $h = D_b$, i.e., when the distributed cracking zone encompasses essentially the whole depth of plate, the moment at failure can be determined as the plastic bending moment, which may be approximately taken as $1.5\times$ larger than the elastic bending moment for the same material strength. This condition and the asymptotic properties for $h \gg D_b$ are satisfied by replacing (4.7) with the criterion:

$$\frac{6M_c}{h^2}\frac{h+D_b}{h+2D_b} + \frac{N}{h} = f'_t \qquad (4.8)$$

Indeed, this is evidenced by the large-size approximation

$$\frac{h+D_b}{h+2D_b} = \frac{1+D_b/h}{1+2D_b/h} \approx \left(1 + \frac{D_b}{h}\right)\left(1 - \frac{2D_b}{h}\right) \approx 1 - \frac{D_b}{h} \qquad (4.9)$$

This means that, for large enough h, (4.8) and (4.7) are equivalent asymptotically, up to the first order in $1/h$.

In Eq. (4.5), we may substitute $M = -Ne = N\mu_e h$ where the normal force N is defined to be positive when tensile, although the actual value of N is negative (compression); and $\mu_e = e/h =$ dimensionless parameter whose value at maximum load may be assumed as approximately constant. This assumption is indicated by the numerical simulation of Bažant and Kim (1998), from which it further transpires that $\mu_e \approx 0.45$, as a consequence of the fact the average crack depth a at maximum load is about $0.8h$ (in any case, $\mu_e < 0.5$, and so a possible error in μ_e cannot have a large effect). The value 0.45 approximately corresponds to the correct number of cracks in the star pattern; if there were more cracks, the depth would be smaller, if fewer, larger.

Next we may express M_c from (4.5) and substitute it into (4.8). Then, taking into account (4.6), we obtain the equation:

$$\frac{3(h+D_b)\mu_w}{2h(h+2D_b)\mu_R}\sigma_N = \left(\frac{6\mu_e(h+D_b)}{h^2(h+2D_b)} + \frac{1}{h}\right)\sqrt{\frac{E'G_f h}{g(\alpha)}} + f'_t \qquad (4.10)$$

which may be rearranged as

$$\sigma_N = \frac{2\mu_R}{3\mu_w}\left\{\left[6\mu_e + \frac{h+2D_b}{h+D_b}\right]\sqrt{\frac{E'G_f}{hg(\alpha)}} + f'_t\frac{h+2D_b}{h+D_b}\right\} \qquad (4.11)$$

Finally, one needs to decide how the values of α at maximum load should change with ice thickness h. Since ice is a quasibrittle material, a finite fracture process zone (FPZ) of a certain characteristic depth $2c_f$ that is a material property must exist at the tip of the vertically propagating radial crack. This zone was modeled in the numerical simulations of Bažant and Kim (1998) as a yielding zone. The tip of the equivalent LEFM crack lies approximately at the middle of the FPZ, i.e., at a distance c_f from the actual crack tip (Bažant and Planas 1998), whose location is denoted as a_0. The value of $\alpha_0 = a_0/h$ may be expected to be approximately constant when ice plates of different thicknesses h are compared. Thus, denoting $g'(\alpha_0) = dg(\alpha_0)/d\alpha_0$, one may introduce the approximation $g(\alpha) \approx g(\alpha_0) + g'(\alpha_0)(c_f/D)$. Substituting this into (4.11) and rearranging, one gets for the size effect the formula

$$\sigma_N = \frac{4\mu_R}{\mu_w}\left[\mu_e + \frac{h+2D_b}{6(h+D_b)}\right]\sqrt{\frac{E'G_f}{hg(\alpha_0) + c_f g'(\alpha_0)}} + \frac{\mu_R}{3\mu_w}\frac{h+2D_b}{h+D_b}f'_t \quad (4.12)$$

The results of numerical simulations of Bažant and Kim (1998) were found to be quite well represented by the simple size effect law in Eq. (2.9) with finite large-size residual strength σ_r (Bažant 1987b), which reads

$$\sigma_N = \sigma_0\left(1 + \frac{h}{h_0}\right)^{-1/2} + \sigma_r \quad (4.13)$$

σ_r, however, appeared to be negliglible in Bažant and Kim's (1998) numerical simulations, in which case this formula reduces to the original size effect law (Bažant 1984a). Equation (4.12) reduces to this law when $D_b = 0$, in which case

$$\sigma_0 = \frac{4\mu_R\mu_e}{\mu_w}\sqrt{\frac{E'G_f}{c_f g'(\alpha_0)}}, \quad h_0 = c_f\frac{g'(\alpha_0)}{g(\alpha_0)}, \quad \sigma_r = \frac{\mu_e}{3\mu_w}f'_t \quad (4.14)$$

Furthermore, $\sigma_r = 0$ if $f'_t = 0$, which seems to be a reasonable simplification. Anyway, the values of D_b and f'_t are probably too small to have an appreciable effect.

The terms in (4.12) containing D_b anyway decrease with increasing h much more rapidly than (4.13)—for large h as $1/h$ compared to $1/\sqrt{h}$. Consequently, they must become negligible for sufficiently large h regardless of the value of D_b. As in Eq. (4.13), equation (4.12) plotted as $\log \sigma_N$ versus $\log h$ approaches for large h a downward inclined asymptote of slope $-1/2$ and represents the large-size form of the size effect law in (4.12).

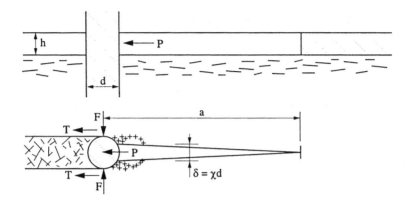

Figure 4.7: Cleavage crack in ice plate pushing against a fixed structure

Force Applied by Moving Ice on a Fixed Structure

Prediction of a force applied by a moving ice plate on a fixed structure, such as an ocean oil platform, is a very complex problem. Several different types of ice break-up mechanism can take place. One mechanism is the global buckling of the ice plate as a plate on elastic foundation. It can occur only for sufficiently thin plates and leads to a reverse size effect of ice thickness, as already discussed in this book. Another mechanism is the compression fracture of ice plate in contact with the structure, which leads to a size effect of ice thickness. Here the general discussion of compression fracture in this book is pertinent. The third possible mechanism is the overall fracture of a finite-size ice floe impacting the structure, which is similar to the fracture problems already discussed and leads to a size effect of the floe size. The fourth possible mechanism consists of a long cleavage crack in the ice plate, propagating against the direction of ice movement (Fig. 4.7). It leads to a size effect of the effective diameter d of the structure which is different from those already discussed. Therefore, it will be considered in some detail.

The resistance of the crack to opening causes the ice to exert on the structure a pair of transverse force resultants F and a pair of tangential forces T in the direction of movement; $T = F \tan \varphi$ where φ may be regarded as the friction angle. Forces T have no effect on the stress intensity factor K_I at the crack tip. Considering the ice plate as infinite, we have

$$K_I = \frac{F}{h}\sqrt{\frac{2\pi}{a}} \qquad (4.15)$$

(Murakami 1987). To determine the crack length a (Fig. 4.7), we need to calculate the crack opening δ caused by F. To this end, one may recall the well-known calculation of the energy release rate:

$$\mathcal{G} = \frac{1}{h}\left[\frac{\partial \Pi^*}{\partial a}\right]_F = \frac{1}{h}\frac{d}{da}\left[\tfrac{1}{2}C(a)F^2\right] = \frac{F^2}{2h}\frac{dC(a)}{da} \qquad (4.16)$$

where $C(a)$ is the load-point compliance of the pair of forces F (Fig. 4.7). Upon using (4.15) and Irwin's relation, we have at the same time

$$\mathcal{G} = \frac{K_I^2}{E} = \frac{2\pi F^2}{Eh^2 a} \qquad (4.17)$$

Equating (4.16) and (4.17), we thus get

$$\frac{dC(a)}{da} = \frac{4\pi}{Eha} \qquad (4.18)$$

This expression is now integrated from $a = d/2$ (surface of structure, considered as circular (Fig. 4.7) to a (note that integration from $a = 0$, which would give infinite C, would be meaningless because a cannot be less than d). In this manner we obtain $C(a)$, and from it the opening deflection δ:

$$\delta = C(a)F = \frac{4\pi F}{Eh}\ln\left(\frac{2a}{d}\right) \qquad (4.19)$$

If cleavage fracture were the only mode of ice breaking, we would have $\delta = d$. However, as will be discussed later, there is likely to be at least some amount of local crushing at, and ahead, of the structure. Consequently, the relative displacement between the two flanks of the crack is no doubt less than d. We denote it as χd where χ is a coefficient less than 1. Setting $\delta = \chi d$, we solve from (4.19):

$$a = \frac{d}{2}\exp\left(\frac{Eh\chi d}{4\pi F}\right) \qquad (4.20)$$

(note that a/d is not constant but increases with d; hence, the fracture modes are not geometrically similar, and so the LEFM power scaling cannot be expected to apply). Substituting (4.20) into (4.15), setting $K_I = K_c = \sqrt{EG_f}$ (Irwin's relation, K_c = fracture toughness of ice), and solving for F, we obtain

$$\frac{2\sqrt{\pi}F}{h\sqrt{EG_f d}} = \exp\left(\frac{Eh\chi d}{8\pi F}\right) \qquad (4.21)$$

The pair of forces F is related to load P on the structure ($P = 2T$, Fig. 4.7) by a friction law, which may be written as $P = 2F\tan\varphi$ where φ is the friction angle. Substituting $F = P/2\tan\varphi$ and $P = \sigma_N hd$ into (4.21), and solving the resulting equation for d, we obtain, after rearrangements,

$$\frac{d}{d_c} = \frac{1}{\tau^2}e^{1/\tau}, \qquad \tau = \frac{\sigma_N}{\sigma_c} \qquad (4.22)$$

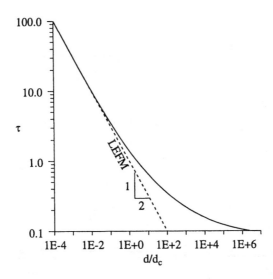

Figure 4.8: Size effect associated with cleavage fracture

in which τ is the dimensionless nominal strength, and d_c and σ_c are constants defined as

$$d_c = \frac{4\pi G_f}{\chi^2 E}, \qquad \sigma_c = \frac{\chi \tan \varphi}{2\pi} E \qquad (4.23)$$

Equation (4.22), plotted in Fig. 4.8, represents the law of cleavage size effect in an inverted form. The small-size asymptotic behavior is the LEFM scaling for similar structures with similar cracks:

$$\text{for } d \ll d_c: \quad \sigma_N \approx \sqrt{d_c/d} \qquad (4.24)$$

The plot of (4.22) in Fig. 4.8 shows that the size effect is getting progressively weaker with increasing structure diameter d (although no horizontal asymptote is approached by the curve). The reason for this is that the cracks of various lengths are dissimilar, i.e., the ratio, a/d, of crack length to structure diameter is not the same for different sizes but increases with the structure size.

4.2. Size Effect on Softening Inelastic Hinges in Beams and Plates

Post-peak softening of inelastic hinges in beams or plates (Fig. 4.9a) can be engendered by both tensile and compression fracture (or by localization of distributed cracking damage). While the effective length of the plastic zone in plastic hinges in steel beams is proportional to the characteristic size D, which

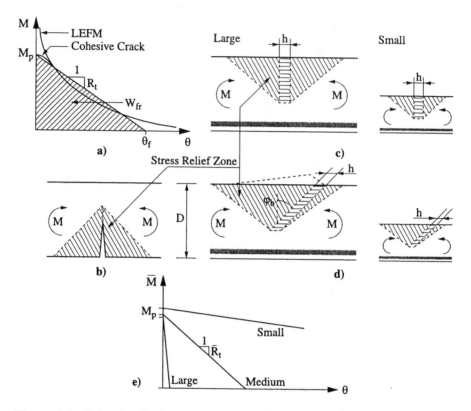

Figure 4.9: Softening inelastic hinge; (a) its moment rotation diagram, (b) hinge failing by tensile crack, (c) and (d) hinge failing by compression fracture band, and (e) size effect on moment-rotation diagram

may be taken as the beam depth D, the effective hinge length of fracturing hinges (Fig. 4.9b,c) is, under certain plausible hypotheses, approximately constant, roughly equal to the characteristic length of the material. The cause is the localization instability of strain softening (Bažant and Cedolin 1991, Chapter 13), and one consequence is the size effect.

Hinge softening due to tensile fracture occurs for example in plain (unreinforced) concrete beams or plates or in floating ice plates, due to propagation of a crack from the tensile face (Fig. 4.9b). The energy dissipated by the crack per unit area of the cross section is G_f.

Hinge softening due to compression fracture occurs for example in reinforced concrete beams if the bending moment M is accompanied by a sufficient axial compression force N (which is the case for prestressed concrete beams, for columns with a large enough axial force, and for frames or arches with a large enough horizontal thrust). It also occurs if there is strong enough tensile reinforcement (Fig. 4.9c,d), for instance, if the beam is overreinforced because of a retrofit with a fiber laminate bonded to the tensile face.

As discussed later in this treatise, the compression fracture consists of an inclined band of axial splitting microcracks (a crushing band, which is often regarded as a shear failure in compression although shear slip becomes possible only after the axial stress gets reduces well below the peak). The plane of the band, inclined by angle ϕ_b with respect to the orthogonal cross section, can intersect the cross section in a line that is vertical (Fig. 4.9c where the inclination would be seen by looking from top down) or horizontal (Fig. 4.9d). The energy, G_b, dissipated by formation of the band, per unit area of the cross section (unit area of the projection of the compression fracture onto the critical cross section) plays the role of the fracture energy of the band, considered to be a material constant.

The assumption of constant G_b is indicated theoretically by studies of localization of softening damage. However, it must be pointed out that this assumption has not yet been systematically tested in the laboratory and lacks at present clear experimental support. On the other hand, it is clear that an assumption of the softening damage in compression remaining distributed, with no localization at all, would definitely be wrong. Only for such an assumption the size effect would be absent (such an assumption would imply the area under the stress-strain curve, including its softening tail, representing energy per unit volume, W_f, to be a constant, while full localization implies G_b, the energy per unit area, to be a constant; the reality might be somewhere in between, but definitely closer to a constant energy per unit area of cross section).

As explained in a previous section, the nominal strength at the initiation of a fracture due to bending moment M_p (peak moment in Fig. 4.9a) exhibits a size effect of the the type described by formula (2.39) or (2.46); therefore, according to formula (2.39),

$$M_p = M_0 q(D), \qquad q(D) = \left(1 + \frac{rD_b}{D}\right)^{1/r} \qquad (4.25)$$

M_0 is a constant representing the bending moment at fracture initiation in a cross section of infinite size, and function $q(D)$, introduced before, describes the size effect at fracture initiation. Since $f_{r\infty} = M_0 D/2I = 6M_0/bD^2$ (I = moment of inertia of cross section), we may write

$$M_0 = f_{r\infty} bD^2/6 \qquad (4.26)$$

For a reinforced concrete cross section failing by concrete crushing, the same expression can be derived except that $f_{r\infty}$ is multiplied by a constant depending on the shape of cross section and the reinforcement ratio.

According to linear elastic fracture mechanics, the diagram of the softening moment-rotation relation for the inelastic hinge is a curve descending from infinity, as sketched in Fig. 4.9a. If the nonlinear cohesive nature of a tensile crack, or the finite size of the fracture process zone of the tensile crack or compression crushing band is taken into account, the diagram starts its descent from a finite value, M_p.

For the sake of simplicity, we will idealize the moment-rotation diagram as linear (triangular, Fig. 4.9a), i.e., $M = R_t(\theta_f - \theta)$ where R_t is the tangent stiffness of the hinge (representing the slope ot the M-θ diagram) and θ_f is the hinge rotation at complete break (Fig. 4.9a). The energy W_{fr} dissipated by a total break of the cross section is given by the area under this diagram (Fig. 4.9a);

$$W_{fr} = \tfrac{1}{2} M_p \theta_f \tag{4.27}$$

Attention will now be restricted to rectangular cross sections and plates (although generalization to arbitrary cross sections would not be difficult). The energy dissipated over the whole cross section upon reaching a complete break may then alternatively be written as

$$W_{fr} = G_b b D \tag{4.28}$$

where b is the width of a rectangular cross section (in the case of a hinge in a plate, we consider a unit width $b = 1$). In view of energy conservation, both expressions for W_{fr} may be set equal, which provides

$$\theta_f = \frac{2 G_b b D}{M_0 q(D)} = \frac{12 G_b}{f_{r\infty}} \frac{1}{D q(D)} \tag{4.29}$$

To bring to light the scaling, it is helpful to introduce the dimensionless bending moment and the dimensionless tangential softening stiffness of the softening inelastic hinge:

$$\bar{M} = \frac{M}{E b D^2}, \qquad \bar{R}_t = \frac{R_t}{E b D^2} \tag{4.30}$$

where the tangent stiffness is defined as $R_t = M_p/\theta_f$. For non-softening elasto-plastic materials, the diagram of $\bar{M}$ versus θ is independent of the beam depth (e.g. Bažant and Jirásek 2001), and so any change in this diagram as a function of structure size reveals a size effect. For our softening material and for a rectangular cross section, we find, upon substituting (4.29) into (4.30) with $R_t = F_{r\infty} b D^2 g(h)/6\theta_f$, that

$$\bar{R}_t = \frac{q^2(D) D}{72 l_{fr}}, \qquad l_{fr} = \frac{E G_b}{f_{r\infty}^2} \tag{4.31}$$

where l_{fr} represents an Irwin-type characteristic length of the material (e.g., Bažant and Planas 1998).

As we see from (4.31) and (4.29), if D is large enough for $q(D)$ to become almost constant, the dimensionless softening stiffness of the hinge increases in proportion to size D, and the rotation at full break decreases at inverse proportion to D (Fig. 4.9). As the structure size approaches infinity, $\bar{R}_t$ becomes infinite (a vertical drop). As the structure size tends to zero, $\bar{R}_t$ becomes zero (a horizontal line), and thus the softening hinge becomes equivalent to a plastic hinge. These are size effects with important consequences for structures, which we explore next. In plasticity, by contrast, one cannot speak of any size effect because $\bar{R}_t = R_t = 0$ for any size, $\theta_f \to \infty$, and $q(D) = $ const. for any D.

Note: Instead of $\bar{M}$ and $\bar{R}_t$ we could work with the nominal bending moment $M_N = M/D^2$ and the nominal tangent stiffness $R_{tN} = R_t/bD^2$. This would be more in line with the entrenched usage of nominal strength σ_N in lieu of the dimensionless strength $\bar{\sigma} = \sigma_N/E$. But there is no precedent for using M_N and R_{tN}, and for the analysis of size effect if makes no difference.

4.3. Size Effect in Beams and Frames Failing by Softening Hinges

The size effects of softening hinges are important for:

- the energy absorption capability of structures (which governs the resistance to earthquake, blast, shock and impact), and

- the strength of redundant brittle beams and frames, as well as the strength of plates.

We now turn attention to the latter (whereas the former will be discussed in a later section).

Consider a statically indeterminate (redundant) beam structure which requires N inelastic hinges to form in order to collapse (e.g., four hinges in Fig. 4.10). Let the inelastic hinges be numbered as $j = 1, 2, ...N$ in the sequence in which they form as the load-point displacement w is increased. Let K_i be the stiffness associated with P if hinges $j = 1, 2, ...i - 1$ have completely softened (i.e., $M = 0$ and $\theta \geq \theta_f$ in these hinges) and hinges $j = i, i + 1, ...N$ have not yet started to form (i.e., $M \leq M_p$ and $\theta = 0$ in these hinges). Obviously, $K_1 > K_2 > K_3 > ... > K_N > 0$.

If all the hinges $j = 1, 2, ...i - 1$ have softened to a zero moment and hinge i has not yet started to form, the load-point deflection w is decided solely by K_i and the start of softening in the next hinge i is decided by a critical stress,

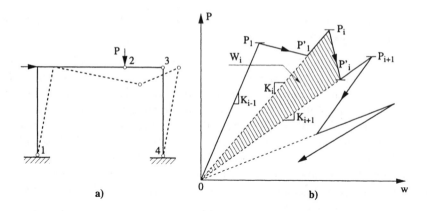

Figure 4.10: (a) Frame failing by softening inelastic hinges, and (b) load-deflection diagram when one and only one hinge is softening at a time

σ_i, representing the stress at the tensile face in the case of tensile fracture or at the compression face in the case of compression fracture; therefore

$$P = K_i w, \qquad \sigma_i = S_i w \qquad (4.32)$$

where S_i are constants. Restricting again attention to rectangular cross sections, one may introduce dimensionless structure stiffness $\bar{K}_i$ and dimensionless critical stress $\bar{S}_i$ in hinge i, such that

$$K_i = \bar{K}_i E b, \qquad S_i = \bar{S}_i E / D \qquad (4.33)$$

where D is the characteristic size of the structure, considered as two-dimensional, and b is the characteristic thickness in the third dimension (K_i has been normalized by b because it increases in proportion to b, while S_i has not, because the stress in a rectangular cross section depends only on the bending moment per unit width).

Assume now that the hinges form and fully soften one by one, i.e., no two hinges are softening at the same time (it will be seen that a large enough structure always fails this way). Then the load-deflection diagram must look as shown in Fig. 4.10b, where the slope of each ray emanating from the origin is K_i and the maximum load on each ray marked P_i $(i = 1, 2, ...)$, corresponds to the start of softening of the next hinge. The trough P'_i on each ray is the load at which the softening of each hinge gets completed (i.e., $M = 0$ and $\theta = \theta_f$). Since we assumed the moment-rotation diagram of the hinges to be linear, the load-deflection diagram from P_i to P'_i must be a straight line.

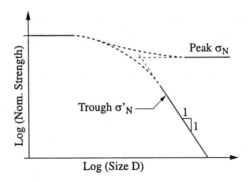

Figure 4.11: Size effects on nominal strengths at a peak and at a trough, matched to small-size plastic limit

The load peaks P_i are determined from the condition $\sigma_i = f_r = f_{r\infty}q(D)$. Noting that $\sigma_i = PS_i/K_i$, we find:

$$P_i = \frac{\bar{K}_i}{\bar{S}_i} f_{r\infty} b D q(D), \qquad (4.34)$$

The corresponding nominal stresses are

$$\sigma_{Ni} = \frac{P_i}{bD} = \frac{\bar{K}_i}{\bar{S}_i} f_{r\infty} q(D) \qquad (4.35)$$

As may have been expected, peaks σ_{Ni} exhibit only the size effect of fracture initiation (Fig. 4.11), the same as that in the modulus of rupture.

To determine the troughs P'_i, consider the area of the shaded triangle in Fig. 4.10b between the rays of slopes K_i and K_{i+1}; the area is

$$W_i = \frac{P_i P'_i}{2}\left(\frac{1}{K_{i+1}} - \frac{1}{K_i}\right) \qquad (4.36)$$

W_i represents the work dissipated when hinge i softens from M_p to 0. Since no other hinge is assumed to be softening simultaneously with hinge i, energy conservation requires W_i to be equal to the work dissipated by hinge i;

$$W_i = G_b b_i D_i = G_{fr}\beta_i \delta_i D^2 \quad \text{with} \quad \beta_i = b_i/D, \quad \delta_i = D_i/D \qquad (4.37)$$

where b_i and D_i are the width and depth of the cross section at hinge i, and β_i and δ_i are constants for structures geometrically similar in three dimensions. Setting this equal to (4.36), we can solve for P'_i and obtain:

$$\frac{\sigma'_{Ni}}{\sigma_{Ni}} = \frac{P'_i}{P_i} = \frac{2EG_b}{\sigma_{Ni}^2}\left(\frac{1}{\bar{K}_{i+1}} - \frac{1}{\bar{K}_i}\right)^{-1}\frac{1}{D} = \frac{\text{const.}}{D} \qquad \text{(for large } D\text{)} \quad (4.38)$$

Factor $1/D$ represents the large-size asymptotic size effect on the troughs (Fig. 4.11). It is a very strong size effect, much stronger than the LEFM size effect of large cracks, which is of the type $1/\sqrt{D}$.

Fig. 4.12 shows a sequence of typical load-deflection diagrams for geometrically similar structures of various sizes D. These diagrams are plotted in dimensionless coordinates w/D and σ_N/E, and the size effect on the peaks stemming from crack initiation, described by function $q(D)$, is ignored in this plot. This has the advantage that (1) the slopes corresponding to K_i (and, due to considering $q(D) = 1$, also the points corresponding to the peaks P_i) are independent of the size, and that (2) the diagrams for structures of different sizes would be identical if the material were elasto-plastic (in which case the size effect is known to be absent). Thus, any change in these diagrams with the structure size signifies a size effect.

The scaling of the troughs P'_i, as described by (4.38), indicates that when the structure size increases, the response approaches a series of narrow spikes of progressively decreasing inclination (Fig. 4.12). The descending part of each spike in Fig. 4.12e is unstable for any type of control of P and w, and is known as the *snapback instability*. This behavior may have been intuitively expected on the basis of the fact that the diagram of $\bar{M}$ versus θ approaches a vertical drop as $D \to \infty$.

When the structure size decreases, the line from P_i to P'_i eventually changes its slope from negative to positive, i.e., P_i ceases to be a peak and P'_i ceases to be a trough. With a further size increase, P'_i eventually becomes coincident with the peak P_i. For still smaller sizes, there exists, during the loading process, a period in which hinges i and $i+1$ soften simultaneously. During such simultaneous softening, the deflection diagram is again linear (lines 67 and 89 in Fig. 4.12), but different for each softening hinge combination and harder to calculate. When the size is decreased further, more and more hinges, or all the hinges, undergo softening at the same time.

Segments 67 and 89 in Fig. 4.12b correspond to simultaneous softening of two subsequently formed hinges. Since areas 0120 and 0340 must be equal to the fracture energies W_{i-1} and W_i of hinges $i-1$ and i, the area 01678950 must be equal to their sum. Therefore the two little shaded triangles of each pair Fig. 4.12b must have equal areas. This property can be exploited for constructing the response graphically.

In the theoretical asymptotic case of a zero size ($D \to 0$), the slope of the diagram of dimensionless moment $\bar{M}$ versus rotation θ tends to horizontal and, consequently, the plastic limit analysis applies. In Fig. 4.9a, the zero-size limiting response consists of line segments parallel to the rays emanating from the origin, which is easy to calculate by well-known methods (note the parallel double dashes marking parallel lines).

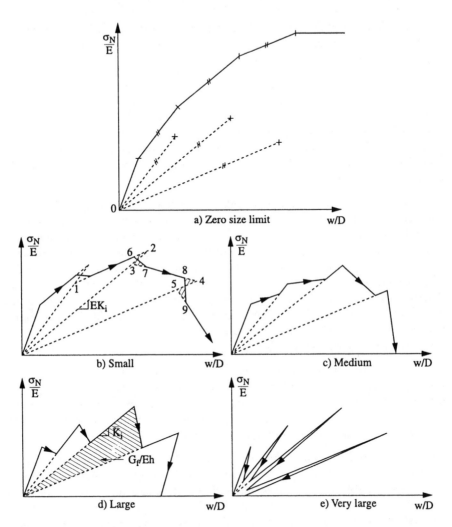

Figure 4.12: Evolution of load-displacement diagrams (in terms of dimensionless nominal strength σ_N/E and relative deflection w/D) with increasing size D of a structure failing by softening hinges (in such coordinates, the secant stiffness slopes, as well as the peak points for $q(D) = 1$, do not change with the size). Parallel dashes in (a) mark parallel lines

The behavior between this asymptotic case and the case of the smallest structure for which only one hinge is softening at a time (in which $P'_i = P_{i+1}$) can be approximately characterized by interpolation, better regarded as asymptotic matching. Matching of the large-size asymptotic size effect in (4.38) to the horizontal small-size asymptote may be achieved by the following simple approximate formulae for all sizes:

$$\sigma'_{Ni} \approx \frac{\sigma'_{0i}}{1 + D/D'_{0i}} \quad \text{or} \quad \frac{\sigma'_{0i}}{(1 + D^s/D'^s_{0i})^{1/s}} \qquad (4.39)$$

where σ'_{0i}, D'_{0i} and s are positive constants. For small sizes, however, the largest size effect is caused by the fact that the overall maximum load for different sizes occurs for different i (compare Fig. 4.12 a and b). This aspect is not reflected in Fig. 4.11.

When the structure is large enough for the load-deflection diagram to involve softening segments, designing for the largest peak $\max P_i$ becomes questionable because, under load control, the structure is unstable during each softening segment. Such a design is even more questionable when the structure is large enough for snapbacks to occur, in which case it is unstable even under load-point displacement control (Bažant and Cedolin 1991). A completely safe design is that for the lowest trough, $\min_i P'_i$, but then again the safety margin would often be unnecessarily high by far. A realistic design load P lies somewhere between $\max P_i$ and $\min_i P'_i$, for a smaller structure closer to $\max P_i$ and for a larger one closer to $\min_i P'_i$.

To decide the proper design load rationally, one should take into account the imperfections and possible dynamic disturbances. In structures with plastic hinges, a dynamic disturbance in which the peak load is reached only temporarily leads merely to a permanent deflection but in the case of post-peak snapback it may trigger stability loss and dynamic failure.

The problem is similar to that of an axially compressed thin cylindrical shell (or a thin spherical dome), for which the critical load of a theoretically perfect shell (unattainable in practice) is followed by a snapback to a very low residual load. It is now well understood (e.g., Bažant and Cedolin 1991) that, due to inevitable imperfections and dynamic disturbances, one must design the shell for that residual load (typically between 1/8 and 1/3 of the theoretical critical load). An effective semi-empirical method has been developed to deal with this problem for shells, and a similar method might have to be developed for large beams and frames with softening hinges.

Finally, it should be noted that brittle materials such as concrete or ice are not the only ones that lead to softening inelastic hinges. As shown by Maier and Zavelani (1970), local buckling of thin flanges in the plastic hinge region can cause sharp softening in the moment curvature diagram. The foregoing analysis applies to that case. Obviously, the local buckling of flanges is much more dangerous in large steel structures than in small ones.

4.4. Size Effect in Floating Ice Subjected to Line Load

It is interesting to extend the foregoing analysis to a floating ice plate loaded by a line load P (force per unit length), which behaves as a beam on elastic foundation. The failure occurs by line cracks propagating gradually through the thickness of ice. As they propagate, the bending moment M in the cracked cross section decreases as a function of the additional rotation θ caused by the cracks. The diagram $M(\theta)$ is nonlinear, but can be simplified as linear (same as Fig. 4.9a), which makes it possible to obtain an exact analytical solution.

The only difference from the preceding section is that the sea water always acts as an elastic foundation, i.e., never reaches plastic response (such response can occur only if water floods the top of ice plate, which is possible only if the deflection exceeds 1/11 of plate thickness and is not considered here). The response of the ice beam between the softening hinges is governed by the differential equation for beams on elastic foundation, whose solution is a sum of complex exponentials. Softening hinges initiate at the location of the maximum bending moment, when $M = M_p$.

The sequence of formation of softening hinges is shown in Fig. 4.13 (top). In the last stage with three hinges, the beam without a foundation would be a mechanism but, for small enough thickness ice h, the solution indicates that the load can increase further. It can be shown that further hinges cannot form; rather the last hinge spreads, creating a continuous hinge segment of a finite length. Such a segment, however, cannot be assumed to transmit a shear force, which means that the load P_2 at the moment of formation of the second hinge must in fact be the maximum load. Similar to the preceding section, the load deflection diagrams evolve as shown in Fig. 4.13 (bottom), and the nominal strengths $\sigma_{Ni} = P_i/h$ corresponding to the first and second peaks and troughs can be shown to be:

$$\sigma_{N1} = Aq(h)h^{1/4}, \quad \sigma'_{Ni} = B\sqrt{12}q(h)h^{-3/4} \qquad (4.40)$$

$$\sigma_{N2} = (e^{\pi/4}/\sqrt{2})Aq(h)h^{1/4}, \quad \sigma'_{N2} = 2.588Bq(h)h^{-1/2} \qquad (4.41)$$

where $A = [3\rho(1-\nu^2)/E]^{1/4}2\sigma_p/3$ and $B = \sqrt{12\rho E/(1-\nu^2)}G_f\sigma_p$, and function $q(h)$ is the same as $q(D)$ defined in (2.39) or (2.46) (since D in plate bending theory usually denotes the cylindrical stiffness, for plates we use h instead of D as the characteristic size).

The difference in scaling between the beams without and with elastic foundation can be easily explained as follows. In the former and latter cases, $K_i \propto 1$

100   Scaling of Structural Strength

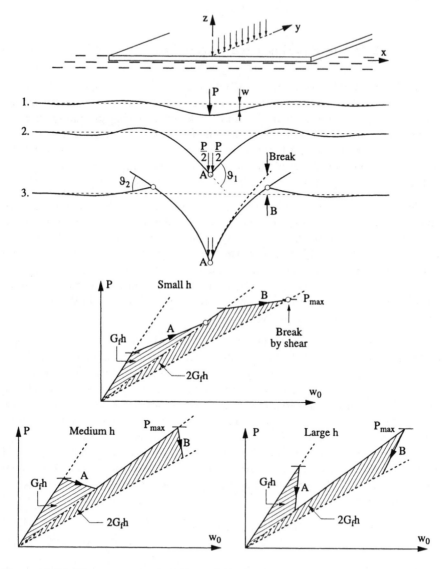

Figure 4.13: Top: sequence of hinge formation in floating ice plate. Bottom: evolution of load-displacement diagrams as ice thickness h increases (the size effect due to crack initiation is not shown)

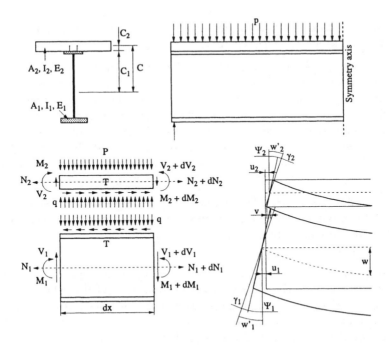

Figure 4.14: Top: composite beam and its cross section. Bottom: internal forces and displacements

and $K_i \propto h^{3/4}$, respectively. So, according to (4.36), $P_i P_i' \propto W_i K_i \propto hq(h)$ and $h^{7/8} q(h)$. Hence, $\sigma_{Ni} \sigma'_{Ni} \propto h^{-1} q^2(h)$ and $h^{-1/4} q^2(h)$, respectively. Since $\sigma_{Ni} \propto q(h)$ and $q(h) h^{1/4}$, we have $\sigma'_{Ni} \propto h^{-1} q^2(h)/\sigma_{Ni} \propto q(h)/h$ without a foundation, and $\sigma'_{Ni} \propto h^{-1/4} q^2(h)/\sigma_{Ni} \propto q(h)/\sqrt{h}$ with a foundation.

4.5. Steel-Concrete Composite Beams and Compound Size Effect

Composite beams consisting of a steel beam and a concrete slab (Fig. 4.14 top) are often used in bridges and buildings. A crucial component are the connectors between the concrete slab and the steel beam. They usually consist of welded steel studs embedded in concrete.

A composite steel-concrete beam of the type used in bridges and building floors (Fig. 4.14) may fail by (1) tensile yielding of the steel beam, (2) compression crushing of the concrete slab, or (3) slip and pullout of the connectors (studs) (see the literature survey in Bažant and Vítek 1999).

Let us now focus on the last mode of failure. As recently demonstrated

(Bažant and Vítek 1999), designing against this failure mode requires consideration of a size effect, which has so far been neglected by structural engineers.

One cause of size effect is the failure of the studs. The studs, always designed strong enough to remain elastic, fail in shear due to formation of large cracks in the concrete slab. This is generally a brittle type of failure exhibiting a strong size effect. The failure is similar to the pullout failure of anchors which is now known to be rather brittle, exhibiting a strong energetic size effect. Recent tests by Kuhlmann and Breuninger (1998) confirm that the shear failure of studs embedded in concrete exhibits a pronounced post-peak softening except when a heavy and dense three-dimensional reinforcing mesh is used.

A second cause of size effect is that, due their softening, the studs cannot reach their maximum shear force simultaneously all along the beam. Rather, the stud failures must propagate along the beam, which is a fracture type behavior calling for energy release analysis.

This problem is instructive for structural designers because it can be solved relying exclusively on the classical beam bending theory. It might be objected by some that the beam bending theory cannot resolve the stress concentrations near the front of the propagating zone of connector failures. Simple though this theory indeed is, it nevertheless does predict satisfactorily the overall energy release from the slender enough beams and is asymptotically exact when the slenderness tends to infinity.

Fig. 4.14 (bottom left) shows an element of the composite beam. Let c_1, c_2 be the distances of the steel–concrete interface from the centroids of the steel beam and the concrete slab, both taken as positive (Fig. 4.14 top); $y = e_1, y = e_2$ are the y-coordinates of the centroids of the beam and slab measured from the centroid of the transformed composite cross section, $e_1 > 0, e_2 < 0$; $Z_i = E_i A_i$ = axial stiffnesses of the concrete and steel parts; $R_i = E_i I_i$ = elastic bending stiffnesses; $S_i = E_i A_{si}$ = elastic shear stiffnesses ($i = 1$ for beam and 2 for slab), and A_{si} = cross section areas A_i after correction for nonuniformity of shear stress distribution.

Let us consider the case of a uniformly loaded simply supported symmetric beam of span L. The slip at midspan vanishes due to the symmetry and only one half of the beam needs to be analyzed. A vertical distributed load p (per unit length of beam) is applied on the concrete slab. The deformations are characterized by deflection w, which is common to both parts of the cross section, slip v between steel and concrete, axial displacements u_1 in the steel beam centroid and u_2 in the concrete slab centroid, and cross section rotations ψ and γ of beam and slab (Fig. 4.14 bottom right). The total bending moment, normal force and shear force transmitted by the whole cross section are denoted as M, N, and V. The interaction of the beam and slab is represented by shear flow F and vertical normal force q distributed along the interface. The steel and concrete parts transmit internal forces M_1, N_1, V_1 and M_2, N_2, V_2, respectively, and the shear flow (shear force per unit length) in the interface is denoted as T.

With regard to design practice, it makes sense to consider geometric similarity in three dimensions, defining the nominal strength of the composite beam as $\sigma_N = P/D^2$ where P = maximum load, with D taken as the cross section depth. In the case of a uniformly distributed load p, P is defined as the load resultant pD, and then $\sigma_N = p/D$.

Generally, the beam may be subjected to an arbitrary distributed or concentrated load with a single load parameter P. The bending moment and shear force in the composite beam may then be expressed in the form $M(x) = PDq(\xi)$ and $V(x) = Pq'(\xi)$ where P = load parameter, x = coordinate measured from the left end of the beam, $\xi = x/D$, $q(\xi)$ = some size-independent dimensionless function, $q'(\xi) = dq(\xi)/d\xi$, and $D = h$ = beam depth.

First we assume, for the sake of simplicity, that there is a sharp (pointwise) transition from connectors that do not slip to connectors that carry the residual shear flow T_r. In other words, the zone of smeared connector failures (analogous to the fracture process zone) is assumed to have a negligible length, as in LEFM. So we consider that there are two symmetrically located regions of length a (Fig. 4.15) such that for $x \leq a$ the shear flow T in the studs has been reduced to the residual shear flow value, T_r (Fig. 4.16) and for $x > a$ there is no slip. In the sense of continuum smearing of the studs, these regions may be regarded as two symmetric sharp interface cracks subjected on their faces to tangential tractions T_r (Fig. 4.16 bottom left).

The interface shear force per unit length of beam (called the *shear flow*) at points just ahead of the tip $x = a$ of the interface crack (i.e., just ahead of the zone of slipped connectors) is given by

$$T(a, c_1) = \frac{V(a)\, Q_{tr}(c_1)}{I_{tr}} \qquad (4.42)$$

$T(a, c_1)$ = shear flow in the interface $(y = c_1)$ just ahead of point $x = a$, I_{tr} = centroidal moment of inertia of the transformed cross section, and $Q_{tr}(c_1)$ = first (static) moment of inertia of the transformed area of concrete about the centroid of the transformed cross section.

Since $V(a) = Pq'(a)$, we may solve from (4.42) the load P_r for which $T(a, c_1) = T_r$ = residual value of the shear flow in the connectors at large slip;

$$P_r = \frac{T_r\, I_{tr}}{Q_{tr}(c_1)\, q'(a)} \qquad (4.43)$$

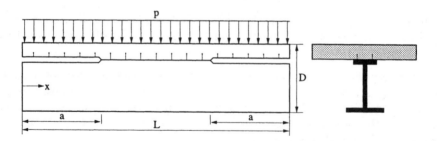

Figure 4.15: Composite beam with studs failed in symmetric crack-like regions of length a

When the load has this value, the shear flow in the connection is continuous through the point $x = a$ and the energy release rate $\mathcal{G}$ due to extending the smeared equivalent interface cracks vanishes.

It is now convenient to imagine the solution for load P as a superposition of the solutions for two loading cases (Fig. 4.16 bottom):

A. the case with load value P_r (or corresponding distributed load p_r), for which the energy release rate $\mathcal{G} = 0$, and

B. the case with load value $P - P_r$ (or distributed load $p - p_r$), for which the tangential tractions on the faces of the smeared equivalent interface cracks vanish.

In loading case A there is obviously no energy release into the tip of the equivalent crack ($\mathcal{G} = 0$) and thus no size effect, and $P = P_r$. In loading case B the energy release rate is non-zero and must be analyzed, which is what we do next. All the size effect arises from this loading case.

The concrete and steel ahead of the crack tip act as a composite beam of bending stiffness
$$R = R_1 + R_2 + A_1 e_1^2 + A_2 e_2^2 \tag{4.44}$$
Behind the crack tip, the steel and concrete in loading case B behave as two separate beams forced to deflect equally (we assume the slab not to lift above the steel beam). Therefore, $M = M_1 + M_2$, and since the curvatures of the steel beam and the slab are equal, their bending moments (for $x < a$) are
$$M_1 = \frac{M R_1}{R_1 + R_2}, \qquad M_2 = \frac{M R_2}{R_1 + R_2} \tag{4.45}$$
respectively, and the bending energies per unit length of beam are $M_1^2/2R_1$ and $M_2^2/2R_2$.

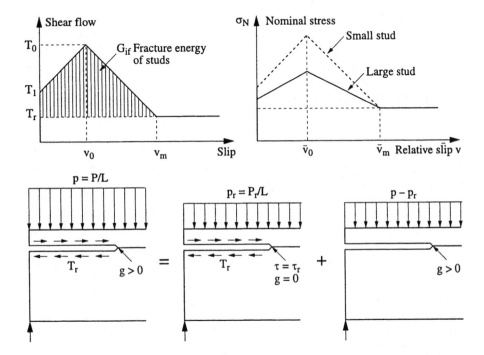

Figure 4.16: Top left: idealized shear force–slip diagram of deformable connectors. Top right: Size effect on nominal stress. Bottom: beam with equivalent interface cracks subjected to residual tractions, and decomposition into two loading cases according the principle of superposition ($p_r = P_r/L$ = distributed load corresponding to P_r)

Small though the contribution of shear strains may be, it is easy to take them into account. Although the Bernoulli-Navier hypothesis of cross sections remaining plane and normal implies zero shear strains, it is known that deflection predictions of deep beams are improved by calculating the shear angles from the shear force V. This is a well known, generally accepted paradox of the classical beam bending theory. For the region without stud failures (no interface crack), the shear angle is V/S where $S =$ shear stiffness of the composite cross section. This means that the complementary shear strain energy $V^2/2S$ per unit length of beam should be included for the region without interface crack. In the region with stud failures (i.e., with interface crack), both parts of the beam undergo the same deflection w. Since $V = V_1 + V_2$ and the shear forces in parts 1 and 2 are $V_1 = R_1 d^3 w/dx^3$ and $V_2 = R_2 d^3 w/dx^3$, one finds that, for the classical bending theory and loading case B, the total shear force $V = V_1 + V_2$ distributes as

$$V_1 = V \frac{R_1}{R_1 + R_2}, \qquad V_2 = V \frac{R_2}{R_1 + R_2} \qquad (4.46)$$

The complementary energies of the shear stresses per unit length of the cracked region of beam are $V_1^2/2S_1$ and $V_2^2/2S_2$.

Integration of the foregoing energy expressions yields the complementary potential energy of the left half of the composite beam:

$$\Pi^* = \int_0^a \left(\frac{M_1^2}{2R_1} + \frac{M_2^2}{2R_2} + \frac{V_1^2}{2S_1} + \frac{V_2^2}{2S_2} \right) dx + \int_a^{L/2} \left(\frac{M^2}{2R} + \frac{V^2}{2S} \right) dx \qquad (4.47)$$

where $a =$ length of the crack (stud failure region). The energy release rate due to growing a is obtained by differentiation with respect to a at constant load. Because the distribution of M and V does not depend on a (as the beam is statically determinate),

$$\left[\frac{\partial \Pi^*}{\partial a} \right]_{P=\text{const.}} = \left[\frac{M_1^2}{2R_1} + \frac{M_2^2}{2R_2} + \frac{V_1^2}{2S_1} + \frac{V_2^2}{2S_2} - \frac{M^2}{2R} - \frac{V^2}{2S} \right]_{x=a} \qquad (4.48)$$

$$= \left[\frac{M^2}{2(R_1 + R_2)} + \frac{V^2}{2S_a} - \frac{M^2}{2R} - \frac{V^2}{2S} \right]_{x=a} \qquad (4.49)$$

in which we inserted (4.45) and (4.46), and made the notation

$$\frac{1}{S_a} = \frac{R_1^2/S_1 + R_2^2/S_2}{(R_1 + R_2)^2} \qquad (4.50)$$

Introducing now dimensionless deflections by $M = (P - P_r) D q(\xi)$, $V = (P - P_r) q'(\xi)$, and noting that the energy release rate must be equal to the energy consumed and dissipated by stud failures per unit length of the interface, $b G_{if}$, one gets:

$$\left[\frac{\partial \Pi^*}{\partial a}\right]_{P=\text{const.}} = \frac{[(P-P_r)Dq(\alpha)]^2}{2E_s D^4 \hat{R}} + \frac{[(P-P_r)q'(\alpha)]^2}{2E_s D^2 \hat{S}} = b\, G_{if} \quad (4.51)$$

with the notations:

$$\frac{1}{\hat{R}} = E_s D^4 \left(\frac{1}{R_1+R_2} - \frac{1}{R}\right), \quad \frac{1}{\hat{S}} = E_s D^2 \left(\frac{1}{S_a} - \frac{1}{S}\right) \quad (4.52)$$

These are positive size-independent dimensionless parameters of the cross-section geometry. The energy dissipation rate, analogous to the fracture energy, may be expressed as $bG_{if} = G_{stud}n/s =$, where $G_{stud} =$ energy dissipated by slip of one stud (representing the cross-hatched area above the residual shear flow line in Fig. 4.16 top), $s =$ longitudinal spacing of studs, and $n =$ number of rows of studs across the width of the interface strip.

Now we can turn attention to the size effect. From (4.51) we may calculate the nominal stress corresponding to $P - P_r$ in loading case B:

$$\sigma_N - \sigma_r = \frac{P-P_r}{D^2} = \frac{1}{Dq(\alpha)}\sqrt{\frac{2E_s G_{stud}\, n}{\{[q(\alpha)]^2 \hat{R}^{-1} + [q'(\alpha)]^2 \hat{Q}^{-1}\}s}} \quad (4.53)$$

From experience with many kinds of fracture, it may now be assumed that, for a limited but significant range of sizes, the failures are geometrically similar, i.e., the values of a/D at maximum load are the same for various sizes D (this assumption is validated a posteriori by the fact that approximately the same size effect ensues by solving the differential equations of the problem; Bažant and Vítek 1999). In this regard note that no size effect would occur only if a were either vanishingly small or constant for various sizes, which is certainly not the case (or if all stud failures were simultaneous rather than propagating, which however cannot be the case since the studs exhibit postpeak softening).

The zone in which the studs are already slipping but their shear flow has not yet been reduced to T_r does not have a vanishing length (i.e., it is not a point). In reality, it occupies a certain finite length, which is denoted as $2c_0$. The behavior may be approximated by an effective (equivalent) sharp LEFM crack in the steel-concrete interface reaching roughly into the middle of the zone. This effective interface crack has the length $a = \alpha_0 D + c_0$ where $\alpha_0 = a_0/D$ and a_0 now represents the length in which the shear flow of the studs has been reduced to T_r.

By analogy with the size of the fracture process zone in quasibrittle materials, we may consider c_0 and α_0 to be approximately constant, i.e., size independent. Now, substituting $\alpha = \alpha_0 + (c_0/D)$ into (4.53) we may conveniently

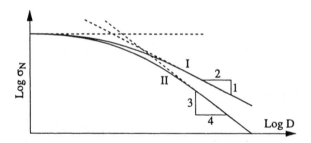

Figure 4.17: Size effects predicted by energy release analysis for geometrical scalings of Types I and II

introduce Taylor series expansions:

$$q(\alpha) = q_0 + q_1(c_0/D) + (q_2/2)(c_0/D)^2 + \ldots \quad (4.54)$$
$$q'(\alpha) = q_1 + q_2(c_0/D) + (q_3/2)(c_0/D)^2 + \ldots \quad (4.55)$$

where constants $q_0 = q(\alpha_0), q_1 = q'(\alpha_0), q_2 = q''(\alpha_0), \ldots$. Thus we obtain

$$\sigma_N - \sigma_r = \frac{1}{D}\sqrt{\frac{2(n/s)E_s G_{stud}}{[q_0 + q_1(c_0/D) + \ldots]^2 \hat{R}^{-1} + [q_1 + q_2(c_0/D) + \ldots]^2 \hat{S}^{-1}}} \quad (4.56)$$

Consider now two basic geometrical types of scaling.

Type I. The composite beam is scaled in proportion to D while the connection characteristics per unit area of steel-concrete interface remain constant (which would be the case for a glued interface with a crack). In that case, n/D, s and G_{stud} are constant, and so are D/n and the transverse spacing of stud rows, b/n. Then, if the series expansions are truncated after the first (linear) term, (4.56) can be rearranged to the form of the usual size effect law with residual stress (Bažant 1987b; see curve I in Fig. 4.17b):

$$\sigma_N = \frac{\sigma_N^0}{\sqrt{1 + (D/D_0)}} + \sigma_r \quad (4.57)$$

in which

$$D_0 = c_0 \frac{A_1}{A_2}, \quad \sigma_N^0 = \sqrt{\frac{2}{s}\left(\frac{n}{D}\right)\frac{E_s G_{stud}}{A_1 c_0}} \quad (4.58)$$

$$A_0 = \frac{q_0^2}{\hat{R}} + \frac{q_1^2}{\hat{S}}, \quad A_1 = 2q_1\left(\frac{q_0}{\hat{R}} + \frac{q_2}{\hat{S}}\right) \quad (4.59)$$

Asymptotically, for very large D, (4.57) indicates that

$$\sigma_N - \sigma_r \propto \frac{1}{\sqrt{D}} \quad (4.60)$$

which is the scaling of LEFM.

Type II. In the second type of scaling, not only the composite beam but also the connectors and their spacing are geometrically scaled. In this case, representing the complete geometric scaling of the entire structure on both the macroscale of beam and the mesoscale of studs, one must take into account the effect of stud diameter d on the nominal strength of the stud. Since d is now proportional to beam depth D, this may be done by expressing the energy required for failure of the studs per unit length of the beam (the shaded area in Fig. 4.16) as follows:

$$G_{stud} = \pi d^2 (\sigma_{N,stud} - \sigma_{r,stud}) v_{stud} = (\sigma_{N,stud} - \sigma_{r,stud}) D^2 \bar{v}_{stud} \quad (4.61)$$

where $\bar{v}_{stud} = \pi (d/D)^2 v_{stud}$ and $d/D = $ constant, and $\sigma_{r,stud} = T_r/\pi d^2 = $ residual nominal strength of stud. The constant v_{stud}, independent of stud sizes, represents the effective slip displacement of stud characterizing its dissipated energy. The part of nominal strength of the stud that exceeds $\sigma_{r,stud}$ is subjected to size effect:

$$\sigma_{N,stud} - \sigma_{r,stud} = \frac{\sigma_{s0}}{\sqrt{1 + (d/d_{s0})}} = \frac{\sigma_{s0}}{\sqrt{1 + (D/D_{s0})}} \quad (4.62)$$

where $d_{s0}, D_{s0} = $ constants. From tests and finite element studies (e.g. Eligehausen and Ožbolt 1992), the pullout nominal strength is known to exhibit a very strong size effect, closely approaching the LEFM size effect $\sigma_{N,stud} \propto d^{-1/2}$. Therefore it may be expected that, for practical stud sizes, $d/d_{s0} = D/D_{s0} \gg 1$, which means that $\sigma_{N,stud} \approx \sigma_{s0} \sqrt{D_{s0}/D}$ and that the residual nominal strength $\sigma_{r,stud}$ is negligible.

Using (4.61) and (4.62), (4.56), one obtains for size effect the law:

$$\sigma_N = \sqrt{\sigma_{s0} \sigma_{b0}} \left(1 + \frac{D}{D_0}\right)^{-1/2} \left(1 + \frac{D}{D_{s0}}\right)^{-1/4} + \sigma_r \quad (4.63)$$

in which D_0 and σ_{b0} are constants;

$$D_0 = c_0 \frac{A_1}{a_0}, \quad \sigma_{b0} = \frac{2n}{A_1 c_0} \left(\frac{D}{s}\right) E_s \bar{v}_{stud} \quad (4.64)$$

For very large sizes D, this expression leads to the asymptotic size effect (curve II in Fig. 4.17b):

$$\sigma_N - \sigma_r \propto D^{-3/4} \quad (4.65)$$

At first it may surprise that this size effect, which may be called the size 'hypereffect', is stronger than the LEFM size effect $D^{-1/2}$. The reason is that this is a compound size effect, in which the size effect due to failure of the beam as a whole (macroscale) is amplified by the size effect in the failure of individual studs (mesoscale).

Obviously, from the viewpoint of the size effect, when a larger composite beam is designed it is better not to increase the size of the studs, if possible. This is the normal design practice anyway. But unfortunately it is not feasible when the beam size is enlarged significantly. In that case, the stud size must be enlarged as well.

4.6. Size Effect Formulae for Concrete Design Codes

Until almost the end of the 1980s, no size effect provisions were present in the concrete design codes of various countries, and the code-making committee generally regarded the size effect as a nuisance that the theoreticians were trying to foist upon them. During the 1990s, fortunately, the attitude changed markedly. Many members of these committees are now convinced that the size effect ought to be introduced in one form or another, and the question now is which form should be adopted and how to calibrate the coefficients of the adopted formula so it optimally reflects the actual experimentally observable behavior of structures.

This has been a healthy trend. However, what is striking is the variety of formulae and the underlying (or absent) concepts. Many proposed formulae ignore the energetic aspects of fracture mechanics. A point to be emphasized in this regard is that the energetic size effect is inevitably present, and so, if some other theory, such as statistical or fractal, is assumed, it could only come on top of the energetic formulation, as its refinement, but not without it, not as a replacement.

The size effect curves according to various formulae so far used in design for the size effect in reinforced concrete structures are plotted in Fig. 4.18. The shaded zone shows the typical range of the existing test data for structural failures. The difficulty in deciding which formula is appropriate is that the scatter of the existing data is too wide for the range of sizes tested.

If the choice should be made strictly empirically, it would be necessary to greatly extend the test data into larger size ranges, and obtain a statistically significant number of test results for geometrically scaled structures. Unfortunately, most large-scale tests have been conducted in the past on structures that were not geometrically scaled (e.g., the bar sizes, cover thickness and bar spacing were not geometrically similar). The effects of the changes of shape (geometry) are known only crudely and introduce additional errors. Therefore, formulae that have the strongest theoretical support ought to be preferred. The

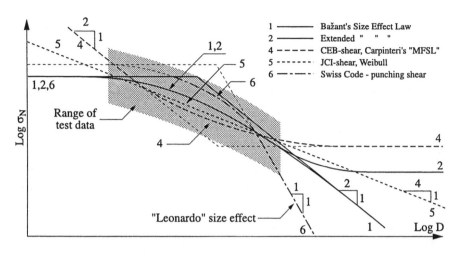

Figure 4.18: Existing 'zoo' of size effect formulae for concrete structures and the range of existing data from structural tests

theory itself may be verified by checks other than extension of the size range of the tests of the given particular failure. Unless such an approach is adopted, the choice between the formulae will be random, depending solely on the voting of committee members.

For the diagonal shear failure of reinforced concrete beams and punching shear failure of slabs, Japan Concrete Institute, based on unprecedented failure tests of very large beams and slabs up to several meters in depth, adopted the Weibull size effect formula – a power law with exponent $n/m = -1/4$ (curve 5 in Fig. 4.18. Although the comparison with tests is acceptable, thanks to a wide random scatter and limited number of data, it is however incorrect to ignore the energetic size effect, which is inevitably present. Anyhow, the power law implies the structure possesses no characteristic dimension (i.e., exhibit complete self-similarity), yet a characteristic dimension must exist due to the size of aggregate as well as the spacing and size of reinforcing bars.

The CEB (European) concrete design code and the German code DIN adopted Carpinteri's MFSL law based on a fractal geometric viewpoint, $\sigma_N = \sigma_0\sqrt{A_1 + A_2/D}$, despite the critique of the underlying concept, already discussed. Anyway, this formula would not be entirely unjustified for the case of bending strength of plain concrete, but not for reinforced concrete failing only after large stable crack growth. Still another size effect formula for the diagonal shear failure was proposed by M. Collins of the University of Toronto, based on some arguments about crack opening width in small and large beams (which have been challenged by the writer).

Swiss code SIA 162 introduced a curious formula, namely

$$\sigma_N = \min\left(\frac{\sigma_0}{1 + D/D_0},\ \tau_c\right) \qquad \text{("Leonardo" size effect)} \qquad (4.66)$$

(curve 6 in Fig. 4.18), to describe the size effect in punching shear failure of slabs (σ_0, D_0, τ_c = constants). The formula was based strictly on test results, but its form is theoretically unacceptable. For sufficiently large sizes D it gives an impossibly strong size effect; it approaches the "Leonardo" size effect (da Vinci 1500s), namely σ_N being inversely proportional to D, which is thermodynamically impossible. Nevertheless, the Swiss code deserves praise for being the first, and so far the only one, to recognize that there indeed is a strong size effect in punching shear.

The LEFM size effect (asymptote of slope $-1/2$ of curve 1 in Fig. 4.18) for anchor pullout was introduced into German and ACI Code Recommendations, based mainly on the tests of Eligehausen. This size effect is excessive for anchors that are very small, although such anchors might not be of great concern.

There are other provisions in various codes which, if scrutinized, imply a size effect without stating it explicitly. For example, the code ACI 318 R-89 implied a huge size effect for the failure of splices, shown graphically in Fig. 4.19. This was in fact the "Leonardo" size effect of slope 1, which is thermodynamically impossible. The figure shows for comparison also the size effect in ACI 318 R-95, in which the discontinuous jump is objectionable (note that these diagrams are plotted assuming the cover thickness to be proportional to the bar diameter, or else one could not speak of size effect). The enormous sudden change between the two plots of two subsequent ACI code provisions shown in this figure, a "U-turn" made in absence of any new revolutionary finding, is striking indeed.

No provisions exist so far in any design code for the size effect in reinforced concrete columns, to the writer's knowledge. Yet this is a case where the size effect is most important, and a strong size effect has been experimentally demonstrated in reduced size test (Bažant and Kwon 1994). Failure of one beam usually does not make a building collapse, but failure of one column often does.

A difficult problem of the various proposed formulae is the prediction of its constants, for example D_0 and σ_0. Formulae for this purpose need to be worked out for many cases. Due to complexity of various reinforced concrete structures, these coefficients may have to be determined partly empirically for each type of failure. Considerable testing might be needed for this purpose.

At present it is clear that the size effect needs to be taken into account in the provisions ensuring safety against all types of brittle failures of concrete structures. They include:

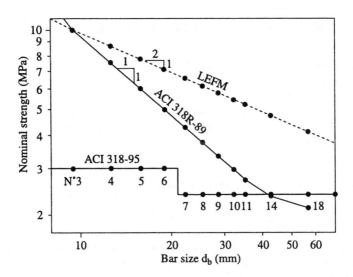

Figure 4.19: Comparison of size effects implied by two subsequent provisions of ACI Code for the failure of splices

1. Diagonal shear failure of reinforced concrete beams.

2. Punching shear failure of reinforced concrete slabs (Bažant and Cao 1987).

3. Torsional failure of reinforced or plain concrete beams (Bažant and Şener 1997; Bažant, Şener and Prat 1988).

4. Failure of reinforced concrete columns (except very strong filled tubes); Bažant and Kwon (1994).

5. Pullout of anchors and of studs in composite beams.

6. Pullout and embedment length of reinforcing bars.

7. Failure of splices of reinforcing bars (Şener, Bažant and Becq-Giraudon 1999).

8. Failure of some types of frame joints.

9. Fracture of gravity dams, e.g., in earthquake (Bažant 1996b).

10. Bending failure of arch dams, unreinforced beams, foundation plinths, retaining walls, etc.

4.7. Size Effect Hidden in Excessive Dead Load Factor in Codes

Since we have discussed the design codes, it is important to point out that the dead load factor values in the current codes are excessive and that this implies a hidden size effect. The currently used dead load factors have recently been criticized by structural engineering statisticians as unjustifiably large, and proposals for a reduction have been made. However, without simultaneously incorporating the size effect into the code provisions, such a reduction would be dangerous.

The point to note is that, the larger the structure, the higher is the percentage of the own weight contribution to the ultimate load $\hat{U}$. Thus, if the load factor for the own weight is excessive, structures of large sizes are overdesigned from the viewpoint of strength theory or plastic limit state design—the theory underlying the current building codes. However, such an overdesign helps to counteract the neglect of size effect in the current codes, which is inherent to plastic limit analysis concepts (Bažant and Frangopol 2001). Doubtless it is the reason why the number of structural collapses in which the size effect was a contributing factor has not been much larger than we have seen so far.

Let $\hat{L}$ and $\hat{D}$ be the internal forces caused by the live load and the dead load, and U the internal force caused by ultimate loads, i.e., the loads magnified by the load factors. Using the load factors currently prescribed by the building code (ACI Standard 319, 1999), one has

$$\hat{U} = 1.4\,\hat{D} + 1.7\,\hat{L} \tag{4.67}$$

Consider now that the dead load factor 1.4 is excessive and that a realistic value, justified by statistics of dead load, should be μ_D. Then the ratio of the required ultimate design value of the internal force to the realistic ultimate value, which may be called the *overdesign ratio* (Bažant 2000), is

$$R = \frac{\hat{U}_{\text{design}}}{\hat{U}_{\text{real}}} = \frac{1.4\,\hat{D} + 1.7\,\hat{L}}{\mu_D\,\hat{D} + 1.7\,\hat{L}} \tag{4.68}$$

Let us limit consideration to dead loads caused by the own weight of structures, which for example dominate the design of large span bridges. For a bridge of very large span, the dead load may represent 90% of the total load, and the live load 10%. In that case, the overdesign ratio is

$$R = \frac{1.4 \times 0.9 + 1.7 \times 0.1}{\mu_D \times 0.9 + 1.7 \times 0.1} \tag{4.69}$$

For small scale tests which were used to calibrate the present code specifications, the own weight may be assumed to represent less that 2% of the total load. In that case, the overdesign ratio is

$$R_0 = \frac{1.4 \times 0.02 + 1.7 \times 0.98}{\mu_D \times 0.02 + 1.7 \times 0.98} \qquad (4.70)$$

Although a precise value is debatable and should be determined by extensive statistics, it seems reasonable to assume that the own weight of a very large structure cannot be underestimated by more than 5%. This means that $\mu_D = 1.05$. So,

$$R \approx 1.28, \qquad R_0 \approx 1.00 \qquad (4.71)$$

It follows that, compared to the tests used to calibrate the code, a structure of a very large span is overdesigned, according to the current theory, by about 28% (Bažant 2000). Such overdesign compensates for a size effect in the ratio 1.28. This is approximately the size effect for very large spans that is unintentionally hidden in the current code specifications.

To give a simple example, consider geometrically similar bridge girders of different sizes D, where D may for example be taken as the depth of the girder at the support. For the sake of simplicity and as an an extreme case, assume that all the dead load is the own weight of a large structure such as a bridge. The weight per unit length of the girder may be written as

$$q = g_o \rho D^2 \qquad (4.72)$$

Here ρ is the average weight of the material per unit volume of the structure and g_o = dimensionless factor characterizing the geometry of the cross section. The bending moment in the critical cross section is expressed as

$$M = g_m q L^2 \qquad (4.73)$$

where L = span and g_m = dimensionless factor characterizing the structural system (for a simply supported beam of a pair of two cantilevers connected by a hinge, for instance, $g_m = 1/8$). The nominal stress σ_{N_o} due to own weight, which may for example be defined as the critical stress σ due to own weight to be compared with material strength σ_0, may be expressed as

$$\sigma_{N_o} = g_s \rho D \qquad (4.74)$$

where g_s = dimensionless geometry factor. If, for instance, a homogeneous beam of rectangular cross section with equal compressive and tensile yield limits is failing by plastic bending, then

$$g_s = 4 g_m g_o (L/D)^2 \qquad (4.75)$$

where L/D = constant when the size effect for geometrically similar structures is considered; for a concrete girder, unprestressed or prestressed, the expression is of course more involved although Eq. (4.74) remains valid. The nominal stress due to the live load is

$$\sigma_{NL} = g_L \hat{L} \tag{4.76}$$

where g_L is again a dimensionless geometry factor.

Now, noting that, in Eq. (4.68) for the ultimate state, the unfactored dead load contribution to the ultimate internal force is proportional to $g_s \rho \hat{D}$ and the unfactored live load contribution is proportional to $g_L \hat{L}$, one concludes that the size effect implied by excessive dead load factor is approximately

$$\frac{\sigma_N}{\sigma_{N_o}} = \frac{1}{R} = \frac{\mu_d\, a\, D + b}{1.4\, a\, D + b}, \qquad a = g_s\, \rho, \quad b = 1.7\, g_L\, \hat{L} \tag{4.77}$$

where a and b are constants if geometrically similar structures are considered.

Using an excessive dead load factor as a substitute for size effect, however, is inadequate, for various reasons:

1. For some brittle failures, 28% as the maximum capacity reduction due to size effect is too small by far, for others excessive.

2. Eq. (4.77) does not approach its final asymptotic value as $1/\sqrt{D}$ but as $1/D$ which is too fast (and in fact thermodynamically impossible). Besides, this equation implies a finite residual strength for $D \to \infty$, which cannot be justified if the residual stress in the critical state is not zero.

3. The hidden size effect implied by the current codes is the same for brittle failures (such as the diagonal shear failure), which do exhibit a size effect, and ductile failures (such as the bending failure due to tensile steel yielding), which do not.

4. Even for brittle failures alone, the own weight is very poorly correlated to the brittleness number which controls the size effect. For example, a very tall column or pier might not be protected by the excessive dead load factor because the own weight might cause no significant bending moment; yet brittle failure due to compression crushing in flexure, which exhibits a size effect, may be caused mainly by horizontal loads such as wind or earthquake, for which the load factor is not excessive.

5. For prestressed concrete structures, which are generally lighter than unprestressed ones, the size effect implied by the code is generally weaker. Yet these structures are more brittle than unprestressed ones and thus exhibit stronger size effects.

6. Eq. (4.77) can never substitute for the Weibull statistical size effect, which is important for bending of unreinforced cross sections thicker than

about 1 m (e.g. for fracture of an arch dam due to flexure in a horizontal plane).

Further it should be noted that a hidden size effect also exists in various proposals for reliability-based codes. This is due to the fact that the reliability implied in the code increases with the contribution of the dead load to the overall gravity load effect (in detail, see Bažant and Frangopol 2000).

The problem is not limited to simplified code formulae. Even when finite element codes with a realistic cracking and fracture model are used, they, too, generate systematic overdesign and overreliability for large structures when combined with the currently prescribed load factors. A much lower dead load factor, justified by statistics, should be used in conjunction with such realistic computational approaches exhibiting size effect. To develop a rational procedure for the design based on computer analysis of structures, the researchers in finite element analysis of concrete structures and in structural reliability must work in synergy.

In conclusion, the question of a possible reduction in the dead load factor cannot be separated from the question of size effect, and vice versa. The reliability experts and fracture experts must collaborate.

4.8. No-Tension Design of Concrete or Rock from the Size Efffect Viewpoint

Design of unreinforced concrete structures such as dams or retaining walls, as well as rock structures such as tunnels, caverns, excavations and rock slopes, has commonly been made under the hypothesis that the material has no tensile strength. The same approach has traditionally been used for masonry—for example stone arches, domes or pillars. During the 1960's, the 'no-tension' hypothesis was introduced into finite element analysis and implemented as the limit case of plasticity in which the tensile yield strength is reduced to zero (as in Rankine yield criterion).

The simple no-tension concept was widely believed to be always on the safe side, which would mean that safe designs could be obtained without bothering about the complexities of fracture mechanics. Although this is usually true, it is, however, not always true. The case where the no-tension approach may fail to be on the safe side is the case of very large structures (Bažant 1996b).

Since the no-tension approach is a plasticity approach, it exhibits no size effect. Fracture mechanics, on the other hand implies size effect, and if the cracks at failure are large, as is the case for a dam overloaded by a horizontal

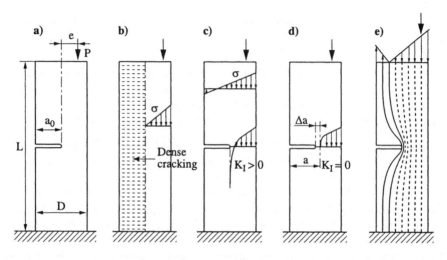

Figure 4.20: (a) Specimen analyzed, (b) distributed cracking implied by no-tension design, (c,d) stress distributions when the crack has the same depth $a_0 = D/2$ as the no-tension zone depth ($K_I > 0$) and when it grows enough to make $K_I = 0$, (e) approximate principal stress trajectories for $a = a_0$

force, the nominal strength approaches zero as the structure size tends to infinity. Thus, the size effect plot of σ_N versus D for the fracture solution must, at some sufficiently large size, cross the horizontal line for the no-tension plasticity solution. Numerical examples of such situations have been given (Bažant 1996b).

These examples imply that by replacing a zero tensile strength with a finite resistance to tensile failure in the form of a finite fracture energy (finite fracture toughness), one can obtain a lower load capacity of the structure. In plasticity, this is impossible.

One simple example is shown in Fig. 4.20. Considering load eccentricity $e = D/3$, the left half of the specimen is, according to the no-tension assumption, stress free. But if the tensile strength is finite, only one crack will form and its stress intensity factor K_I will be positive (which may be expected upon noting that the principal stress trajectories from the tensile side must crowd together in passing aroung the crack tip; Fig. 4.20). So this crack will propagate (precisely to the length $a = 0.549D$ if $K_I = 0$ is assumed). Since the uncracked ligament becomes shorter, the load capacity for subsequently applied horizontal loads will be diminished. Demonstrations of a reduced capacity for the vertical load have been given for a specimen loaded by water pressure on crack faces (Bažant 1996b).

The no-tension concept nevertheless remains valid and useful for the design of dams (as shown by Elices). But the design should be checked by fracture mechanics, especially when the dam is very large.

Chapter 5

Energetic Scaling of Compression Fracture and Further Applications to Concrete, Rock and Composites

5.1. Propagation of Damage Band Under Compression

The fracture of quasibrittle materials due to compressive stress is one of the most difficult aspects of fracture mechanics. In compression fracture, one must distinguish two distinct phenomena: (1) micromechanics of initiation of compression fracture, and (2) mechanics of global compression fracture causing failure.

The first problem has been investigated much more than the second, and various micromechanical mechanisms that initiate fracture under compressive stresses have been identified; e.g., the growth of axial splitting cracks from voids (Cotterell 1972, Sammis and Ashby 1986, Kemeny and Cook 1991, Steif 1984, Wittmann and Zaitsev, 1981, Zaitsev 1985, Fairhurst and Cornet 1981, Ingraffea and Heuzé 1980, Nesetova and Lajtai 1973, Carter 1992, Yuan et al. 1993) or near inclusions, the creation of axial splitting cracks by groups of hard inclusions, and the formation of wing-tip cracks from sliding inclined surfaces (Hawkes and Mellor 1970, Ingraffea 1977, Ashby and Hallam 1986, Horii and Nemat-Nasser 1982, 1986, Sanderson 1988, Schulson 1990, Costin 1991, Schulson and Nickolayev 1995, Lehner and Kachanov 1996, and a critique by Nixon 1996).

It must be realized, however, that these mechanisms do not explain the global failure of the structure. They can cause only a finite extension of the axial splitting cracks whose length is of the same order of magnitude as the size of the void, the inclusion, or the inclined microcrack. Each of these mechanisms can produce a zone of many splitting cracks approximately parallel to the uniaxial compressive stress or, under triaxial stress states, to the compressive principal stress of the largest magnitude.

Biot (1965) proposed that the cause of compression failure may be three-dimensional internal buckling which can occur either in the bulk of specimen or within an inclined band. However, he considered only elastic behavior and did not conduct any energy analysis. Finite strain analysis of compression failure caused by internal buckling of an orthotropically damaged material or orthotropic laminate was presented by Bažant (1967). Kendall (1978) showed that, with the consideration of buckling phenomena under eccentric compressive loads, the energy balance condition of fracture mechanics yields realistic predictions of compression fracture of test cylinders loaded only on a part of the end surface.

The global compression fracture has been analyzed (Bažant 1993, Bažant and Xiang 1997) under the hypothesis that some of the aforementioned micromechanisms create a band of axial splitting cracks (shown in Fig. 5.1), which propagates laterally, in a direction either inclined or normal to the direction of the compressive stress of the largest magnitude. In the post-peak regime, the axial splitting cracks interconnect to produce what looks like a shear failure although there is no shear slip before the post-peak softening (in fact, shear failure *per se* is probably impossible in concrete). The energy analysis of the propagating band of axial splitting cracks shows that, inevitably, there ought to be a size effect. Let us discuss it for the prismatic specimen shown in Fig. 5.1.

Formation of the axial splitting cracks causes a narrowing of the band and, in an approximate sense, a buckling of the slabs of the material between the splitting cracks as shown in the figure (alternatively, this can be modeled as internal buckling of damaged continuum). This causes a reduction of stress, which may be considered to occur approximately in the shaded triangular areas.

For the calculation of the energy change within the crack band one needs to take into account the fact that the slabs of material between the axial splitting cracks ought to undergo significant post-buckling deflections corresponding to the horizontal line 3-5. Thus, the energy change in the splitting crack band is given by the difference of the areas 0120 and 03560 (the fact that there is a residual stress σ_{cr} in compression fracture is an important difference from a similar analysis of tensile crack band propagation). The energy released must be consumed and dissipated by the axial splitting cracks in the band. This is one condition for the analysis.

The second condition is that the narrowing of the band due to microslab buckling must be compatible with the expansion of the adjacent triangular areas due to the stress relief. One needs to write the condition that the shortening

Energetic Scaling of Compression Fracture

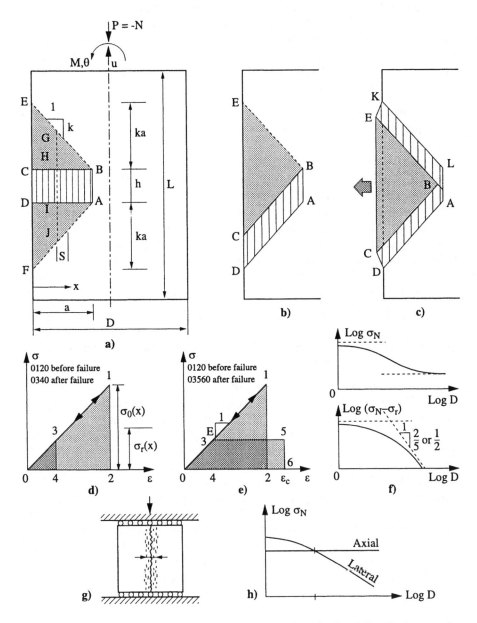

Figure 5.1: (a,b,c) Sideways propagations of a band of axial splitting cracks, with energy release zones, (d,e) reduction of strain energy density outside and inside the band, and (f) resulting approximate size effect curve

of segment HI in Fig. 5.1 on top left is compensated for by the extension of segments GH and IJ, which is a compatibility condition. The energy release from the crack band is given by the change of the areas under the stress-strain diagrams in the middle of Fig. 5.1 (bottom), caused by the drop of stress from the initial compressive stress σ_0 to the final compressive stress σ_{cr} carried by the band of splitting cracks.

The resulting size effect on the nominal strength of large structures failing in compression has, according to this analysis, the form:

$$\sigma_N = C_1 D^{-2/5} + C_0 \tag{5.1}$$

where C_1, C_0 = constants. Mathematical formulation of the foregoing arguments (Bažant 1993; Bažant and Xiang 1997) provided for the compression failure a formula which exhibits a size effect. This size effect is plotted in Fig. 5.1(f), with the logarithm of size D as a coordinate and either $\log \sigma_N$ or $\log(\sigma_N - \sigma_r)$ as the ordinate. In the latter plot (Fig. 5.1f), the size effect is shown to approach an asymptote of slope $-2/5$. This is another interesting feature, which results from the fact that the spacing of the axial splitting cracks is not constant but depends on the overall energy balance.

The solution of the nominal strength σ_N was obtained under the assumption of arbitrary spacing s. It was noted that σ_N exhibits a minimum for a certain spacing s, which depends on size D. It is this condition of minimum which causes the asymptotic slope to be $-2/5$ instead of $-1/2$.

Why do small uniaxial compression specimens fail by an axial splitting crack and exhibit no size effect? In a uniform uniaxial stress field, a sharp planar axial crack does not change the stress field and thus releases no energy. Therefore a damage band of finite width (Fig. 5.1g) must precede the formation of an axial splitting crack. The energy is released only from this band but not from the adjacent undamaged solid. Therefore, the energy release is proportional to the length of the axial splitting crack, which implies that there is no size effect (Fig. 5.1h). Thus, the lateral propagation of a band of splitting cracks, which involves a size effect, must prevail for a sufficiently large specimen size (Fig. 5.1h, Bažant and Xiang 1996). The reason that the axial splitting prevails for a small enough size is that the overall fracture energy consumed (and dissipated) by a unit axial extension of the splitting crack band is smaller than that consumed by a unit lateral extension, for which new cracks must nucleate.

5.2. Size Effect in Reinforced Concrete Columns

The results of the aforementioned approximate analysis in terms of a band of axial splitting cracks propagating sideway have been compared to the test results (Bažant and Kwon 1994) on size effect in eccentrically loaded reduced-scale tied reinforced concrete columns of three different sizes (in the ratio 1:2:4) and three different slendernesses, $\lambda = 19.2$, 35.8 and 52.5 (Fig. 5.2). The columns were made of concrete with reduced aggregate size.

The test results indicated a size effect which is seen in Fig. 5.2 (and is ignored by the current design codes). The formulas obtained by the foregoing approximate energy analysis of the propagation of a band of axial splitting cracks are shown by the solid curves in the figures, indicating a satisfactory agreement.

Recently, Bažant and Kwon's (1994) tests of eccentrically loaded reduced-scale reinforced concrete columns (Fig. 5.2) have been analyzed using three-dimensional finite elements (Brocca and Bažant 2000). The meshes are shown in deformed and undeformed configurations in Figs. 5.3. The steel bars were assumed not to slip against concrete at the mesh nodes.

The stress-strain relations for concrete was simulated by the latest version M4 of the microplane model (Bažant and Novák 2000a) which was generalized to finite strain as described in Bažant and Novák (2000b) (the use of finite strain automatically captured the second-order bending moments due to axial load). The characteristic length of the material, necessary for simulating the quasibrittle size effect, was introduced through the crack band model.

Looking at mesh deformations seen in Fig. 5.3, one can discern inclined crack bands developing during failure, which agrees with observations made during the tests. Fig. 5.4 shows the computed load-displacement curves, which exhibit postpeak softening due to the combined effect of softening damage of concrete and of geometrical nonlinearity.

Fig. 5.5 shows the computed nominal strength values in the bi-logarithmic size effect plots. For comparison, the test data as well as the results of the aforementioned simplified analysis are shown as well. The comparison with test data is seen to be satisfactory.

In comparison with the simplified analysis, the slope of the computed size effect plot is somewhat smaller. This small discrepancy might have two explanations: (1) the neglect of bond slip of steel bars at mesh nodes, and (2) the fact that the crack band model tends to be too stiff for damage propagation in directions inclined to the mesh lines.

The finite element results confirm that the current design procedures based on plastic limit analysis, for which the size effect plot is a horizontal line, are questionable. Their safety margin for very large columns is systematically less than it is for normal size columns. A change is needed. However, further size effect tests ought to be caried out on full-size columns with normal aggregate and normal reinforcing bars.

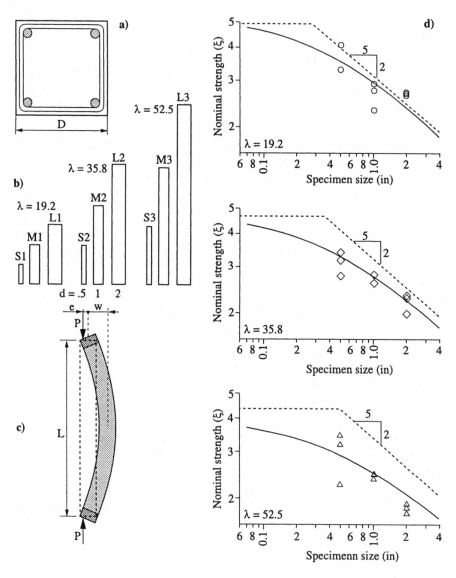

Figure 5.2: (a,b,c) Reduced-scale reinforced concrete columns of different sizes and slendernesses, tested by Bažant and Kwon (1994); (d) Measured nominal strength versus column size, and fits by formula (Bažant and Xiang 1997)

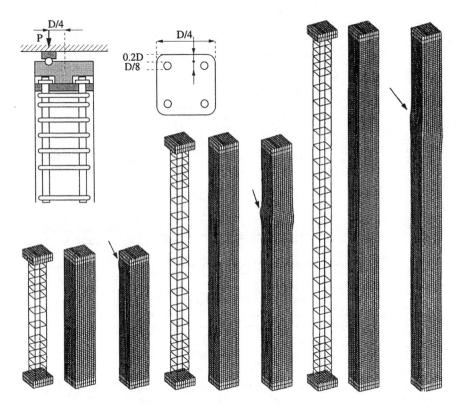

Figure 5.3: Undeformed and deformed meshes used for the computations of failure of columns (after Brocca and Bažant 2000)

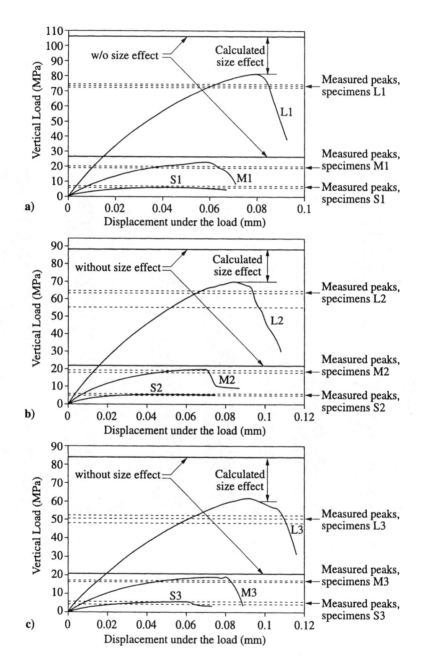

Figure 5.4: Load-displacement curves of reinforced concrete columns of various slendernesses and sizes computed by Brocca and Bažant (2000)

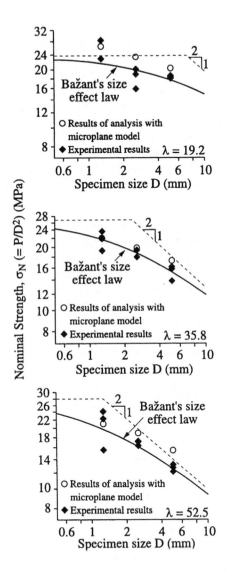

Figure 5.5: Size effect data computed by finite elements (Brocca and Bažant 2000), compared to measured data and curves obtained with the simplified model

5.3. Fracturing Truss (Strut-and-Tie) Model for Shear Failure of Reinforced Concrete

It appears that compression failure is also the final failure mechanism in shear failures of reinforced concrete beams, such as diagonal shear of beams and torsion of beams, punching of plates, pullout of anchors, failure of corbells and frame connections, etc. The importance of the size effect in shear failure of beams has been experimentally documented by many investigators (Leonhardt and Walther 1962; Kani 1967; Kupfer 1964; Leonhardt 1977; Walraven 1978, 1995; Iguro et al. 1985; Shioya et al. 1989; Shioya and Akiyama 1994; Bažant and Kazemi 1991a; Walraven and Lehwalter 1994; see also Bažant and Cao 1986a, 1986b, 1987; Bažant and Sun 1987; Bažant, Şener and Prat 1988). Let us briefly outline the mechanics (Bažant, 1996) of the size effect in the diagonal shear failure of reinforced concrete beams.

According to the truss model of Ritter (1899) and Mörsch (1922), refined by Nielsen and Braestrup (1975); Thürlimann (1976); Collins (1978); Collins et al. (1996); Marti (1980); Collins and Mitchell (1980); Hsu (1988); Schlaich et al. (1987) and others, and recently called the *strut-and-tie model*, a good approximation is to assume that a system of inclined parallel cracks forms in the high shear zone of a reinforced concrete beam before the attainment of the maximum load (Fig. 5.6). The cracks are assumed to be continuous and oriented in the direction of the principal compressive stress (which is, of course, an approximation). This assumption implies that there is no shear stress on the crack planes and that the principal tensile stress has been reduced to 0.

According to this simplified picture, the beam acts as a truss. The truss consists of the longitudinal reinforcing bars, the vertical stirrups (which are in tension), and the inclined compression struts of concrete between the cracks. If the reinforcing bars and stirrups are designed sufficiently strong, there is only one way the truss can fail—by compression of the diagonal struts.

In the classical approach, the compression failure of the struts has been handled according to the strength concept, which cannot capture the localization of compression fracture and implies the compression fracture to occur simultaneously everywhere in the inclined strut. In reality, the compression fracture, called *crushing*, develops within only a portion of the length of the strut (in a region with stress concentrations, as on the top of beam in Fig. 5.6). Then it propagates across the strut.

For the sake of simplicity, the band of axial splitting cracks forming the crushing zone may be assumed to propagate as shown in Fig. 5.6 and reach, at maximum load, a certain length c. The depth of the crushing band may be expected to increase initially but later to stabilize at a certain constant value h governed by the size of aggregate.

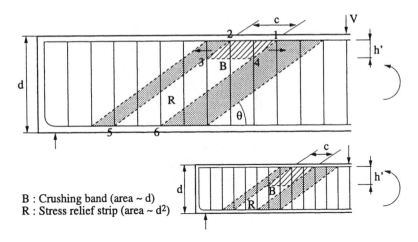

Figure 5.6: Fracture adaptation of truss model for diagonal shear failure of reinforced concrete beams: Compression crushing zone and energy release zone in beams of different sizes (after Bažant 1996a)

It is now easy to explain how the size effect arises. Because of the existence of parallel inclined cracks at maximum load, the formation of the crushing band reduces stress in the entire inclined white strip of width c and depth d (beam depth shown in Fig. 5.6). The area of the white strip is cd or $(c/d)d^2$ and its rate of growth is $(c/d)2d\dot{d}$, in which c/d is approximately a constant when similar beams of different sizes are compared.

So, the energy release rate is proportional to $\sigma_N^2 d\dot{d}/E$, where the nominal strength is defined as $\sigma_N = V/bd$ = average shear stress, V = applied shear force and b = beam width. The energy consumed is proportional to the area of the crushing band, ch or $(c/d)hd$, that is, to $G_f d/s$, and its rate of $G_f \dot{d}/s$ where G_f = fracture energy of the axial splitting cracks (s = crack spacing). This expression applies asymptotically for large beams because for beams of a small depth d the full width h of the crushing band cannot develop. Equating the derivatives of the energy release and energy dissipation expressions, i.e., $\sigma_N^2 d\dot{d}/E \propto G_f \dot{d}/s$, we conclude that the asymptotic size effect ought to be of the form:

$$\sigma_N \propto s^{-1}\sqrt{EG_f d} \qquad (5.2)$$

The complete size effect represents a transition from a horizontal asymptote to the inclined asymptote in the size effect plot given by this equation. Relatively simple design formulas are obtained in this manner (Bažant, 1996a). The analysis can also be done in a similar way for the diagonal shear failure of beams with longitudinal reinforcement but without vertical stirrups, and further for torsion, etc.

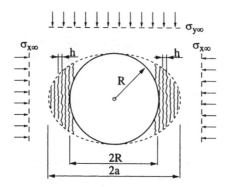

Figure 5.7: Borehole in rock and growth of an elliptical zone of axial splitting cracks (after Bažant, Lin and Lippmann 1993)

5.4. Breakout of Boreholes in Rock

A size effect is known to occur also in the breakout of boreholes in rock, as experimentally demonstrated by Nesetova and Lajtai (1973), Carter (1992), Carter et al. (1992), Yuan et al. (1992), and Haimson and Herrick (1989). It is known from the studies of Kemeny and Cook (1987, 1991) and others that the breakout of boreholes occurs due to the formation of splitting cracks parallel to the direction of the compressive stress of the largest magnitude, $\sigma_{y\infty}$. An approximate energy analysis of the breakout was conducted under the simplifying assumption that the splitting cracks occupy a growing elliptical zone (although in reality this zone is narrower and closer to a triangle).

The assumption of an elliptical boundary permitted the energy release from the surrounding infinite solid to be easily calculated according to Eshelby's theorem for eigenstrains in ellipsoidal inclusions (Bažant, Lin and Lippmann, 1993). According to the theorem, the energy release from the infinite rock mass can be approximated as

$$\Delta \Pi = -\pi[(a+2R)R\sigma_{x\infty}^2 + (2a+R)a\sigma_{y\infty}^2 - 2aR\sigma_{x\infty}\sigma_{y\infty}$$
$$-2a^2\sigma_{cr}^2](1-\nu^2)/2E \tag{5.3}$$

in which R = borehole radius, a = principal axis of the ellipse (Fig. 5.7), $\sigma_{x\infty}$ and $\sigma_{y\infty}$ = remote principal stresses, E = Young's modulus of the rock, and ν = Poisson ratio.

A similar analysis as that for the propagating band of axial splitting cracks, already explained, has provided a formula for the breakout stress. This formula has a plot similar to those in Fig. 5.1(f). Its asymptotic behavior is described by Eq. (5.1).

5.5. Asymptotic Equivalent LEFM Analysis for Cracks with Residual Bridging Stress

In the case of compression fracture due to lateral propagation of a band of axial splitting cracks, a residual stress given by the critical stress for internal buckling in the band remains. Lumping the fracturing strains in the band into a line, one may approximately treat such a fracture as a line crack in which interpenetration of the opposite faces is allowed and the softening compressive stress-displacement law terminates with a plateau of residual constant stress σ_Y. Likewise, a constant residual stress σ_Y may be assumed for characterizing the tensile stress-displacement law for a crack in a fiber-reinforced composite (e.g. fiber-reinforced concrete).

The asymptotic formulae (2.53)–(2.55) for the case of many loads can be applied to this case because the uniform pressure σ_Y along the crack can be regarded as one of two loads applied on the structure. We write the stress intensity factors due to the applied load P and the uniform crack pressure σ_Y as $K_I^2 = \sigma_N^2 D g(\alpha_0 + \theta)$ (with $\theta = c_f/D$), and $K_I^2 = \sigma_Y^2 \gamma(\alpha_0 + \theta)$, respectively, where g and γ are dimensionless functions taking the role of g_1 and g_2 in the preceding formulae. In this manner, (2.53) and (2.54) yield, after rearrangements, the following formula for the size effect (and shape effect) in the case of a large crack (Bažant 1997a; 1998a,b):

$$\sigma_N = \frac{\sqrt{EG_f} + \sigma_Y \sqrt{\gamma'(\alpha_0)c_f + \gamma(\alpha_0)D}}{\sqrt{g'(\alpha_0)c_f + g(\alpha_0)D}} \tag{5.4}$$

For geometrically similar structures and size-independent α_0, this formula yields a size effect curve that terminates, in the log D scale, with a horizontal asymptote in the manner shown in Fig. 2.12 (top right) and Fig. 2.15, but has also a horizontal asymptote on the left.

In the case of initiation of a crack with uniform residual stress σ_Y, equations (2.53) and (2.55) can be reduced to the following size (and shape) effect formula:

$$\sigma_N = \frac{\sqrt{EG_f} + \sigma_Y \sqrt{\gamma'(0)c_f + \frac{1}{2}\gamma''(0)\frac{c_f^2}{D}}}{\sqrt{g'(0)c_f + \frac{1}{2}g''(0)\frac{c_f^2}{D}}} \tag{5.5}$$

(Bažant 1997a; 1998a,b), whose logarithmic plot also terminates with a horizontal asymptote as in Fig. 2.12 (top right).

If the residual stress is compressive and is determined by internal buckling in a band of axial splitting cracks of arbitrary spacing, then σ_Y in the foregoing equations is not constant. As already explained, minimization of σ_N with respect to the crack spacing s shows that the crack spacing in the band should vary with D. Then, in the foregoing equations (5.4) and (5.5):

$$\sqrt{EG_f} \quad \text{must be replaced by} \quad \sqrt{EG_f}D^{1/10} \qquad (5.6)$$

Furthermore, the σ_Y value will also depend on the crack spacing, according to the formula for the critical buckling load. The overall trend will be well approximated by (5.1), and in particular σ_N will approach the large-size asymptotic limit as $D^{-2/5}$.

Arguing in favor of his MFSL law (to be discussed later), Carpinteri made the point that some measured size effect plots (of $\log \sigma_N$ versus $\log D$) exhibit a positive curvature and approach a horizontal asymptote (as in Fig. 2.12). However, as we have seen by now, this can have any of the following four (deterministic nonfractal) causes:

1. In unnotched structures, the relative length α_0 of traction-free crack might not be constant but may decrease with increasing size D.

2. There may be a residual cohesive (crack-bridging) stress σ_Y in the crack.

3. The failure may occur at the initiation of macroscopic crack growth.

4. Above a certain size D, there may be a transition to some plastic failure mechanism.

5.6. Application to Compression Kink Bands in Fiber Composites

Equations (5.5) and (5.6), after some refinements (Bažant et al. 1999), have been shown applicable to the compression kink bands in composites reinforced by parallel fibers, and in wood. This problem has so far been treated by elastoplasticity, and solutions of failure loads which give good agreement with the existing test data have been presented (Rosen 1965, Argon 1972, Budianski 1983, Budianski et al. 1997, Budianski and Fleck 1994, Kyriakides et al. 1995, Christensen and DeTeresa, 1997).

Measurements of the size effect over a broad size range, however, appear to be unavailable at present, yet there is a good reason to suspect that a size effect exists. This is indicated by observing that (1) the shear slip and fracture along the fibers in the kink band probably exhibits softening, i.e., a gradual reduction of the shear stress to some final asymptotic value, and (2) the kink band does not form simultaneously along the entire kink band but has a front that propagates, in the manner of the band of parallel compression splitting cracks.

5.7. Effect of Material Orthotropy

In the case of fiber composites, as well as some rocks, LEFM must be generalized to take into account the orthotropy of the material. The stress intensity factor of a sharp crack with a negligibly small FPZ may always be written in the form:

$$K_I = \sigma_N \sqrt{\pi D \alpha}\, F(\alpha) \qquad (\alpha = a/D) \tag{5.7}$$

where σ_N = nominal stress, considered here at maximum load, D = characteristic dimension, a = crack length, α = relative crack length, and $F(\alpha)$ = function characterizing structure geometry and material orthotropy.

The energy release rate $\mathcal{G}$ may be related to K_I using Bao et al.'s (1992) generalization of Irwin's (1958) relation for orthotropic materials:

$$\mathcal{G} = \frac{K_I^2}{\bar{E}} = \frac{D}{\bar{E}}\sigma_N^2 g(\alpha), \qquad g(\alpha) = \pi\alpha[F(\alpha)]^2 \tag{5.8}$$

where $g(\alpha)$ = dimensionless energy release function, characterizing the structure geometry and material orthotropy, and

$$\bar{E} = \frac{1}{Y(\rho)^2} \frac{(E_2/E_1)^{1/4}}{\sqrt{(1+\rho)/2E_1 E_2}} \tag{5.9}$$

$$\rho = \frac{\sqrt{E_1 E_2}}{2G_{12}} - \sqrt{\nu_{12}\nu_{21}} \tag{5.10}$$

and

$$Y(\rho) = [1 + 0.1(\rho - 1) - 0.015(\rho - 1)^2 + 0.002(\rho - 1)^3][(\rho + 1)/2]^{-1/4} \tag{5.11}$$

Subscripts 1 and 2 refer to Cartesian axes $x_1 \equiv x$ and $x_2 \equiv y$; x_2 coincides with the fiber direction; E_1, E_2, G_{12}, and ν_{12} are the orthotropic elastic constants; and parameters E_2/E_1 and ρ characterize the degree of orthotropy. The formula is valid when the crack propagates in the direction x_1 orthogonal to the fibers, but it is used here as an approximation even for propagation directions forming a small angle with x_1.

For fracture specimens in the form of long notched strip or slender notched beams, function $g(\alpha)$ or $F(\alpha)$ may be taken approximately the same as for isotropic specimens.

The results obtained by Bažant, Daniel and Li (1996) with the size effect law modified for orthotropy are shown in Fig. 2.6.

Chapter 6

Scaling via J-Integral, with Application to Kink Bands in Fiber Composites

6.1. J-Integral Analysis of Size Effect on Kink Band Failures

It is instructive to show now the application of the J-integral to the derivation of the basic scaling properties of kink band failures, as presented in Bažant et al. (1999). This represents the most fundamental approach to fracture. Let us analyze the specimen with unidirectional (axial) fiber reinforcement shown in Fig. 6.1. The kink band has length a which can be long or short compared with the specimen width D taken as the characteristic dimension. The width of the kink band, considered to be small, is denoted as w, and its inclination as β (Fig. 6.2a, 6.3). Although tractable, the bending stiffness of the fibers is neglected, for the sake of simplicity.

The loading is assumed to produce cohesive shear cracks that are parallel to the fibers and have a certain characteristic spacing s. The axial normal stress transmitted across the kink band (band-bridging stress) is denoted as σ (Fig. 6.3). Although Fig. 6.2a depicts an in-plane fiber inclination, the behavior is similar for the out-of-plane fiber inclination in the test specimen used because what matters for the analysis is the reduction of axial stress across the kink band, which is the same for both cases.

The diagram of the shear stress τ transmitted across the shear cracks versus the slip displacement η_{fr} on these cracks must exhibit post-peak softening (Fig. 6.4 top left). This is confirmed by two important recent experimental findings. First, Fleck and Shu (1995) placed strain gauges at the flanks of the kink band and, as the kink band grew, observed the strain in the gages to decrease, rather than remain constant (see also Fleck and Hutchinson 1997). Second,

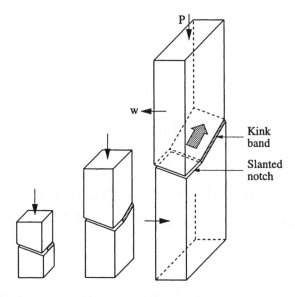

Figure 6.1: Geometrically similar single-edge notched carbon-PEEK (poly-ether-ether-keton) specimens tested (Bažant et al. 1999)

Moran et al. (1995) recently discovered the phenomenon of band broadening (see also Sutcliffe et al. 1996), which implies that the relative displacement across the band increases as the band grows, and thus indicates that the kink band plays a role similar to a crack (whose opening width grows with the distance from the front and the transmitted stress decreases), rather than to a line of dislocations (on which the relative displacement as well as the transmitted stress remains constant).

For the sake of simplicity, the stress-displacement diagram of the axial shear cracks is considered to be bilinear, as shown in Fig. 6.4 (top right) where τ_p = peak stress or shear strength = shear stress parallel to fibers at which the cohesive crack initiates, and τ_r = the residual shear strength, representing the final yield plateau. According to the analysis of mode II slip bands by Palmer and Rice (1973), the area of the diagram above the yield plateau is known to play the role of shear (Mode II) fracture energy, G_f (see the shaded triangle in Fig. 6.4 top left) (the critical value J_{cr} of the J-integral also includes the rectangle below the triangle). The fracture energy of the kink band, that is, the energy dissipated by fracture per unit length of the band is

$$G_b = G_f w/s \qquad (6.1)$$

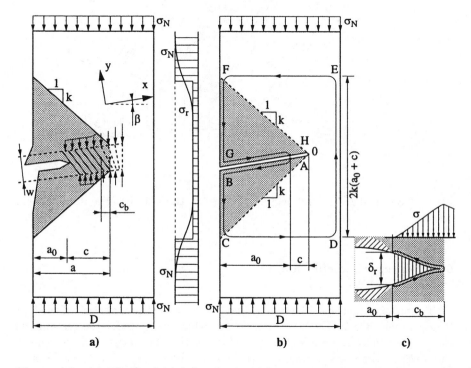

Figure 6.2: (a) Idealized kink band of width w in a notched specimen, with a fracture process zone of effective length c_b; (b) path of J-integral, with energy release (stress relief) zones OFGO, OBCO; (c) fracture process zone of equivalent cohesive crack

140   Scaling of Structural Strength

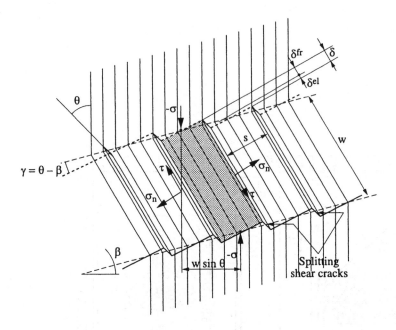

Figure 6.3: Idealized microbuckling of fibers in the kink band and axial shear cracks

6.2. J-integral Calculations

To approximately calculate the energy release due to propagation of the kink band, we use Rice's (1968a) J-integral, for which we consider the rectangular closed path $ABCDEFGH$ shown in Fig. 6.2b. The start and the end of this path at the crack surfaces must lie at the boundary of the FPZ because the residual stress across the band does work (for Mode II cracks this was shown by Palmer and Rice, 1973). The top, bottom and right sides of this rectangular path, $CDEF$, are sufficiently remote from the crack band for the initially uniform stress state to remain undisturbed.

On the left downward sides of the rectangular path, FG and BC, the distribution of the axial stress has some kind of a curved profile sketched on the left of Fig. 6.2b. The precise shape of this profile is not important but it is important that asymptotically, for large sizes $D \gg w$, the profiles must become geometrically similar. This observation is the basic idea of the asymptotic size effect analysis via the J-integral.

For the sake of simplicity, we may replace this profile by the stepped piecewise constant profile shown, in which the stress drops abruptly from the initial

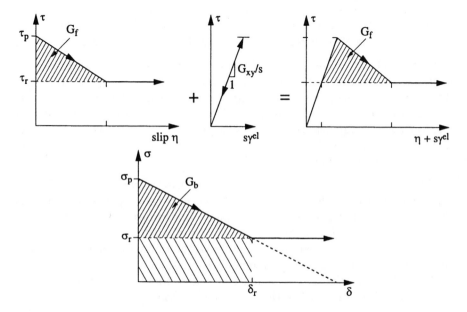

Figure 6.4: Top left: assumed bilinear diagram of shear stress versus slip displacement on the axial cracks of spacing s, crossing the kink band. Top right: superposition of elastic deformation between the cracks to obtain the diagram of shear stress versus total shear displacement accumulated over distance s between cracks. Bottom: Diagram of the axial normal stress σ versus axial displacement δ across the kink band, and area representing the kink band fracture energy G_b

stress σ_N to the residual stress σ_r which is transmitted across the band after the band contracts sufficiently. An important point again is that, for large enough geometrically similar specimens ($D \gg w$), the locations of the stress steps in this replacement profile must also be similar, that is, points F and C, must lie on inclined rays of a certain constant slope k shown dashed in Fig. 6.2b. These rays may be imagined to emanate from the tip of the equivalent crack of length $a = a_0 + c_b$ (Fig. 6.2b) where c_b characterizes the length of the FPZ of the kink band and represents approximately the distance from the center of the FPZ of kink band to the point where the stress is reduced to its residual value σ_r (Fig. 6.4 bottom; the length of the FPZ is about $2c_b$). Slope k depends on the structure geometry and on the orthotropic elastic constants.

The area between these rays and the kink band roughly represents the zone of stress relief caused by the drop of axial stress transmitted by the kink band. The strain energy contained within this area is released and is dissipated by the axial shear cracks forming at the front of the kink band. Noting that this area, and thus the energy release, increases in proportion to D^2, while the energy dissipated at the kink band front increases linearly with D, one immediately concludes that there must be size effect.

The zone at kink band front in which the axial shear cracks are forming represents the FPZ of the kink band. Its length c_b may be regarded as a material property, almost independent of the specimen dimensions and geometry. It may be considered to be of the same order of magnitude as the width w. Throughout this zone, the fiber inclination increases from the initial misalignment angle $\bar{\varphi}$ up to the value $\bar{\varphi} + \varphi$ corresponding to the residual cracks. To make test evaluation simple, the specimens must be notched and the FPZ at maximum load must still be attached to the notch (i.e. $c = c_b$).

Referring to the sketch in Fig. 6.2b, the crack band of length $a_0 + c_f$ is approximately equivalent to a mode I crack whose faces are imagined to interpenetrate. The length of this crack is $a_0 + c$ where $c = c_f + (w/2k)$, which may again be assumed to be approximately a constant when the size is varied. Consequently, the height FC of the rectangular path in Fig. 6.2b is approximately $2k(a_0 + c)$, as labeled in the figure.

In view of these considerations, the first part of the J-integral may be approximately expressed as follows (Bažant et al. 1999):

$$\oint \overline{W} dy = 2k(a_0 + c) \left(\frac{\sigma_N^2}{2E_y} - \frac{\sigma_r^2}{2E_y} \right) \tag{6.2}$$

in which $\overline{W}$ = strain energy density, and y = coordinate normal to the direction of propagation (Fig. 6.2b), and E_y = effective elastic modulus of the orthotropic fiber composite in the fiber direction y (with different values for plane strain and plane stress). In (6.2) we have considered that the parts of the integral over the horizontal segments are 0, and that the stress on the vertical segment DE may be assumed undisturbed by the kink band, i.e., equal to σ_N. The portions of the integral over the crack surface segments GH and AB are, likewise, 0.

The second part of the J-integral may be calculated in a similar manner as that introduced by Palmer and Rice (1973) for the propagation of Mode II shear fracture with residual friction;

$$\oint \vec{\sigma} \cdot \frac{\partial \vec{u}}{\partial x} ds = \int_{AB} \sigma_r \frac{d}{dx}\left[\frac{1}{2}\delta(x)\right] dx - \int_{GH} \sigma_r \frac{d}{dx}\left[\frac{1}{2}\delta(x)\right] dx$$

$$= \int_{x=0}^{a_0} \sigma_r d\delta(x) = \sigma_r \int_{x=0}^{a_0} d\delta(x) = \sigma_r \delta_{BG} \quad (6.3)$$

in which $\vec{\sigma}$ = stress vector acting from the outside on the domain enclosed by the path, $\vec{u}$ = displacement vector, s = length coordinate along the path, δ = relative displacement across the band, and δ_{BG} = relative displacement between points B and G. That displacement can be estimated as the difference between the changes of length $\overline{ED}$ and length $\overline{FC}$;

$$\delta_{BG} = \Delta\overline{ED} - (\Delta\overline{FG} + \Delta\overline{BC}) \quad (6.4)$$

$$= 2k(a_0+c)\frac{\sigma_N}{E_y} - 2k(a_0+c)\frac{\sigma_r}{E_y} = 2k(a_0+c)\frac{\sigma_N - \sigma_r}{E_y} \quad (6.5)$$

Now the J-integral may be readily evaluated as follows:

$$J = \oint\left(\overline{W}dy - \vec{\sigma}\cdot\frac{\partial \vec{u}}{\partial x}ds\right) = \frac{k}{E_y}(a_0+c)\left[\sigma_N^2 - \sigma_r^2 - 2(\sigma_N - \sigma_r)\sigma_r\right]$$

$$= \frac{k}{E_y}(a_0+c)(\sigma_N - \sigma_r)^2 \quad (6.6)$$

The energy consumed may be calculated again with the help of the J-integral. Similar to Rice (1968b) and Palmer and Rice (1973), the integration path that runs along the equivalent crack surface and around the crack tip (Fig. 6.2c) may be used;

$$J_{cr} = \oint \vec{\sigma} \cdot \frac{\partial \vec{u}}{\partial x} dx \quad (6.7)$$

This represents the critical value, J_{cr}, of the J-integral required for propagation. This critical value may be subdivided into two terms:

$$J_{cr} = G_b + \sigma_r \delta_r \quad (6.8)$$

where G_b is the fracture energy, i.e., the energy required to produce the axial shear cracks across the kink band, and $\sigma_r \delta_r$, represents the plastic work that is done by the residual stresses σ_r within the FPZ of the kink band and is leaving the FPZ in its wake. This work corresponds in Fig. 6.4 (top left) to the shaded rectangle lying under the shaded triangle. Following the way shown by Rice (1968b) and Palmer and Rice (1973) for shear bands, J_{cr} may be evaluated (Fig. 6.2c) as follows:

$$J_{cr} = \oint \vec{\sigma} \cdot \frac{\partial \vec{u}}{\partial x} dx$$

$$= -\int_{x=a_0}^{a_0+c} f[\delta(x)] \frac{d}{dx}\left[\frac{1}{2}\delta(x)\right] dx + \int_{x=a_0+c}^{a_0} f[\delta(x)] \frac{d}{dx}\left[\frac{1}{2}\delta(x)\right] dx$$

$$J_{cr} = -\int_{x=a_0}^{a_0+c} f[\delta(x)] \frac{d\delta(x)}{dx} dx = \int_0^{\delta_r} f[\delta(x)] d\delta(x) \qquad (6.9)$$

(Bažant et al. 1999a,b). This means that J_{cr} represents the sum of the shaded triangle and shaded rectangle in the stress-displacement diagram of Fig. 6.4 (top left). Therefore, according to (6.8), fracture energy G_b is represented by the area under the descending stress-displacement curve and above the horizontal line for the residual stress.

6.3. Case of Long Kink Band

Setting (6.6) equal to (6.8), and solving for the nominal strength σ_N of the specimen, we obtain (Bažant et al. 1999):

$$\sigma_N = \sigma_R + \sqrt{\frac{E_y(G_b + \sigma_r \delta_r)/kc}{1 + D/D_0}} = \sigma_R + \frac{\sigma_0}{\sqrt{1 + D/D_0}} \qquad (6.10)$$

in which

$$D_0 = \frac{c}{\alpha_0}, \quad \sigma_0 = \sqrt{\frac{E_y(G_b + \sigma_r \delta_r)}{kc}}, \quad \alpha_0 = \frac{a_0}{D}, \quad \sigma_R = \sigma_r \qquad (6.11)$$

(for other geometries, σ_R need not be equal to σ_r). The resulting formula (6.10) has the same form as that proposed by Bažant (1987a) for the general case of quasibrittle failures with a residual plastic mechanism, and subsequently verified for several applications to concrete structures (Bažant and Xiang 1997). This formula is valid when a long enough kink band transmitting constant residual stress σ_r develops in a stable manner before the maximum load is reached. Because of σ_r, such stable propagation can happen even in specimens of positive geometry (i.e., for increasing $g(\alpha)$). Stable propagation is helped by rotational restraint of specimen ends.

6.4. Failure at the Start of Kink Band from a Notch or Stress-Free Crack

In the case of notched test specimens (of positive geometry), the maximum load is achieved while the FPZ of the kink band is still attached to the notch. Except for the sign of the band-bridging stresses, the situation is analogous to tensile fracture of notched specimens. From experiments on concrete as well as analytical studies based on the cohesive crack model, it is known that only a short initial portion of the softening stress-displacement curve of the cohesive crack comes into play. It is only the initial downward slope of this curve which matters for the maximum load (the tail of the postpeak load-deflection diagram, of course, depends on the entire stress-displacement curve of the cohesive crack); see Bažant and Li (1995) or Bažant and Planas (1998).

A similar situation must be expected for kink bands in notched specimens. Since the shape of the softening stress displacement curve of the cohesive crack model is irrelevant for the maximum load, except for the initial downward slope of the curve, the maximum load must be the same as that for a linear stress-displacement diagram, shown by the descending dashed straight line shown in Fig. 6.4 (bottom).

It follows that in this case the residual stress σ_r should be disregarded and the fracture energy G_B that mathematically governs the kink band growth at maximum load of a notched specimen corresponds to the entire area under the extended descending straight line in Fig. 6.4. Obviously, $G_B > G_b$ if $\sigma_r > 0$. Consequently, setting $\delta_r = 0$ in (6.11) and replacing G_b by G_B, we have the size effect law:

$$\sigma_N = \frac{\sigma_0}{\sqrt{1 + D/D_0}} \qquad (6.12)$$

with

$$D_0 = \frac{c_0}{\alpha_0}, \qquad \sigma_0 = \sqrt{\frac{E_y G_B}{k c_0}}, \qquad c_0 = c_b + \frac{w}{2k} \qquad (6.13)$$

(Bažant et al. 1999). This coincides with the approximate size effect law proposed in Bažant (1983, 1984); Fig. 6.5 (left, for $\sigma_R = 0$).

From experience with other materials, the length (at maximum load) of the crack band up to the beginning of the FPZ, a_0, may often be considered to be roughly proportional to the specimen size D, within a certain range of sizes. In other words, the ratio D/a_0 at maximum load of geometrically similar structures is often approximately constant. So is the value of D_0 in (6.12), provided that the specimens are geometrically similar.

Similar results are obtained via equivalent LEFM (Bažant et al. 1999). That approach relies on some stronger simplifications but has the advantage that effect of structure geometry is also captured.

146   Scaling of Structural Strength

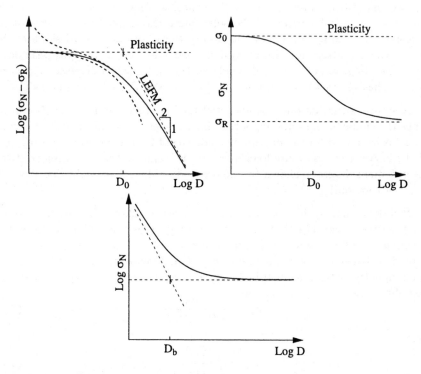

Figure 6.5: Top left: size effect law (solid curve) for specimens with a long kink band or notch (Eq.2.8) and asymptotic formulas (dashed curves). Top right: same but with σ_N instead of $\log(\sigma_N - \sigma_R)$ as the ordinate. Bottom: size effect law for when P_{max} occurs at kink band initiation

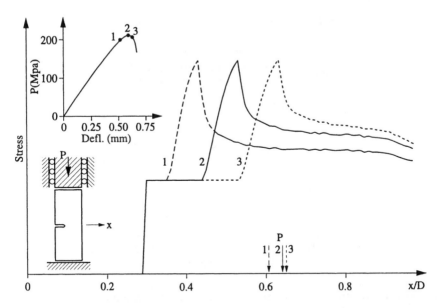

Figure 6.6: Stress profiles across the ligament of the carbon-PEEK specimens before, at and after the maximum load (after Bažant et al. 1999). Note the shift of the compression resultant P which makes a stable kink band growth possible

6.5. Comparison with Size Effect Tests of Kink Band Failures

To demonstrate the existence of a size effect in kink band failure and justify the present analysis, tests of relatively large carbon fiber-PEEK specimens of three different sizes (Fig. 6.1) have been carried out at Northwestern University (Bažant et al. 1999). Slanting of the notches was found to achieve that no axial shear-splitting would precede or accompany the kink band growth (Fig. 6.1). Rigid restraint against rotation at the ends made it possible for the kink band to grow stably for a considerable distance before attaining the maximum compression load. Attainment of this goal was verified experimentally and was also demonstrated theoretically by the subsequent stress profiles along the kink band calculated with the cohesive crack model (Fig. 6.6).

The results of individual tests are shown in Fig. 6.7. Despite high scatter, which is probably inevitable in the case of fiber composites, one can see that the present theory does not disagree with the test results (Bažant et al. 1999).

The present fracture theory exhibiting size effect may also be verified by comparison with the test results of Soutis, Curtis and Fleck (1993). They used rectangular prisms with circular holes. Although they did not vary the

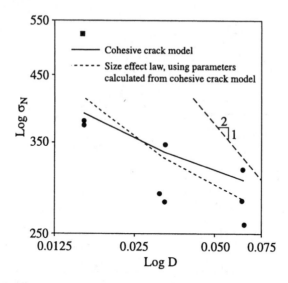

Figure 6.7: Optimum fits by size effect formulae and by numerical analysis with the cohesive crack model, after Bažant et al. (1999)

specimen size, they varied the hole size, which represents a combination of size effect and shape effect. Thanks to the fact that the formulation based on the equivalent LEFM captures also the shape effect, the asymptotic formula derived in Bažant et al. (1999) could be used to fit the data. The comparison showed that the theory exhibiting size effect allows much better fits than a theory lacking the size effect (Fig. 6.8).

The results make it clear that the present theories of kink-band failure, which are based on plasticity or strength criteria, are adequate only for small structural parts. For large ones, the size effect must be taken into account.

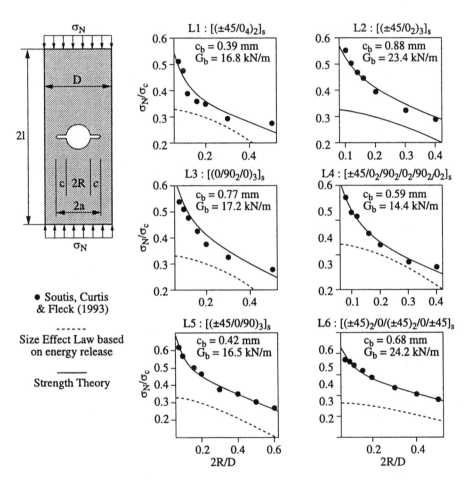

Figure 6.8: Soutis, Curtis and Fleck's (1993) test results for quasi-isotropic and orthotropic carbon–epoxy laminates of six different layups, with holes of various radii R (data points) and constant width D. Solid curves: optimum fits by size effect law. Dashed curves: predictions of strength theory exhibiting no size effect

Chapter 7

Time Dependence, Repeated Loads and Energy Absorption Capacity

7.1. Influence of Loading Rate on Size Effect

Strictly speaking, fracture is always a time dependent phenomenon. In polymers, strong time dependence of fracture growth is caused primarily by viscoelasticity of the material (see the works of Williams and others beginning with the 1960s). In rocks and ceramics, the time dependence of fracture is caused almost exclusively by the time dependence of the bond ruptures that cause fracture. In other materials such as concrete, both sources of time dependence are very important (Bažant and Gettu 1992, Bažant and Wu 1993; Bažant and Li 1997; Li and Bažant 1997). Both sources of time dependence have a significant but rather different influence on the scaling of fracture.

Consider first the rupture of an interatomic bond, which is a thermally activated process. The frequency of ruptures is given by the Maxwell-Boltzmann distribution, defining the frequency f of exceeding the strength of atomic bonds, $f \propto e^{-\mathcal{E}/RT}$, where T = absolute temperature, R = gas constant and $\mathcal{E}$ = energy of the vibrating atom. When a stress is applied, the diagram of the potential energy surface of the interatomic bonds is skewed as sketched in Fig. 7.1a. This causes the activation barrier for bond breakages to be reduced from Q to a smaller value $Q - c\sigma$, and the activation barrier for bond restorations to be increased from Q to $Q + c\sigma$, where Q = activation energy = energy barrier at no stress, and c = constant. This causes that the frequency of bond ruptures, f^+, becomes greater than the frequency of bond breakages, f^-, with the net

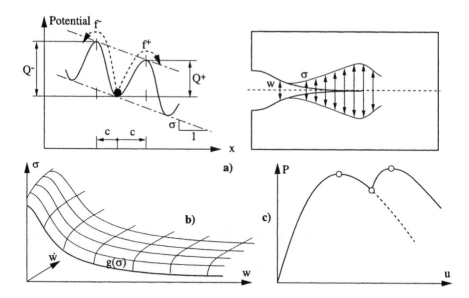

Figure 7.1: (a) Skewing of the potential surface of interactomic bond caused by applied stress, with corresponding reduction of activation energy Q^+; (b) dependence of cohesive stress on crack opening and cohesive stress; (c) response change after a sudden increase of the loading rate

difference

$$\Delta f = f^+ - f^- \propto e^{-\overbrace{(Q - c\sigma)}^{Q^+}/RT} - e^{-\overbrace{(Q + c\sigma)}^{Q^-}/RT} \qquad (7.1)$$

which becomes

$$\Delta f \propto \sinh(c\sigma/RT) e^{-Q/RT} \qquad (7.2)$$

The rate of the opening w of the cohesive crack may be assumed approximately proportional to Δf. From this, the following rate-dependent generalization of the crack-bridging (cohesive) law for the cohesive crack has been deduced (Bažant 1993, 1995c; Bažant and Li 1997; Li and Bažant 1997):

$$w = g\left[\sigma - \kappa\, e^{Q/RT} \operatorname{asinh}\left(\frac{\dot w}{c_0}\right)\right] \qquad (7.3)$$

The dependence of the stress displacement curves for the cohesive crack on the crack opening rate $\dot w$ is shown in Fig. 7.1b.

The effect of linear viscoelasticity in the bulk of the structure can be introduced into the aforementioned equations of the cohesive crack model on the

Time Dependence, Repeated Loads and Energy Absorption Capacity   153

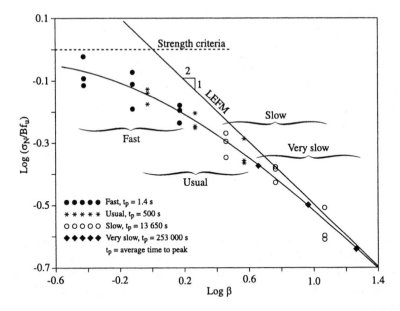

Figure 7.2: Nominal strengths of 4 groups of 3 specimens of different sizes tested at 4 different times to peak, t_p, plotted as as function of size relative size $\beta = D/D_0$ (after Bažant and Gettu 1992)

basis of elastic-viscoelastic analogy (correspondence principle). Numerical solutions of fracture specimens show that viscoelasticity in the bulk (linear creep) causes the points in the size effect plot to shift to the right, toward increasing brittleness. This explains the observations of Bažant and Gettu (1992), which show the data points on the size effect plot for groups of similar small, medium and large notched specimens tested at various rates of crack mouth opening displacement (Fig. 7.2; Bažant and Li 1997). These rates are characterized by the time t_p to reach the peak. As revealed in Fig. 7.2, the groups of data points move to the right with an increasing t_p.

The fact that the brittleness of response is increasing with a decreasing rate of loading or increasing load duration may at first be surprising but can be explained (as revealed by calculations according to the time dependent cohesive crack model) by relaxation of the stresses surrounding the fracture process zone, which cause the process zone to become shorter. This behavior is also clarified by the plot of the nominal strength (normalized with respect to the material strength f'_t) versus the crack mouth opening displacement (normalized with respect to the critical crack opening w_c). For a specimen in which the only source of time dependence is creep, the peaks of these stress displacement curves shift with an increasing rate of loading to the left and the softening

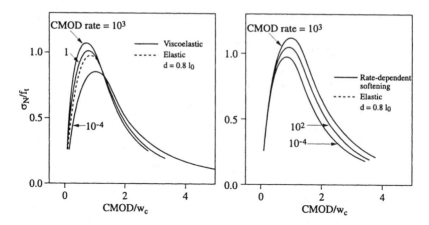

Figure 7.3: Curves of nominal stress versus relative crack mouth opening displacement (CMOD) for different CMOD rates, calculated by cohesive crack model under the assumption that the material exhibits only viscoelasticity in the bulk (left) or only rate-dependent crack opening (right) (Li, Hongh and Bažant 1995)

curves cross (Fig. 7.3 left). On the other hand, when the rate dependence is caused only by the bond breakages, the peaks shift to the right, as seen in Fig. 7.3 (right), and in that case there is no shift of brittleness of the kind seen in Fig. 7.2. It must be emphasized that these results are valid only in the range of static loading, that is, in absence of inertia forces and wave propagation effects. The behavior becomes more complicated in the dynamic range.

7.2. Size Effect on Fatigue Crack Growth

Related to the time dependence is the influence of fatigue on fracture (Paris and Erdogan, 1967). The rate of growth of a crack caused by fatigue loading is approximately given by the Paris law (or Paris-Erdogan law) which reads: $\Delta a/\Delta N = \kappa(\Delta K_I/K_{Ic})^n$, in which a = crack length, N = number of cycles, ΔK_I = amplitude of the applied stress intensity factor; κ, n = dimensionless empirical constants; and K_{Ic} = fracture toughness introduced only for the purpose of dimensionality. The interesting point is that the rate of growth does not depend on the maximum and minimum values of K_I, as a good approximation.

This law has found wide applicability for fatigue growth of cracks in metals. If similar structures with similar cracks are considered, this equation implies the size effect of LEFM, which is however too strong for not too large quasibrittle

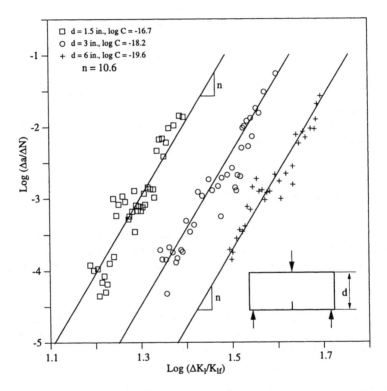

Figure 7.4: Crack growth per cycle versus amplitude or relative stress intensity factor for three different sizes of concrete specimens (after Bažant and Xu 1991)

structures. It was shown (Bažant and Xu, 1991; Bažant and Schell, 1993) that the Paris law needs to be combined with the size effect law for monotonic loading, yielding the following generalization of Paris law in which the effect of structure size D is taken into account:

$$\frac{\Delta a}{\Delta N} = \kappa \left(\frac{\Delta K_I}{K_{Ic}} \sqrt{1 + \frac{D_0}{D}} \right)^n \tag{7.4}$$

in which D_0 is the same exponent as in Paris law, and K_{Ic} is a constant denoting the fracture toughness of an infinitely large structure.

The necessity of the size correction is demonstrated by the test results of Bažant and Xu (1991) for concrete in Fig. 7.4. At constant size D, the logarithmic plot of the crack growth rate versus the amplitude of K_I should be approximately a straight line. This is clearly verified by Fig. 7.4. However, for different specimen sizes, different lines are obtained. The spacing of these straight lines is well predicted by Eq. (7.4), while for the classical Paris law these three lines would have to be identical.

7.3. Wave Propagation and Effect of Viscosity

The effective length c_f of the fracture process zone may be regarded as the characteristic length of the material. In dynamics, the existence of a material length further implies the existence of a characteristic time (material time):

$$\tau_0 = c_f/v \tag{7.5}$$

This represents the time for a wave of velocity v to propagate the distance c_f. Obviously, for times shorter than τ_0, the material length cannot get manifested and the damage localization that leads to a crack band or a sharp crack has not enough time to happen.

In dynamic problems, any type of viscosity η of the material (present in models for creep, viscoelasticity or viscoplasticity) implies a characteristic length, even if there is no characteristic length for very slow static response. Indeed, since η = stress/strain rate $\sim$ kg/ms, and the Young's modulus E and mass density ρ have dimensions $E \sim$ kg/ms^2 and $\rho \sim$ kg/m^3, the material length associated with viscosity is given by

$$\ell_v = \frac{\eta}{v\rho}, \qquad v = \sqrt{\frac{E}{\rho}} \tag{7.6}$$

where v = wave velocity. Consequently, any rate dependence in the constitutive law implies a size effect of quasibrittle type (and a nonlocal behavior as well).

There is, however, an important difference. Unlike the size effect associated with a real material length, the viscosity-induced size effect (as well as the width of damage localization zones) is not time independent. It varies with the rates of loading and deformation of the structure, and vanishes as the rates drop to zero. For this reason, an artificial viscosity or rate effect can approximate the nonviscous size effect and localization only within a narrow range of time delays and rates, but not generally.

7.4. Ductility and Energy Absorption Capacity of Structures

For dynamic loads such as blast or earthquake, the most important characteristic of a structure is its energy absorption capability, which represents the energy corresponding to the area under the complete load-deflection diagram (with its entire post-peak softening tail). This area depends on the ductility of the structure, which diminishes with increasing structure size.

To illustrate the size effect on structural ductility, consider the diagrams of load P versus load-point displacement u_f caused by fracture when the energy release rate is constant, being equal to fracture energy G_f of the material. For a zero crack length ($a = 0$), these diagrams descend from infinity ($P \to \infty$, Fig. 7.5), which means that a crack cannot start from a smooth surface according to LEFM. These responses can be of two types:

- *Type I*, for which the $P(u_f)$-curve has always a negative slope (Fig. 7.5a), and

- *Type II*, for which the slope of the $P(u_f)$-curve reverts at a certain point (called the *snapback point*) to positive (Fig. 7.5b).

(for calculations of such curves from LEFM, see Bažant and Cedolin, 1991, Fig. 12.17). It can be shown that the type of response is decided by the following criterion:

$$\text{for } \lim_{a \to a_1} ND/M = 0 \;\ldots\; \text{Type I}, \qquad > 0 \;\ldots\ldots\; \text{Type II} \qquad (7.7)$$

where a_1 = crack length for full break of cross section; M, N = bending moment and normal force transmitted across the ligament.

When the structure has a notch, of length a_0 (or a pre-existing traction-free crack), the response first follows a straight line emanating from the origin (lines 01 in Fig. 7.5), and when the $P(u_f)$ curve for a propagating crack is reached, there is an abrupt slope change. If $R(c)$ varies according to a smooth R-curve,

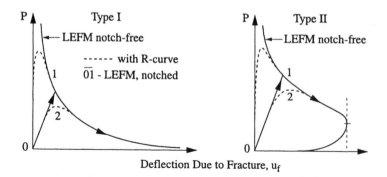

Figure 7.5: Two types of LEFM fracture response—with and without a snapback

the slope change is not abrupt but smooth, i.e. the peak is rounded (see the dashed curves in Fig. 7.5).

The elastic deflection of a crack-free structure, the additional deflection caused by the crack, and the deflection of the spring are additive. The loads causing these three deflections are equal. So the coupling of the corresponding parts may be imagined as a series coupling; see Fig. 7.6 where adding the segments a, b, and c on any horizontal line yields the horizontal coordinate $a + b + c$ of the load-deflection curve of the whole system. The inverse slopes (compliances) at the points with the same P (lying on a horizontal line) are also added.

The larger the structure, the lower its stiffness. As clarified by Fig. 7.6, a sufficient enlargement of structure size, corresponding to a sufficiently soft spring (i.e., a large enough C_s or C_0), will cause the total load-deflection diagram to exhibit a snapback—a point at which the descending load-deflection curve ceases to have a negative slope. If the curve is smooth, it is a point with a vertical tangent ($d\Delta/dP = 0$), which represents a point of (locally) maximum deflection (or an inflexion point); Fig. 7.7. This is known to represent the stability limit under displacement control (Bažant and Cedolin 1991).

For type I fracture, no vertical tangent nor snapback will occur if the spring and the crack-free specimen are sufficiently stiff (i.e, if C_s and C_0 are small enough). Since an increase of slenderness L/D can cause C_0 to exceed any given value, a snapback occurs if (and only if) the structure is sufficiently slender. But this is different for type II structures, for which a snapback occurs for any slenderness.

From the geometrical construction in Fig. 7.6, it now transpires that, on the curve of P versus the crack-produced deflection u_f, the point of snapback instability occurs at the point at which the tangent to the curve has the slope

Time Dependence, Repeated Loads and Energy Absorption Capacity 159

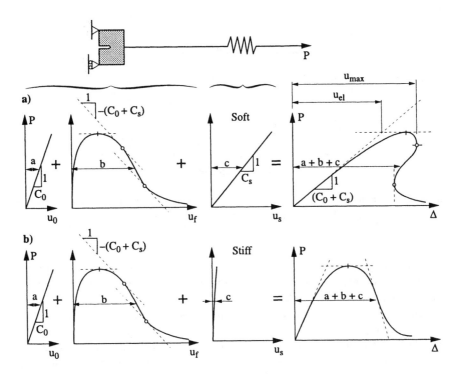

Figure 7.6: Superposition of deflections due to elasticity of structure with no crack (a) and of spring (b), and deflection due to crack (c)

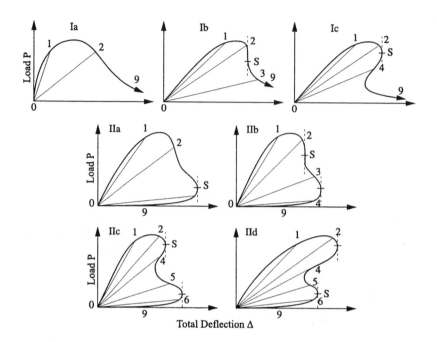

Figure 7.7: Various kinds of load-deflection curves for crack growth characterized by an R-curve, for monotonic loading and reloading

$1/C_{cr} = -1/(C_0 + C_s)$ (shown in Fig. 7.6). From this observation one can readily deduce various type I or II combinations of the $P(u_f)$ curve with the combined elastic deflection characterized by the combined compliance $C_f + C_0$ (the additional effect of the R-curve behavior, which is not shown in Fig. 7.7, is to 'round off' the sharp corners on these curves in the manner shown in Fig. 7.5). These combinations lead to seven different kinds (Ia,...IId) of the overall load-deflection curves $P(\Delta)$, illustrated in Fig. 7.7. For the types possessing more than one point at which the line of slope $1/C_{cr}$ is tangent, the failure under static displacement-controlled loading will occur at the first such point.

If the structure is pre-cracked or notched, the initial loading follows an inclined straight line until the tip of the existing crack or notch becomes critical ($\mathcal{G} = R$); see the line segments 01, 02, 03,... in Fig. 7.7. So the load-deflection diagrams can follow any of the paths 019, 029, 039,..., 0190, 0290, 0390,... identified in Fig. 7.7. Thus, as seen in Fig. 7.7, an enormous variety of responses can be encountered in an elastic structure with one growing crack.

The concept of ductility has often been hazy in practice. A distinction must be made between material ductility, characterizing the strain at which a plastically yielding material will fail due to a microcrack, and structure duc-

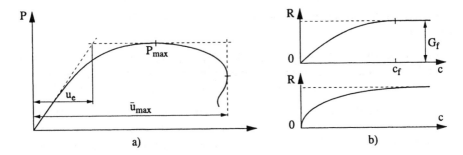

Figure 7.8: (a) Definition of ductility; (b) Two types of R-curve considered: $R(c) = G_f\sqrt{c/(c_0 + c)}$ (upper curve), and geometry dependent R-curve (lower curve)

tility. We consider only the latter concept of ductility. A rational definition must be based on the stability loss of the structure, as proposed for fracturing or damaging structures in Bažant (1976), and studied in more detail in Bažant, Pijaudier-Cabot and Pan (1987a,b). For load control conditions (i.e., for gravity loads), the stability loss occurs when the maximum load is reached.

Ductility is a different concept from the maximum load or strength of the structure. It characterizes the deformation capability under the most stable type of loading, which is the loading under displacement control. In that case, stability is lost at the snapback point (the first point of vertical tangent, if it exists) of the overall load-deflection curve of the structural system.

Therefore, we define ductility (Fig. 7.8a) as (Bažant 1976) $\lambda = \Delta_{max}/\Delta_{el}$ where Δ_{max} = total deflection of the system (cracked structure together with the spring) at the first point of vertical tangent (snapback), and Δ_{el} = elastic (recoverable) part of the total deflection at maximum load (Fig. 7.8a). We can say that the ductility is infinite (or unbounded, $\lambda = \infty$) when no point of vertical tangent exists, as shown in Fig. 7.7, case Ia.

The loss of stability under displacement control, i.e. the ductility limit, corresponds in Fig. 7.5 and 7.6 to the first point of vertical tangent. For type Ia in Fig. 7.7 there is no stability loss under displacement control, which means the ductility is unbounded.

Would it make sense to define ductility by some post-peak point with a certain finite softening slope? It would not. Stability loss does not occur at such a point if the total system is considered. If the structure is considered without the spring through which is it loaded, and the slope is equal to the negative of the spring stiffness, then of course such a point does indicate stability loss (Bažant and Cedolin 1991, Sec. 13.2), but such a point must then correspond to a point of vertical tangent on the load-deflection curve of the complete system with the spring.

If a structure of infinite ductility exhibits postpeak softening, no matter how mild, its sufficient enlargement will make the ductility finite, i.e., will produce a snapback. A further enlargement will make that finite ductility smaller.

The relative energy absorption may be defined as the area under the complete load-deflection diagram divided by the elastic strain energy recovered upon unloading from the maximum load state. Obviously, its value decreases with increasing structure size.

Detailed examples of calculation of effects of size as well as its slenderness on the ductility of cracked three-point-bend beams loaded through springs of various stiffnesses are found in Bažant and Becq-Giraudon (1999). Some of the results are plotted in Fig. 7.9. It should be noted that the family of the curves of ductility versus structure size at various spring stiffnesses is characterized by a certain critical spring stiffness that represents a transition from bounded single-valued functions of D to unbounded two-valued functions of D. The flexibility (force) method has been adapted to extend the ductility analysis to arbitrary structural assemblages for which the stress intensity factor of the cracked structural part considered alone is known.

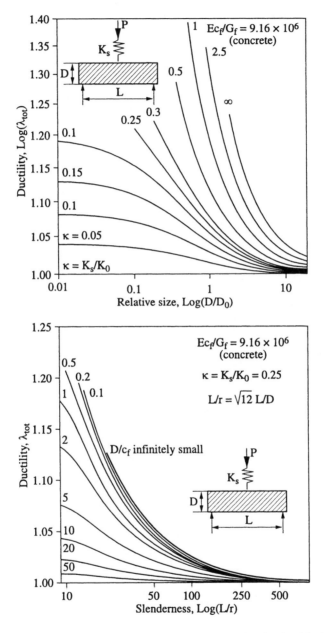

Figure 7.9: Top: ductility of concrete vs. specimen size in double logarithmic plot. Bottom: ductility of concrete vs. slenderness in semi-logarithmic plot. Both obtained for the geometry dependent R-curve corresponding to Bažant's size effect law (after Bažant and Becq-Giraudon 1999)

Chapter 8

Computational Approaches to Quasibrittle Fracture and Its Scaling

8.1. Eigenvalue Analysis of Size Effect via Cohesive (Fictitious) Crack Model

Computationally the most efficient model capable of resolving the stress profile along the fracture process zone is the cohesive crack model (as well as the crack band model which is essentially equivalent). There is more than one way to calculate the size effect for this model. One primitive way is to solve repeatedly by finite elements the complete response history of the structure for a sequence of increasing sizes and then to collect the maximum load values. There exists, however, a direct way which is much more efficient. It relies on converting the integral equation of the cohesive crack model to an eigenvalue problem (Bažant and Li 1995). The solution of the response history becomes unnecessary.

According to the cohesive crack model, introduced for concrete under the name *fictitious crack model* by Hillerborg et al. (1976), the crack opening in the fracture process zone (cohesive zone) is assumed to be a unique decreasing function of the crack-bridging stress (cohesive stress) σ; $w = g(\sigma)$. The basic equations of the cohesive crack model imply the condition that the crack opening calculated from the bridging stresses must be compatible with the elastic deformation of the surrounding structure, and the condition that the total stress intensity factor K_I^{tot} at the tip of the cohesive crack must be zero in order for the stress to be finite. They read:

$$g[\sigma(\xi)] = -\int_{\alpha_0}^{\alpha} D\, C(\xi,\xi')\sigma(\xi')d\xi' + D\, C_P(\xi)P \qquad (8.1)$$

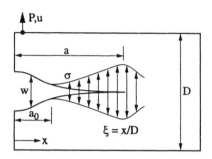

Figure 8.1: Cohesive crack and distribution of bridging stress

$$K_I^{tot} = -\int_{\alpha_0}^{\alpha} \kappa(\xi)\sigma(\xi)Dd\xi + P\kappa_P = 0 \qquad (8.2)$$

in which $\xi = x/D$, $x =$ coordinate along the crack (Fig. 8.1), $\alpha = a/D$, $\alpha_0 = a_0/D$, $a, a_0 =$ total crack length and traction free crack length, $C(\xi, \xi')$, $C_P(\xi) =$ compliances of the surrounding elastic structure for loads and displacements at the crack surface and at the loading point (Fig. 8.1), and $\kappa(\xi)$, $\kappa_P =$ stress intensity factors at the tip of cohesive crack $(x = a)$ for unit loads applied at the crack surface or at the loading point.

The usual way to solve the maximum load of a given structure according to the cohesive crack model was to integrate these equations numerically for step-by-step loading (Petersson, 1981). However, recently it was discovered that, under the assumption that there is no unloading in the cohesive crack (which is normally the case), the size effect plot can be solved directly, without solving the history of loading before the attainment of the maximum load. To this end one needs to invert the problem so that one looks for the size D for which a given relative crack length $\alpha = a/D$ corresponds to the maximum load P_{max}.

Then it is found that this size D represents the first eigenvalue of the following integral equation over the crack bridging zone (Bažant and Li 1995a,b):

$$D\int_{\alpha_0}^{\alpha} C(\xi, \xi')v(\xi')d\xi' = -g'[\sigma(\xi)]v(\xi) \qquad (8.3)$$

in which the eigenfunction $v(\xi)$ has the meaning of the derivative $\partial\sigma(\xi)/\partial\alpha$. The maximum load is then given by the following quotient

$$P_{max} = \frac{\int_{\alpha_0}^{\alpha} v(\xi)d\xi}{D\int_{\alpha_0}^{\alpha} C_P(\xi)v(\xi)d\xi} \qquad (8.4)$$

These results have also been generalized to obtain directly the load and displacement corresponding, on the load-deflection curve, to a point with any

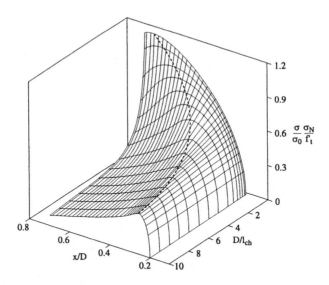

Figure 8.2: Stress profiles along the crack line for the maximum load and for various sizes of similar specimens (the peaks represent the tips of the cohesive crack)

given tangential stiffness, including the displacement at the snap-back point, which characterizes the ductility of the structure.

The cohesive crack model nicely illustrates the transition from failure at a relatively large fracture process zone in the case of small structures to the failure at a relatively small process zone in the case of large structures. See the plot of the profiles of the normal stress ahead of the tip of the traction-free crack length (notch length) shown in Fig. 8.2. The points at the tip of the cohesive zone represent the maximum stress points in these stress profiles. Note how the maximum stress points move, in relative coordinates, closer to the tip of the notch if the structure size is increased. These results of the cohesive crack model confirm that, for large sizes, the size effect of LEFM should be approached.

8.2. Microplane Model

Finite element analysis of failure of quasibrittle material requires a constitutive model for strain-softening damage. To be objective and capable of representing the size effect, the constitutive model must be tied in some way to a characteristic length of the material.

Modifications of classical plasticity, in which the stress strain relation is formulated empirically in terms of the strain and stress tensors and their invariants, have not proven very effective for complex materials such as concrete. It seems more effective to develop first-principles models which try to directly capture the main inelastic phenomena such as microcrack opening and closure and frictional slip, on various planes in the material.

This approach is taken in the microplane model, in which the constitutive law is expressed not in terms of tensors but in terms of vectors of stress and strain acting on planes of various orientations called the *microplanes*. This model, proposed in Bažant (1984c) as a generalization and adaptation of an idea of Taylor (1938), has been developed in detail for concrete (Bažant et al. 2000), although various versions have also been formulated and used for metal plasticity, shape memory alloys, stiff foams and rocks.

The microplane constitutive model is defined by a relation between the stresses and strains acting on a plane in the material, the microplane, having an arbitrary orientation characterized by its unit normal n_i. The basic hypothesis, which ensures stability of postpeak strain softening (Bažant, 1984c), is that the strain vector $\vec{\epsilon}_N$ on the microplane is the projection of strain tensor ϵ, i.e., $\epsilon_{N_i} = \epsilon_{ij} n_j$. The normal strain on the microplane is $\epsilon_N = n_i \epsilon_{N_i}$, that is,

$$\epsilon_N = N_{ij} \epsilon_{ij} \tag{8.5}$$

where $N_{ij} = n_i n_j$ (repetition of the subscripts, referring to Cartesian coordinates x_i, implies summation over $i = 1, 2, 3$).

The shear strains on each microplane are characterized by their components in chosen directions M and L given by orthogonal unit coordinate vectors $\vec{m}$ and $\vec{l}$, of components m_i, l_i, lying within the microplane. The shear strain components in the directions of $\vec{m}$ and $\vec{l}$ are $\epsilon_M = m_i(\epsilon_{ij} n_j)$ and $\epsilon_L = l_i(\epsilon_{ij} n_j)$, where m_j and n_j are the components of two suitably chosen orthogonal unit vectors m and l lying in the microplane. By virtue of the symmetry of tensor ϵ_{ij},

$$\epsilon_M = M_{ij} \epsilon_{ij}, \qquad \epsilon_L = L_{ij} \epsilon_{ij} \tag{8.6}$$

in which $M_{ij} = (m_i n_j + m_j n_i)/2$ and $L_{ij} = (l_i n_j + l_j n_i)/2$.

The static equivalence (or equilibrium) of stresses between the macro and micro levels can be enforced only approximately. This is done by the principle of virtual work (Bažant 1984c) written for surface Ω of a unit hemisphere;

$$\frac{2\pi}{3} \sigma_{ij} \delta \epsilon_{ij} = \int_\Omega (\sigma_N \delta \epsilon_N + \sigma_L \delta \epsilon_L + \sigma_M \delta \epsilon_M) \, d\Omega \tag{8.7}$$

This equation means that the virtual work of macrostresses (continuum stresses) within a unit sphere must be equal to the virtual work of microstresses (microplane stress components) regarded as the tractions on all the surface elements of the sphere. Substituting $\delta\epsilon_N = N_{ij}\delta\epsilon_{ij}$, $\delta\epsilon_L = L_{ij}\delta\epsilon_{ij}$ and $\delta\epsilon_M = M_{ij}\delta\epsilon_{ij}$, and noting that the last variational equation must hold for any variation $\delta\epsilon_{ij}$, one gets the following basic equilibrium relation:

$$\sigma_{ij} = \frac{3}{2\pi} \int_\Omega s_{ij}\, d\Omega \approx 6 \sum_{\mu=1}^{N_m} w_\mu s_{ij}^{(\mu)}, \qquad (8.8)$$

with

$$s_{ij} = \sigma_N N_{ij} + \sigma_L L_{ij} + \sigma_M M_{ij} \qquad (8.9)$$

The integral is in numerical calculations approximated by an optimal Gaussian integration formula for a spherical surface representing a weighted sum over the microplanes of orientations $\vec{n}_\mu$, with weights w_μ normalized so that $\sum_\mu w_\mu = 1/2$.

The constitutive relation on the microplane level is written as a relation between the microplane stress and strain components. These relations can directly reflect friction and slip, as well as cracking and and crack closure, separately for planes of various orientation in the material. For concrete, this approach has proven very powerful and has been used successfully in explicit dynamic analysis of impact, groundshock and blast on systems with up to several million finite elements (Bažant et al. 2000).

8.3. Spectrum of Distributed Damage Models Capable of Reproducing Size Effect

A broad range of numerical methods which can simulate damage localization, fracture propagation and size effect is now available. They can be classified as follows:

1. Discrete fracture, with elastic analysis:
   (a) R-curve model
   (b) Cohesive (fictitious) crack model

2. Distributed cracking damage—nonlinear analysis by:
   (a) Finite elements:
      i. Crack band model

ii. Nonlocal damage model:
   A. averaging type (semi-empirical)
   B. based on crack interactions (micromechanics)
iii. Gradient localization limiter:
   A. 1st gradient
   B. 2nd gradient—explicit
   C. 2nd gradient—implicit

(b) Discrete elements—random particle model:

   i. with axial forces only (random truss model)
   ii. with transmission of both normal and shear forces between particles.

(c) Element-free Galerkin models, using moving least-square interpolants, and partition of unity approach.

8.4. Simple, Practical Approaches

The simplest is the R-curve approach, which can often yield an analytical solution. The cohesive (or fictitious) crack model is efficient if the behavior of the elastic body surrounding the cohesive crack is characterized a priori by a compliance matrix or a stiffness matrix.

A great complication arises in general applications in which the direction of fracture propagation is usually unknown. For such situations, Ingraffea (1977, with later updates) has had great success in developing an effective remeshing scheme (in his computer program FRANC); however, this approach has not yet spread into practice.

The engineering firms and commercial finite element programs (e.g. DI-ANA, SBETA, Červenka and Pukl 1994, ATENA), as it appears, use almost exclusively the crack band model. This model is the simplest form of finite element analysis that can properly capture the size effect.

The basic idea in the crack band model (Bažant 1982; Bažant and Oh 1983) is to describe fracture or distributed cracking by a band of smeared cracking damage that has a single-element width, and to treat the band width, i.e., the element size in the fracture zone, as a material property (as proposed by Bažant 1976). This is the simplest approach to avoid spurious mesh sensitivity and ensure that the propagating crack band dissipates the correct amount of energy (given by the fracture energy G_f).

8.5. Nonlocal Concept and Its Physical Justification

A more general and powerful but also more complex approach is the nonlocal damage approach, in which the stress at a given point of the continuum does not depend only on the strain and that point but also on the strains in the neighborhood of the point. While the crack band model can be regarded as a simplified version of the nonlocal concept, the truly nonlocal finite element analysis involves calculation of the stress from the stress values in the neighboring finite elements.

The simplest and original form of nonlocal approach (Bažant, Belytschko and Chang 1984; Bažant, 1984b) involves an empirical weighted averaging rule. There are many possible versions of nonlocal averaging. But the most realistic results (Jirásek 1996) are apparently obtained with a nonlocal approach in which the secant stiffness matrix for the strain-softening stress-strain relation (which describes the evolution of damage or smeared cracking) is calculated from the spatially averaged strains and the stress is then obtained by multiplying with this matrix the local strain.

Physically a more realistic nonlocal damage model is obtained by continuum smearing of the matrix relations that describe interactions among many cracks in an elastic solid. One type of such a matrix interaction relation, due to Kachanov (1985, 1987), has led to the following field equation (Bažant 1994b):

$$\Delta \overline{S}^{(1)}(\boldsymbol{x}) - \int_V \Lambda(\boldsymbol{x}, \boldsymbol{\xi}) \Delta \overline{S}^{(1)}(\boldsymbol{\xi}) \mathrm{d}V(\boldsymbol{\xi}) = \langle \Delta S^{(1)}(\boldsymbol{x}) \rangle \qquad (8.10)$$

This is a Fredholm integral equation in which V = volume of the structure; $\Lambda(\boldsymbol{x}, \boldsymbol{\xi})$ = crack influence function, characterizing in a statistically smeared manner the normal stress across a frozen crack at coordinate $\boldsymbol{x}$ caused by a unit pressure applied at the faces of a crack at $\boldsymbol{\xi}$; $\langle . \rangle$ is a spatial averaging operator; $\Delta S^{(1)}$ or $\Delta \overline{S}^{(1)}$ = increment (in the current loading step) of the principal stress, labeled by (1), before or after the effect of crack interactions. The integral in this equation is not an averaging integral because its kernel has spatial average 0. The kernel is positive in the amplification sector of crack interactions and negative in the shielding sector.

So, in this nonlocal damage model, aside from an averaging integral, there is an additional nonlocal integral over the inelastic stress increments in the neighborhood. These increments model the stress changes that relax or enhance the crack growth. They reflect the fact that a neighboring crack lying in the shielding zone of a given crack inhibits the crack growth, while another crack lying in the amplification zone enhances the crack growth (Bažant 1994b; Bažant and Jirásek 1994a, 1994b).

This formulation shows that the nonlocality of damage is principally a consequence of the interactions among microcracks and provides a physically based micromechanical model. Application of this concept in conjunction with the microplane constitutive model for damage has provided excellent results for fracture and size effect in concrete (Ožbolt and Bažant 1996). However, the analysis is more complex than with the classical empirical averaging approach to nonlocal damage.

In practical terms, what has been gained from the crack interaction approach is that the failures dominated by tensile and shear fractures could be described by one and the same material model with the same characteristic length for the nonlocal averaging. This proved impossible with the previous models.

If the characteristic length involved in the averaging integral of a nonlocal damage model is at least three times larger than the element size, the directional bias for crack (or damage) propagation along the mesh lines gets essentially eliminated. However, this may often require the finite elements to be too small (although it is possible to adopt an artificially large characteristic length, provided that this is compensated by modifying the post-peak slope of the strain-softening constitutive equation so as to ensure the correct damage energy dissipation).

If the characteristic length is too small, or if the crack band model is used, then it is necessary either to know the crack propagation direction in advance and lay the mesh lines accordingly, or to use a remeshing algorithm of the same kind as developed by Ingraffea (1977) for the discrete crack model.

The earliest nonlocal damage model, in which not only the damage but also the elastic response was nonlocal, exhibited spurious zero-energy periodic modes of instability, which had to be suppressed by additional means, such as element imbrication (Bažant et al. 1984; Bažant 1984b). This inconvenience was later eliminated by the formulation of Pijaudier-Cabot and Bažant (1987) (see also Bažant and Pijaudier-Cabot, 1988), in which the main idea was that only the damage, considered in the sense of continuum damage mechanics (and later also yield limit degradation, Bažant and Lin 1988b), should be nonlocal and the elastic response should be local. The subsequent nonlocal continuum models with an averaging type integral are various variants on this idea.

8.6. Prevention of Spurious Localization of Damage

While from the mechanics viewpoint, the principal purpose of introducing the nonlocal concept is to make it possible to reproduce the quasibrittle size effect, from the viewpoint of finite element analysis it is to prevent arbitrary spurious localization of damage front into a band of vanishing width if the mesh is refined. Because, in the damage models with strain softening, the energy dissipation per unit volume of material (given by the area under the complete stress-strain curve) is a finite value, a vanishing width of the front of the damage band implies the fracture to propagate with zero-energy dissipation, which is obviously physically incorrect. This phenomenon also gives rise to spurious mesh sensitivity of the ordinary (local) finite element solutions according to continuum damage mechanics with strain softening.

From the physical viewpoint, the strain softening, characterized by a non-positive definite matrix of tangential moduli, appears at first sight to be a physically suspect phenomenon because it implies the wave speed to be complex (and thus wave propagation to be impossible), and because it implies the type of partial differential equation for static response to change from elliptic to hyperbolic (Hadamard 1903; Hill 1962; Mandel 1964; Bažant and Cedolin 1991).

To avoid these problems, and to make simulation of quasibrittle size effect feasible, one of two measures may be introduced:

1. some type of a mathematical device, called the localization limiter, which endows the continuum damage model with a characteristic length; or

2. rate-dependence of softening damage.

The conclusion that strain softening causes the wave speed to be complex rather than real, however, is an oversimplification, because of two phenomena:

- First, a strain softening material can always propagate unloading waves, because the tangent stiffness matrix for unloading always remains positive definite, as discovered experimentally in the 1960s (Rüsch and Hilsdorf 1963, Evans and Marathe 1968).

- Second, as revealed by recent tests at Northwestern University (Bažant and Gettu 1992; Bažant, Guo, and Faber 1995; Tandon et al. 1995), a real strain-softening material can always propagate loading waves with a sufficiently steep front.

The latter phenomenon is a consequence of the rate effect on crack propagation (bond breakage), which causes that a sudden increase of the strain rate always reverses strain softening to strain hardening (followed by a second peak); see Fig. 7.1(c). This phenomenon, which is mathematically introduced by Eq. (7.3), is particularly important for the finite element analysis of impact.

Another type of localization limiter are the gradient limiters, in which the stress at a given point of the continuum is considered to depend not only on the strain at that point but also of the first or second gradients of strains at that point. This concept also implies the existence of a certain characteristic length of the material. It appears to give qualitatively reasonable results for various practical problems of damage propagation, as well as the size effect. An effective form that reduces the order of continuity requirements for finite elements is an implicit form in which the nonlocal strain components are obtained as solutions of Helmholtz differential equations (Peerlings et al. 1996, de Borst and Gutiérrez 1999). This approach may be seen as a kind of nonlocal model in which the weight function is a solution of the Helmoltz equation.

It should be kept in mind, however, that the gradient localization limiters have not been directly justified physically. They can be derived in the sense of an approximation to the nonlocal damage model with an integral of averaging type. Indeed, expansion of the kernel of the integral and of the strain field into Taylor series and truncation of these series yields the formulation with a gradient localization limiter, and thus also justifies it physically (provided the integral formulation is based on the smearing of crack interactions).

8.7. Discrete Elements, Lattice and Random Particle Models

The discrete element models for damage and fracture are a fracturing adaptation of the model for granular solids proposed by Cundall (1971) and Cundall and Strack (1979). They are very demanding of computer power but are becoming more and more feasible as the power of computers increases. In these models, the material is represented by a system of particles whose links break at a certain stress.

The typical spacing of the particles acts as a localization limiter, similar to the crack band model, and controls the rate of energy dissipation per unit length of fracture extension (Bažant et al. 1990). The particles can simulate the actual aggregate configurations in a material such as concrete, or may simply serve as a convenient means to impose a certain characteristic length on the model, as in the case of the simulation of sea ice floes (Jirásek and Bažant, 1995a,b).

It must be warned, however, that trying to reduce the number of unknowns by using particles and particle spacings that are larger than the typical inhomogeneities of the material, one imposes an incorrect, excessively large characteristic length. A size effect will still be present, but will be weaker, with the approach to the LEFM asymptote pushed into larger sizes, and the transitional size D_0 too large.

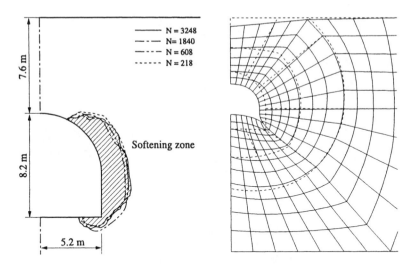

Figure 8.3: Analysis of tunnel excavation using nonlocal yield limit degradation, with deformed mesh (right) (after Bažant and Lin 1988b)

In the case of isotropic materials, it is important that the configuration of particles be random. With a regular particle arrangement there is always a bias for fracture propagation along the mesh lines, even when all the properties of the particle links are randomized (Jirásek and Bažant, 1995b).

In the simplest discrete element model, the interactions between particles are assumed to be only axial. But that causes the Poisson ratio of the homogenizing continuum to be 1/4 for the three-dimensional case, or 1/3 for the two-dimensional case, and so materials with other Poisson ratios cannot be modeled (unless some artifices are used). Another disadvantage is that the damage band appears to be too narrow.

An arbitrary Poisson ratio and a wider damage band can be achieved by a particle model in which the links between particles transmit not only axial forces but also shear forces. This is the case for the model of Zubelewicz and Bažant (1987), as well as the model of Schlangen and van Mier (1992) and van Mier and Schlangen (1993). In the latter, the particle system is modeled as a frame with bars that undergo bending (the bending of the bars is of course fictitious and unrealistic, but it does serve the purpose of achieving a shear force transmission through the links between particles). Van Mier and co-workers have had considerable success in modeling concrete fracturing in this manner.

An example of numerical solutions with nonlocal models and random particle models has already been given in Figs. 2.10 and 2.11. Further two examples are presented in Figs. 8.3 and 8.4, which show applications of a nonlocal finite

176   Scaling of Structural Strength

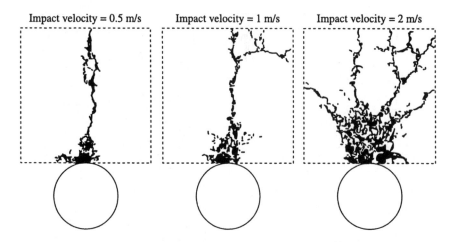

Figure 8.4: Random particle simulation of the breakup of an ice floe travelling at different velocities, after it impacts a rigid obstacle (Jirásek and Bažant 1995b)

element damage model to the analysis of failure of a tunnel excavated without lining, and to the simulation of the break-up of a traveling sea ice floe after it impacts a rigid obstacle.

Chapter 9

New Asymptotic Scaling Analysis of Cohesive Crack Model and Smeared-Tip Method

Let us now attempt a detailed analysis of the asymptotic scaling properties of the cohesive crack model, facilitated by a new smeared-tip approach.

It is proper to begin by commenting on the nature and history of the cohesive crack model, which was already used in (8.1) and (8.2). The cohesive crack model lumps into one line all the inelastic deformations in the fracture process zone of a finite length and width, including distributed microcracking and frictional or plastic slips. The benefit is that all the body volume can be treated as elastic. From nonlocal damage analysis with finite elements it is known (e.g. Bažant and Cedolin 1979) that the lumping causes little error whenever the softening damage localizes into a narrow band, which is true in most practical situations where damage ultimately leads to fracture. In these situations, the cohesive crack model is essentially equivalent (Bažant and Cedolin 1979) to the crack band model of finite element analysis (Bažant 1976; Bažant and Oh 1983; Bažant and Planas 1998). It also represents a limiting case of nonlocal damage formulations.

The basic hypothesis of the cohesive crack model is that there exists a unique relation between the cohesive (crack-bridging) stress σ and the opening displacement v, representing a material property. This model was originated by Barenblatt (1959, 1962), while Dugdale's (1960) work, extended to peak load analysis, led to an analogous model for fracture with a plastic zone. Although Barenblatt considered only the decrease of interatomic forces at separation of surfaces, the model has later been applied on a much larger scale of the material—to quasibrittle fracture of heterogeneous materials. In these materials, the crack opening $2v$ represents the relative displacement that is accumulated across the width of the fracture process zone (FPZ) and is caused by nucleation and growth of microcracks and voids combined with plastic or frictional slip in the FPZ, and the cohesive (crack-bridging) stress is the homogenized continuum stress in the FPZ.

To simplify analysis, the stress-displacement law was, in some later works (Willis 1967, Smith 1974, Reinhardt 1985, Wnuk 1974, Knauss 1973, 1974), replaced with an assumed stress profile along the FPZ. Hillerborg et al. (1976) and Petersson (1981) slightly generalized the cohesive crack model, under the name 'fictitious' crack model, and introduced it into the studies of concrete. A nearly equivalent model, which is more convenient for finite element analysis and yields almost identical results, is the crack band model, in which the FPZ is represented by a band of small finite width h_c with transverse inelastic strain $\epsilon_y^{fr} = 2v/h_c$.

The energetic aspect of the cohesive crack model was clarified by means of the J-integral (Rice 1968a, Kfouri and Rice 1977, Suo et al. 1992). Although the basic characteristic of the original models of Barenblatt and Dugdale is the vanishing of the stress intensity factor (SIF), some recent studies posited a non-zero SIF (Rice 1992) at the tip of cohesive crack. While the traditional assumption has been that the cohesive (crack-bridging) stress begins decreasing with crack opening from a certain finite initial value representing the tensile strength σ_f, some recent models (Needleman 1990, Tvergaard and Hutchinson 1992) assume that the cohesive stress at first increases as the crack begins to open, with the softening to come only later at larger openings (but in that case one must get reconciled to a paradox due to the existence on the crack flanks of stresses exceeding the stress σ_f that triggers the initial opening of a crack). The present analysis will be kept sufficiently general to accommodate any of these assumptions about the stress displacement relation or profile.

In the case of cohesive crack model as well as the almost equivalent crack band model, finite element solutions have demonstrated the existence of size effect and corroborated (e.g. Bažant and Li 1995a,b; Bažant and Planas 1998) the deterministic energetic size effect laws already presented. An important study of the asymptotics of the cohesive crack model was carried out by Planas and Elices (1992, 1993). They established the basic large size asymptotic character of cohesive crack solutions and the asymptotic fields. They also presented highly accurate numerical solutions clarifying the large-size and small-size asymptotic behaviors.

An explicit scaling law, however, has not been analytically derived in the aforementioned studies from the cohesive crack model. The structure geometry effect has neither been captured analytically; nor has the asymptotic matching between the large-size and small-size behaviors of this fundamental fracture model. Filling these gaps is the objective of this section. The geometry effect will be incorporated by introducing into the cohesive crack formulation the energy release function of linear elastic fracture mechanics (LEFM).

Traditionally, fracture with a cohesive or yielding zone has been analyzed on the basis of Green's functions or compliance functions, or distributed dislocations. These approaches, however, appear ineffective for general asymptotic analysis. Another approach named the *smeared-tip method*, developed by Planas and Elices (1986, 1992, 1993) and Bažant (1990a), will be adopted here, but with a new modification that makes it more effective for asymptotic analysis.

In the smeared-tip method, the cohesive fracture solution is represented as a superposition of infinitely many solutions of the given body for the case of sharp (LEFM) cracks whose tips are continuously distributed (or smeared) along the actual cohesive crack (Fig. 9.1). In the original version of the smeared-tip method, named here the P-version, the intensity of the distributed crack-tip singularities was characterized in terms of the density of the load sharing. In the new version (proposed in Bažant 2000), called the K-version, this intensity is characterized by the density of a continuously distributed SIF. Its distribution may be conveniently assumed as the basic material characteristic. For each assumed distribution, analysis of a very large structure yields the corresponding stress-displacement law of the cohesive crack model, and vice versa.

In contrast to many previous solutions based on a chosen fixed stress profile along the FPZ, the choice of fixed SIF (stress intensity factor) density profile has the advantage that the dimensionless SIF function $k(\alpha)$ can be exploited to capture the effects of structure geometry. These previous solutions (e.g. Willis 1967, Smith 1974, Reinhardt 1985, Wnuk 1974, Knauss 1973, 1974) were of course intended only for materials in which the FPZ is so small compared to the structure size that the geometry effect is the same as in LEFM.

In pursuing this approach, a new law will be obtained for the size effect in structures that fail only after the FPZ has moved away from the notch tip or the body surface. This happens in structures with an initially negative fracture geometry that later changes to positive.

9.1. Limitations of Cohesive Crack Model

With regard to a modification to be proposed, it is helpful to mention first the limitations of the cohesive crack model. This model is classically defined by a unique relation between the cohesive (crack-bridging) stress σ and the

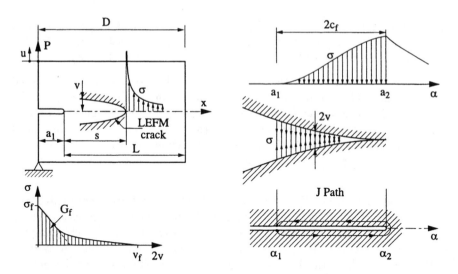

Figure 9.1: Top left: Near-tip opening and stress distributions. Top right: Fracture process zone and bridging stress profile. Bottom left: Stress-displacement curve of cohesive crack model and fracture energy. Bottom right: Rice's J-integral path for calculating fracture energy of cohesive crack model

opening w of the crack faces (or half crack width; Fig. 9.1 bottom left);

$$\frac{\sigma}{f_t} = \phi\left(\frac{w}{w_f}\right) \tag{9.1}$$

where ϕ is a decreasing function assumed to characterize the material; $f_t =$ tensile strength of material = stress σ at which the crack begins to open; $w_f =$ opening width (relative displacement of crack faces) at which the crack-bridging stress is reduced to zero. A rather general smooth expression for the inverse of function ϕ is

$$\frac{w}{w_0} = 1 - \frac{\sigma}{f_t} + \left(\frac{w_f}{w_0} - 1\right)\left(1 - \frac{\sigma}{f_t}\right)^n \qquad (0 \leq \sigma \leq f_t) \tag{9.2}$$

where $n =$ positive constant, and $w_0 =$ intersection (for $n > 1$) of the initial tangent of the softening curve with the w axis (Fig. 9.1 bottom left). For quasi-brittle materials, the softening curve is considered to start along an inclined tangent (whose horizontal intercept is w_0, $n \geq 1$), while for ductile materials it starts horizontally ($n < 1$). The test data for concrete suggest that the area under the complete stress displacement curve is about 2.5 times larger than the area under the initial tangent, and this ratio is achieved by setting $w_f/w_0 = 1 + 3(n+1)/4$ (if $n > 1$).

The standard cohesive crack model is of course a simplification of the reality, and it might not be the best one for all situations. The reality is a tortuous meandering final crack path and an FPZ having in the transverse direction y a certain finite effective width, $2h(x)$ (Fig. 9.2). The opening width (relative displacement of the faces) of a cohesive crack, w, represents the inelastic strain accumulated over width $2h$, i.e.

$$w(x) = \int_{-h(x)}^{h(x)} \epsilon_y''(x,y)\,dy, \qquad \epsilon_y'' = \epsilon_y(x,y) - \epsilon_y^{el}(x,y) \qquad (9.3)$$

where ϵ_y and ϵ_y^{el} are the total normal strain and the elastic normal strain, and ϵ_y'' is the inelastic strain due to microcracking, plastic frictional slip and pullout of fragments or inclusions. The deformations in the FPZ are of course randomly scattered and all strains must be understood in the sense of some type of statistical smoothing, describing the average behavior among many random realizations. The magnitude of inelastic strain ϵ_y'' increases on the average from the margins toward the middle line of the FPZ and its localization becomes progressively sharper as the crack-bridging stress is being reduced.

A unique relationship of the transverse y-component $\sigma(x)$ of stress on the crack line to opening $w(x)$ could be true only if each infinitesimal transverse strip across the FPZ (Fig. 9.2) acted independently, uniaxially. In reality, shear and normal stresses τ_{xy} and σ_x act on the sides of each strip (Fig. 9.2). This means that $\sigma_y(x)$ must depend also on the opening w at the neighboring points x. In other words, a more realistic law would be the following nonlocal cohesive law:

$$\frac{\sigma}{f_t} = \phi\left(\frac{\overline{w}}{w_f}\right), \qquad \overline{w}(x) = \int_{a_1}^{a_2} \Psi(x,x')\,w(x')\,dx' \qquad (9.4)$$

where a_1 and a_2 are the beginning and end of the FPZ, and $\Psi(x,x')$ is a bell-shaped interaction function (or weight function) that, in theory, could be deduced if the mean solution of the stochastic boundary value problem of random stress and strain fields in the FPZ could be obtained. This problem is of course hardly amenable to analytical solutions. Numerical stochastic finite element solutions, which must employ some form of a nonlocal approach, or nonlinear fracturing random lattice models, are currently the only way to model the behavior of the fracture process zone more realistically. So far, such approaches have provided no evidence of a unique stress-displacement curve for the overall response of the FPZ.

The standard cohesive crack model is in fact equivalent to assuming that the FPZ acts as a nonlinear (softening) Winkler foundation. The limitations of Winkler foundation are well known.

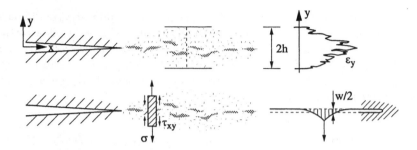

Figure 9.2: Fracture process zone with a strip assumed to characterize the standard cohesive crack model and stresses acting on it

9.2. K-Version of Smeared-Tip Method for Cohesive Fracture

In the smeared-tip method, one superposes the solutions of the given body for various lengths of sharp (LEFM) cracks, the tips of which are continuously distributed (smeared) along the crack path. Such superposition is used to represent the solution of a cohesive crack (Fig. 9.1). Any opening profile and any stress profile can be represented in this manner.

For a single LEFM mode I crack,

$$P = b\sqrt{D}\,\frac{K_I(\alpha)}{k(\alpha)} \qquad (\alpha = a/D) \qquad (9.5)$$

where P applied load or a parameter of a system of distributed and concentrated loads; b body thickness; K_I = SIF for Mode I and crack tip located at α; $k(\alpha) = \sqrt{g(\alpha)}$ = dimensionless SIF = SIF value for $D = b = P = 1$, which is a function of the relative crack length $\alpha = a/D$ for a body of given geometry; $g(\alpha)$ = dimensionless energy release function; a = actual crack length; and D = characteristic size or dimension of the structure (Fig. 9.1). The stresses σ on the crack line ahead of the crack tip and the crack opening width w behind the crack tip are, according to LEFM,

$$\sigma(\xi) = K_I(\alpha)\,S(\xi,\alpha), \qquad \alpha \in (0,\xi) \qquad (9.6)$$

$$S(\xi,\alpha) = \frac{1}{\sqrt{2\pi D(\xi-\alpha)}}\,[1 + b_1(\xi)(\xi-\alpha) + b_2(\xi)(\xi-\alpha)^2 + ...] \qquad (9.7)$$

$$w(\xi) = K_I(\alpha)\,W(\xi,\alpha), \qquad \alpha \in (\xi, L/D) \qquad (9.8)$$

$$W(\xi,\alpha) = \frac{\sqrt{32}}{E'\sqrt{\pi}}\,\sqrt{D(\alpha-\xi)}\,[1 + c_1(\xi)(\alpha-\xi) + c_2(\xi)(\alpha-\xi)^2 + ...] \qquad (9.9)$$

where $E' = E/(1-\nu^2)$ for plane stress, $E' = E$ for plane strain, E = Young's modulus, ν = Poisson's ratio; and $b_1, b_2, ..., c_1, c_2, ...$ are nonsingular dimensionless functions which depend on structure geometry (shape). They can be approximated by polynomials obtained by curve-fitting finite element results.

The applied load P, the crack-bridging (cohesive) stresses σ and the crack opening w obtained by superposition of the solutions (9.5), (9.6) and (9.8) for infinitely many cracks with continuously distributed (smeared) tips (Fig. 9.3) may in general be written as

$$P = \int dP = b\sqrt{D} \int_0^{L/D} \frac{dK_I(\alpha)}{k(\alpha)} \qquad (9.10)$$

$$\sigma(\xi) = \int_0^\xi S(\xi, \alpha)\, dK_I(\alpha) \qquad (9.11)$$

$$w(\xi) = \int_\xi^{L/D} W(\xi, \alpha)\, dK_I(\alpha) \qquad (9.12)$$

where L is the final length of the crack at total break (Fig. 9.1 top left) and $dK_I(\alpha)$ is the SIF of the smeared tips lying between α and $\alpha + d\alpha$.

The relative coordinates at the beginning and the end of the FPZ will be labeled as α_1 and α_2. It will be convenient to introduce a further dimensionless coordinate ρ having size-independent FPZ limits such that $\rho = 0$ and $\rho = 1$ correspond to the beginning and end of the FPZ (Fig. 9.3);

$$\rho = \frac{\alpha - \alpha_1}{2\theta} \qquad \theta = \frac{c_f}{D} = \tfrac{1}{2}(\alpha_2 - \alpha_1) \qquad (9.13)$$

where α_1 is the end of the stress-free crack portion and α_2 is the tip of the cohesive crack (end of FPZ).

The K-profile along the FPZ may be characterized as

$$dK_I(\alpha) = \frac{K_c D}{2c_f} q[\rho(\alpha)]\, d\alpha = K_c\, q(\rho)\, d\rho \qquad (9.14)$$

where K_c = fracture toughness (critical SIF); $q(\rho)$ = K-profile = dimensionless SIF density as a function of dimensionless crack length ρ (Fig. 9.3 bottom). For reasons to become clear later, it will be assumed that function $q(\rho)$ is such that the function $q(\rho)/\sqrt{|\omega - \rho|}$ be integrable for $0 < \omega < 1$, which also guarantees the total SIF at the cohesive crack tip to vanish (boundedness of $q(\rho)$ is sufficient but not necessary to satisfy these conditions). For $D \to \infty$ the fracture process zone in the relative coordinate α becomes a point, and all the solutions being superposed correspond to the same crack tip location in terms of α. Therefore, LEFM must apply, which means that $\int dK_I = K_c$ or

$$I_1 = \int_0^1 q(\rho)\, d\rho = 1 \qquad (9.15)$$

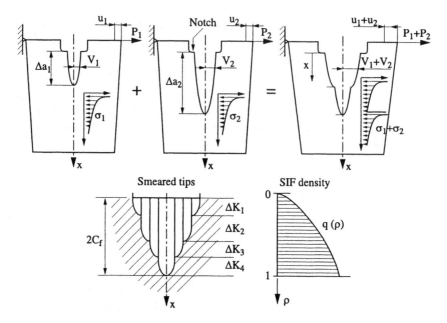

Figure 9.3: Top: Superposition of two LEFM solutions for bodies with different crack tip locations. Bottom: Crack opening profile caused by increments $\Delta K_1, \Delta K_2, \ldots$ of stress intensity factor (SIF) along crack line, and SIF profile $q(\rho)$

The smeared-tip method can be formulated in two versions. In the original version, labeled here the P-version (Planas and Elices 1986, 1992, 1993; Bažant 1990a; Bažant and Beissel 1994; Bažant and Planas 1998), the SIF of the smeared crack tips between a and $a + da$ was associated not with an SIF increment $dK_I(\alpha)$ but with a load contribution $dP = p(\alpha)Dd\alpha$ where $p(\alpha)$ was the load-sharing distribution that had the meaning (Bažant and Planas 1998, Eq. 7.5.65)

$$p(\alpha) = \frac{bK_c\sqrt{D}}{2c_f k(\alpha)} q[\rho(\alpha)] \tag{9.16}$$

The present approach is labeled the K-version since the smeared crack tips are associated with the increments of K_I.

Unlike the asymptotic K-profile $q(\rho)$, the asymptotic distribution $p(\alpha)$ is not size- and shape-independent, and so its asymptotic form is not a material property. Although Planas and Elices (1992, 1993) gave the expression for function $q(\rho)$, they used in their asymptotic analysis the P-version. The difference between the P- and K-versions is of course nothing but a substitution of a new variable, but the K-version is much more convenient.

9.3. Nonstandard Cohesive Crack Model Defined by a Fixed K-Profile

For sufficiently large structures, functions $b_1, b_2, ..., c_1, c_2, ...$ may be neglected. Then (9.11) with (9.14) furnish

$$\sigma(\rho) = \frac{K_c}{2\sqrt{\pi c_f}} S(\rho), \qquad S(\rho) = \int_0^\rho \frac{q(\omega)\, d\omega}{\sqrt{\rho - \omega}} \qquad (9.17)$$

$$w(\rho) = \frac{16\, K_0 \sqrt{c_f}}{E'\sqrt{\pi}} W(\rho), \qquad W(\rho) = \int_\rho^1 q(\omega)\sqrt{\omega - \rho}\, d\omega \qquad (9.18)$$

It will be convenient to write these equations also as

$$\sigma(\rho) = f_t \frac{S(\rho)}{S(1)}, \qquad v(\rho) = w_f \frac{W(\rho)}{W(0)} \qquad (9.19)$$

If K_c, c_f and the profile $q(\rho)$ are known, equation (9.17) provides a parametric description of the stress-displacement $\sigma(w)$ curve of the cohesive crack model. Indeed, choosing a series of values of ρ, the corresponding pairs of w and σ can be obtained by evaluating the integrals $S(\rho)$ and $W(\rho)$, which are independent of structure size and shape. Thus, for each K-profile, there exists a corresponding stress-displacement curve (softening law) $\sigma(w)$ of the cohesive crack model. Vice versa, assuming this relationship to be invertible, one can find for each f_t, w_f and function ψ the values K_c, c_f and function $q(\rho)$. Therefore, defining the cohesive crack model by a stress-displacement curve or by a fixed *asymptotic* K-profile is equivalent.

For the sake of simplicity, we will now introduce a nonstandard form of the cohesive crack model defined by the hypothesis that *the K-profile $q(\rho)$ and its length $2c_f$ are material properties*, i.e., are fixed. In the perspective of the generalized cohesive law (9.4), it seems that this nonstandard form might be no less realistic than the standard form. Both seem to involve roughly equal degrees of simplification.

A cohesive crack model defined by the K-profile is asymptotically equivalent to the standard cohesive crack model defined by the stress-displacement curve. But for finite sizes, it is different (the effect of this difference will be explored later).

If (9.17) and (9.18) are substituted into (9.2), one gets

$$F(\rho) = \int_0^\rho \frac{q(\omega)\, d\omega}{\sqrt{\rho - \omega}} \qquad (9.20)$$

$$F(\rho) = S(1)\, \phi\left(\frac{1}{W(0)} \int_\rho^1 \sqrt{\omega - \rho}\, q(\omega)\, d\omega\right) \qquad (9.21)$$

In particular, for the law in (9.2),

$$\frac{S(\rho)}{S(1)} = 1 - \frac{w_f}{w_0}\frac{W(\rho)}{W(0)} + N(\rho) \qquad (9.22)$$

$$N(\rho) = \left(\frac{w_f}{w_0} - 1\right)\left(1 - \frac{S(\rho)}{S(1)}\right)^n \qquad (9.23)$$

An important point is that the structure size D is absent from (9.20). This shows that our hypothesis of size-independence of the profile $q(\rho)$, which greatly simplifies size effect analysis, is in agreement with neglecting $b_1, b_2, ..., c_1, c_2, ...$.

Eq. (9.20) has a weakly singular kernel and represents an integral equation of the first kind for function $q(\rho)$. This equation, whose equivalent form in terms of $p(\rho)$ was introduced by Planas and Elices (1992) (see also Bažant and Planas (1998), Eq. 7.5.67), represents the well-known Abel integral equation if the right-hand size $F(\rho)$ is known. An explicit solution exists for that case and may be used in successive approximations to find an accurate solution of (9.20). One starts with some assumed function, evaluates the right-hand side, then obtains from Abel's formula a new solution, evaluates an improved right-hand side, etc.

In the special case of a linear (triangular) softening curve ($w_f = w_0$), we have $N(\rho) = 0$, and so equation (9.20) simplifies to the following linear integral equation of the first kind:

$$\frac{S(\rho)}{S(1)} + \frac{W(\rho)}{W(0)} = 1 \qquad (9.24)$$

Function $N(\rho)$, representing the nonlinear part of the integral equation, is negligible (for $n > 1$) when σ is very close to the strength limit f_t. This situation occurs for $\rho \to 1$. Consequently, if the softening curve begins by a downward inclined tangent ($n > 1$), the terminal part of the profile $q(\rho)$ near $\rho = 1$ is governed by the linear integral equation (9.24). This observation agrees with the well-known fact that the maximum load of structures with a notch or preexisting stress-free crack is known to be almost independent of the shape of the softening curve and to depend only on f_t and the ratio w_0/f_t characterizing the downward slope of the initial tangent, shown in Fig. 9.1 (bottom left).

According to (9.14) and (9.10), the nominal strength of the structure, which is a load parameter defined as $\sigma_N = P/bD$, may now be expressed as

$$\sigma_N = \frac{K_c}{\sqrt{D}} \int_0^1 \frac{q(\rho)\,d\rho}{k[\alpha(\rho)]} \qquad (9.25)$$

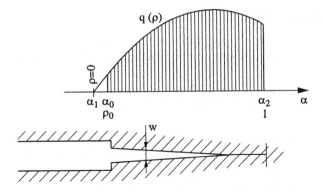

Figure 9.4: Incomplete SIF density profiles when the FPZ is growing from a notch

provided that the FPZ is not shortened by a notch. If it is, then, more generally (Fig. 9.4),

$$\sigma_N = \frac{K_c}{\sqrt{D}} \int_{\rho_0}^{1} \frac{q(\rho)\,d\rho}{k[\alpha(\rho)]} \qquad (9.26)$$

where ρ_0 is the coordinate at the notch tip.

If the K-profile $q(\rho)$ along with c_f and the dimensionless SIF $k(\alpha)$ are known, then σ_N can be evaluated from (9.25). In this manner, the size effect curve of the cohesive crack model can be computed for any given structure geometry provided that function $k(\alpha)$ characterizing the geometry is known. For many geometries, this function is given in handbooks, and for others it can be adequately approximated by curve-fitting elastic finite element results.

If the shape of the softening stress-displacement law is fixed (e.g., if w_0/w_f and n are fixed), the cohesive crack model is characterized by only two material parameters, which may be chosen either as f_t and w_f or as f_t and G_f where G_f = fracture energy = area under the complete curve of stress versus crack opening. According to Irwin's relation,

$$G_f = K_o^2 / E' \qquad (9.27)$$

To relate c_f to the basic parameters of the cohesive crack model, we may utilize Rice's J-integral (Rice 1968a) giving the energy flux J into a fully developed FPZ. The J-integral path must envelop the entire fracture process zone. Following Rice (1968a), we choose a J-integral path (Fig. 9.1 bottom right) that begins at the lower crack face at α_1, runs along this face to α_2, then along an infinitely small circle around the cohesive crack tip, and finally back to α_1

along the opposite crack face;

$$J = \int_{\alpha_1}^{\alpha_2} \sigma(\alpha) \frac{d[-w(\alpha)/2]}{d\alpha} d\alpha + \int_{\alpha_2}^{\alpha_1} \sigma(\alpha) \frac{d[w(\alpha)/2]}{d\alpha} d\alpha \qquad (9.28)$$

$$= -\int_0^1 \sigma(\rho) \frac{dw(\rho)}{d\rho} d\rho$$

which further yields $J = \int_0^{w_f} \sigma dw$. If the FPZ is fully developed (i.e. if $\sigma = 0$ at $\rho = 0$), it can move forward only if $J = G_f$, and so this condition must asymptotically be met. Assuming this, substituting (9.17) and (9.18) or (9.11), and taking into account (9.27), one obtains:

$$J = \frac{2 I_0^2}{\pi} G_f = \frac{f_t w_f I_0^2}{2 S(1) W(0)} \qquad (9.29)$$

with the notation

$$I_0 = \sqrt{\int_0^1 \left(\int_0^\rho \frac{q(\omega)\,d\omega}{\sqrt{\rho-\omega}} \right) \left(\int_\rho^1 \frac{q(\omega)\,d\omega}{\sqrt{\omega-\rho}} \right) d\rho} \qquad (9.30)$$

Since $J = G_f$ asymptotically, in view of (9.29) we must have

$$I_0 = \sqrt{\pi/2}, \qquad 4 S(1) W(0) G_f = \pi f_t w_f \qquad (9.31)$$

To determine the half-length c_f of the fracture process zone, one may now express from (9.17) the condition $\sigma(1) = f_t$. Thus, taking into account the last expression in (9.29) and (9.31), one gets

$$c_f = \frac{S^2(1)}{\pi} l_0 = \frac{\pi}{16\, W^2(0)} \frac{E' w_f^2}{G_f}, \qquad l_0 = \frac{K_c^2}{f_t^2} \qquad (9.32)$$

where l_0 represents Irwin's characteristic length. Now it should be noted that, in view of (9.15) and (9.18), $W(0) \leq 1$, and so a lower bound on the half-length of the fracture process zone, for any shape of the stress-displacement law, is

$$c_{f\,min} = \frac{\pi}{16} \frac{E' w_f^2}{G_f} \qquad (9.33)$$

The fact that neither D nor $k(\alpha)$ appears in (9.32) confirms that the hypothesis of size- and shape-independence of c_f agrees with the neglect of $b_1, b_2, ...,$ and $c_1, c_2, ...$

9.4. Asymptotic Scaling Analysis

Let $\alpha_0 = a_0/D$ denote the relative crack length for $D \to \infty$, which is the case of LEFM. For $D \to \infty$, the relative FPZ size in terms of α shrinks to a point, in which case $\alpha_0 = \alpha_1 = \alpha_2$. In studying the size effect, we consider structures of different sizes that are geometrically similar, and so α_0 is constant.

Depending on the values of $k_0 = k(\alpha_0)$ and $k'_0 = [dk(\alpha)/d\alpha]_{\alpha=\alpha_0}$, three different cases of scaling of the smeared-tip model with a fixed K-profile must be distinguished.

9.4.1. Case 1. Positive Geometry with Notch or Stress-Free Initial Crack, for Fixed K-Density ($g_0 > 0, g'_0 > 0$)

When there is a notch or preexisting stress-free crack (which may be produced by fatigue under previous repeated loads), $k(\alpha_0) = 0$. If the fracture geometry is negative, i.e. $k'(\alpha) < 0$, FPZ moves away from the tip in a stable manner at increasing load. For failure to occur while the FPZ is still attached to the tip of notch or preexisting stress-free crack (Fig. 9.5, Case 1), the geometry must be positive. Thus, for Case 1, we assume:

$$k_0 > 0, \quad k'_0 > 0 \quad (9.34)$$

where $k_0 = \sqrt{g_0}$, $k'_0 = g'_0/2\sqrt{g_0}$; $g_0 = g(\alpha_0), g'_0 = dg(\alpha_0)/d\alpha$. The beginning and end of the FPZ are

$$\alpha_1 = \alpha_0 > 0, \quad \alpha_2 = \alpha_0 + 2\theta < L/D \quad (\theta = c_f/D) \quad (9.35)$$

It will be convenient to introduce a Taylor series expansion of the inverse of the dimensionless SIF function about point α_0;

$$\frac{1}{k[\alpha(\rho)]} = \frac{1}{k_0} - \frac{k'_0}{k_0^2} 2\theta(\rho - \rho_0) + (...)(\rho - \rho_0)^2 + ..., \quad \rho_0 = \frac{\alpha_0 - \alpha_1}{2\theta} \quad (9.36)$$

For geometrically similar notched structures, the LEFM scaling is $\sigma_N \propto 1/\sqrt{D}$. Thus $\sigma_N^2 D = $ const. in LEFM. To obtain deviations from LEFM due to finite FPZ, it is therefore convenient to expand in power series of θ either $\sigma_N^2 D$ or $1/(\sigma_N^2 D)$. The latter will provide a more direct derivation of the size effect law.

Let us first assume, for the sake of simplicity, that the complete K-profile is developed when the maximum load is attained (i.e., the stress continuously approaches 0 at the notch tip). In that case, the notch tip $\alpha_0 = \alpha_1$ or $\rho_0 = 0$.

190   Scaling of Structural Strength

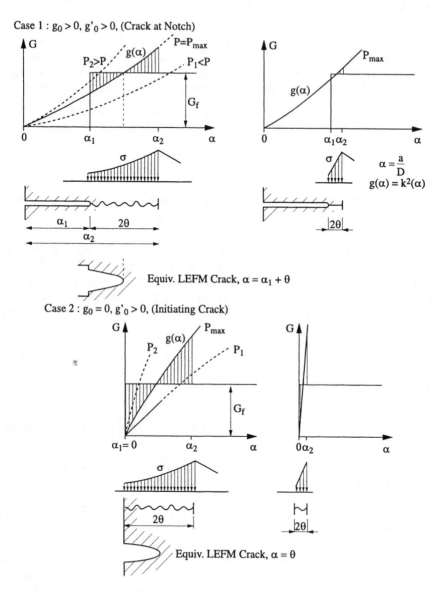

Figure 9.5: Lines of dimensionless energy release function $g(\alpha)$ for increasing values of constant load P_1, P, P_2 for small (left) and large (right) sizes D, along with stress profiles in FPZ for three basic types of failure: Case 1—notch or stress-free crack; Case 2—initiating crack

Substituting expansion (9.36) into (9.25) and taking into account (9.15), one gets

$$\frac{1}{\sigma_N^2 D} = \frac{k_0^2}{K_c^2}\left[1 - \frac{\gamma_0}{2}\theta + (...)\theta^2 + ...\right]^{-2} \approx \frac{k_0^2}{K_c^2}(1 + \gamma_0\theta) \qquad (9.37)$$

with the notations

$$\gamma_0 = \frac{4k_0'}{k_0}I_2, \qquad I_2 = \int_0^1 q(\rho)\,\rho\,d\rho \qquad (9.38)$$

For small θ, one has the approximation

$$\sigma_N = \frac{K_c}{k_0\sqrt{D}}(1 + \gamma_0\theta)^{-1/2} \qquad (9.39)$$

This may be rewritten in the form of the classical size effect law proposed in (Bažant 1983, 1984a), i.e.

$$\sigma_N = \sigma_0\left(1 + \frac{D}{D_0}\right)^{-1/2} \qquad (9.40)$$

in which

$$\sigma_0 = \frac{K_c}{k_0\sqrt{D_0}}, \qquad D_0 = \gamma_0 c_f = \frac{\gamma_0}{\pi}S^2(1)l_0 = 4I_2\frac{k_0'}{k_0}c_f \qquad (9.41)$$

The size effect law (9.40) has been derived in many different ways and received broad and diverse experimental support for many different quasibrittle materials, including concrete, rock, fiber composites, tough ceramics and sea ice (Bažant and Planas 1998; Bažant and Chen 1997; Bažant 1997a, 2000). The asymptotic form of (9.40) for $D \to \infty$ is the LEFM scaling:

$$\sigma_N = \frac{K_c}{k_0\sqrt{D}} \qquad (9.42)$$

In view of the assumptions made, the present solution can be accurate only up to the first two terms, i.e., up to the term with θ. Therefore, any other size effect formula for which the first two terms of the expansion of $\sigma\sqrt{D}$ would coincide with those in (9.39) must be regarded as equally justified. One such formula is

$$\sigma_N = \frac{K_c}{k_0\sqrt{D}}(1 + r\gamma_0\theta)^{-1/2r} \qquad (9.43)$$

or

$$\sigma_N = \frac{\sigma_0}{\sqrt{D}}\left(1 + \frac{rD_0}{D}\right)^{-1/2r} \qquad (9.44)$$

192   Scaling of Structural Strength

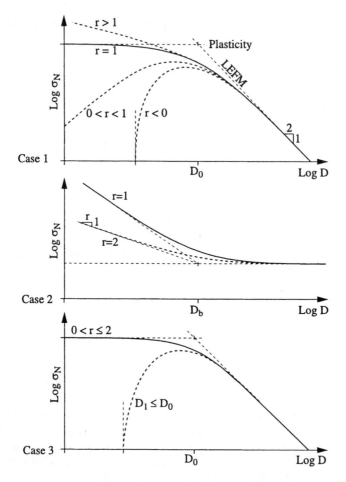

Figure 9.6: Size effect laws for three basic cases at failure: Case 1—notch or stress-free crack, Case 2—initiating crack, Case 3—negative-positive geometry transition

with r being any real number except 0 (Fig. 9.6). Noting the Taylor series expansion $(1+x)^n = 1 + nx + \frac{1}{2}n(n-1)x^2 + ...$, we see that formula (9.43) for $r = \frac{1}{2}$ agrees with (9.39) up to the first two terms in powers of θ. Therefore, (9.43) is no less justified than (9.39) as far as the large-size asymptotic behavior is concerned.

The question now is what is the correct or optimum value of r. The answer depends on whether we seek a size effect law that is optimum only for very large D or prefer a size effect formula approximately applicable also for medium and small sizes D. The latter is much more useful.

The plots of (9.44) for various r are shown in Fig. 9.6. For $r < 1$ the plots of σ_N versus D are non-monotonic. For $r \geq 1$ the plot is monotonic but for $r > 1$ the small-size asymptotic value of σ_N is infinite, which seems unjustified. Moreover, for $r < 0$, the σ_N values below a certain size D are imaginary.

Obviously, σ_N must be real for all D. Moreover, σ_N for $D \to 0$ must be finite (which is demonstrated for the cohesive crack model in what follows). So, to have a size effect formula applicable through the entire size range, r must be 1, which gives a monotonic size effect plot and a finite small-size asymptotic value.

The physically most realistic way to demonstrate the finiteness of σ_N for $D \to 0$ is not the cohesive crack model because for very small sizes the width of the FPZ is not negligible, and in fact the structure width is less than the width of a fully developed FPZ observed in large bodies. A more realistic model seems to be the nonlocal continuum damage mechanics. Numerical solutions with this approach have established the finiteness of the small size limit of σ_N.

The foregoing arguments may be seen in the context of the so-called asymptotic matching between $D \to \infty$ and $D \to 0$. The technique of asymptotic matching, originated with the boundary layer theory by Prandtl (1904), has been used with great success in fluid mechanics (Barenblatt 1979, 1987; Sedov 1959). It is a general approach for obtaining approximate solutions to problems that are very tough on the practical scale but become much easier for both much larger and much smaller scales. Rather sophisticated techniques of asymptotic matching have been found necessary in fluid mechanics. In the present problem, however, the right asymptotic matching formula has been obtained almost trivially.

One might wonder whether more than the first two terms should not be taken in the Taylor series expansion (9.36). This would of course improve accuracy for very large D; however, it would impair the representation of size effect for normal sizes and deprive the size effect formula of its asymptotic matching character.

Formula (9.44) differs from the generalization

$$\sigma_N = \sigma_0 \left[1 + \left(\frac{D}{D_0} \right)^r \right]^{-1/2r} \qquad (9.45)$$

proposed long ago in order to allow better fitting in the mid-range of computer results obtained with the cohesive crack model (Bažant 1985a; Bažant and Planas 1998). For this formula, the value $r = 0.45$ was found optimum for a notched beam (Bažant 1985b) while a value of r greater than 1 was optimum for a center cracked panel loaded on the crack faces. Eq. (9.44), however, has only the first order asymptotic accuracy for large D; the second term of the expansion of $\sigma_N \sqrt{D}$ in terms of θ differs from the second term obtained from (9.43). Nevertheless, this formula might provide better asymptotic matching, with a better description of the size effect in mid-range.

Eq. (9.43) is not the only simple formula with the required asymptotic properties for $D \to \infty$ and $D \to 0$. For example, another such formula is

$$\sigma_N = \sigma_0 \sqrt{1 - e^{-D_0/D}} \qquad (9.46)$$

Indeed, its large size asymptotic expansion is

$$\sigma_N = \sigma_0 \sqrt{D_0/D}[1 - \tfrac{1}{4}(D_0/D) + \tfrac{5}{96}(D_0/D)^2 - ...] \qquad (b)$$

However, unlike (9.43), this formula is not amenable to a linear regression plot for identifying D_0 and σ_0 from size effect data. Moreover, the approach of this formula to the small-size horizontal asymptote in the size effect plot of $\log \sigma_N$ versus $\log D$ appears too abrupt compared with the test data for concrete.

Effect of stress cutoff at notch tip

Our assumption that the K-profile at notch tip is fully developed has been of course an approximation. Let us now check whether it makes any difference asymptotically when we consider that the K-profile is cut off by the notch, i.e., the notch tip, of coordinate α_0, corresponds to some coordinate $\rho_0 > 0$. In that case one must calculate σ_N from (9.26) rather than (9.25). The value of ρ_0 must be found from the condition of maximum load (stability limit under load control), which may be written as

$$d\sigma_N/d\rho_0 = 0 \qquad (9.47)$$

Here one must substitute the integral (9.26) with a variable lower limit ρ, and note that α in function $k(\alpha)$ depends on ρ_0 because $\alpha = \alpha_0 + 2c_f(\rho - \rho_0)/D$. Differentiation according to the Leibnitz rule yields, after some rearrangements, the maximum load condition

$$\frac{q(\rho_0)}{2\theta \, k[\alpha(\rho_0)]} = \psi(\rho_0) \qquad \text{where} \qquad \psi(\rho_0) = \int_{\rho_0}^{1} \frac{k'[\alpha(\rho)]}{k^2[\alpha(\rho)]} q(\rho) \, d\rho \qquad (9.48)$$

Let us now expand this condition into a power series in ρ_0.

For $D \to \infty$ (or $\theta \to 0$), it is necessary that $q(\rho_0) \to 0$ (Fig. 9.4). This condition characterizes the tail end of the K-profile $q(\rho)$ (this fact is intuitive, since the density of SIF must vanish when the material is almost torn apart). Since $q(0) = 0$, the first two nonzero terms of the expansion of $q(\rho)$ for finite but small θ are

$$q(\rho) = q'_0 + \tfrac{1}{2}q''_0 \rho_0 + ... \qquad (9.49)$$

where $q'_0 = dq(0)/d\rho$ and $q''_0 = d^2q(0)/d\rho^2$. Then we calculate the first derivative of integral $\psi(\rho_0)$ with respect to ρ_0 and evaluate it at $\rho_0 = 0$. Because $q(0) = 0$, we get

$$\left[\frac{d\psi(\rho_0)}{d\rho_0}\right]_{\rho_0=0} = -2\theta \int_0^1 \frac{k''[\alpha(\rho)]k[\alpha(\rho)] - 2k'^2[\alpha(\rho)]}{k^3[\alpha(\rho)]} q(\rho) d\rho \quad (9.50)$$

So the condition (9.48) has the expansion

$$\psi(\rho_0) = \psi_0 + \psi'_0 \rho_0 + \ldots = (q'_0 \rho_0 + \tfrac{1}{2} q''_0 \rho_0 + \ldots)/k_0 \quad (9.51)$$

where $k_0 = k[\alpha(\rho_0)] = $ constant. Neglecting the higher-order small terms with ρ_0 and solving for ρ, one thus gets the approximation

$$\rho_0 = \psi_a \theta, \qquad \psi_a = \frac{2}{q'_0} k_0 \psi_0 \quad (9.52)$$

where ψ_a and $\psi_0 = \psi(0)$ are constants. This ρ_0-value must now be substituted into the lower limit of the integral (9.26). Using also the expansion (9.36), one gets again equation (9.37) but, instead of (9.38),

$$\gamma_0 = \frac{4k'_0}{k_0} \frac{I'_2}{I'_1}, \quad I'_1 = \int_{\psi_a \theta}^1 q(\rho) d\rho, \quad I'_2 = \int_{\psi_a \theta}^1 q(\rho)(\rho - \psi_a \theta) d\rho \quad (9.53)$$

For small θ, these integrals may be approximated as

$$I'_1 = I_1 - \tfrac{1}{2} q'_0 \psi_a^2 \theta^2 \approx I_1, \quad I'_2 = I_2 - \psi_a I_1 \theta + \left(\tfrac{1}{2} q'_0 - \tfrac{2}{3}\right) \psi_a^3 \theta^3 \approx I_2 - \psi_a I_1 \theta \quad (9.54)$$

Consequently, instead of (9.39) we have

$$\sigma_N = \frac{K_c}{k_0 \sqrt{D}} \left(1 + \frac{4k'_0 (I_2 - \psi_a \theta)}{k_0} \theta\right)^{-1/2} \quad (9.55)$$

Now we see that the correction to the size effect law is higher-order small in terms of θ or $1/D$. This proves what might have been intuitively expected—namely that all the size effect law forms given before are asymptotically correct up to $1/D$.

9.4.2. Case 2. Fracture Initiation from Smooth Surface, for Fixed K-Density ($g_0 = 0, g'_0 > 0$)

Structures may also fail at the initiation of fracture from a smooth surface of the body or in the interior (Fig. 9.5, Case 2). A typical example is the standardized modulus of rupture tests of concrete beams. For $D \to \infty$, i.e. when the FPZ is negligible compared to D, the failure occurs when the relative crack length vanishes; therefore,

$$k_0 = 0, \quad k'_0 > 0 \tag{9.56}$$

But for finite dimensions, failure can occur only after a FPZ of a finite size $2c_f$ gets formed (Fig. 9.5). So we have

$$\alpha_1 = \alpha_0 = 0, \quad \alpha_2 = 2\theta \tag{9.57}$$

At fracture initiation, the dimensionless SIF has in general the form $k(\alpha) \propto \sqrt{\alpha}$, which means that $k'_0 \to \infty$. Therefore, we now need to expand into Taylor series the dimensionless energy release function $g(\alpha) = [k(\alpha)]^2$. Since $g(0) = 0$, the expansion, unlike before, cannot be truncated after the linear term, i.e., the quadratic term must now be included; so

$$g(\alpha) = g'_0 \alpha + \tfrac{1}{2} g''_0 \alpha^2 + \ldots \tag{9.58}$$

where $g'_0 = dg(\alpha)/d\alpha$ and $g''_0 = d^2 g(\alpha)/d\alpha^2$ at $\alpha = 0$. Setting $\alpha = 2\theta\rho$ and substituting $k(\alpha) = \sqrt{g(\alpha)}$ into (9.25), one obtains

$$\sigma_N = \frac{K_c}{\sqrt{D}} \int_0^1 \frac{q(\rho)}{\sqrt{2\theta \rho g'_0}} \left(1 + \frac{g''_0}{g'_0}\theta\rho\right)^{-1/2} d\rho \tag{9.59}$$

This equation shows that $\lim \sigma_N$ for $D \to \infty$ is finite.

Typically, $g''_0 < 0$ (as in beam bending). So, equation (9.59) gives imaginary σ_N for small enough D, and must therefore be modified to serve as a size effect formula for the entire range. Since this equation gives $\sigma_N \sqrt{D}$ with only second order accuracy in θ, it may now be noted that the formula

$$\sigma_N = \frac{K_c}{\sqrt{D}} \int_0^1 \frac{q(\rho)}{\sqrt{2\theta \rho g'_0}} \left(1 - \frac{r g''_0}{2 g'_0}\theta\rho\right)^{1/r} d\rho \tag{9.60}$$

for any real r (except 0) gives the same terms of expansion up to the second order in θ, and is therefore equally justified. The original formula (9.59) corresponds to $r = -2$. Any positive r is acceptable.

Consider now that $r = 1$. Eq. (9.60) may be rearranged as follows:

$$\sigma_N = K_c \left(\frac{I_3}{\sqrt{2 c_f g'_0}} + \langle -g''_0 \rangle \sqrt{\frac{c_f}{8 g'^3_0}} \frac{I_4}{D} \right) \tag{9.61}$$

in which

$$I_3 = \int_0^1 \frac{q(\rho)}{\sqrt{\rho}} d\rho, \quad I_4 = \int_0^1 q(\rho)\sqrt{\rho}\, d\rho \tag{9.62}$$

(where $I_3 \geq 1$ and $I_4 \leq 1$). These integrals are constants which can be evaluated if the K-profile is known or chosen. The Macauley brackets $\langle . \rangle$ in (9.61) signify the positive part of the argument. They have been inserted because for $g_0'' > 0$ there is no (deterministic) size effect (as is the case for direct tension of unnotched bar).

Eq. (9.61) may be simply rewritten as

$$\sigma_N = \sigma_\infty \left(1 + \frac{D_b}{D}\right) \qquad (9.63)$$

in which

$$\sigma_\infty = \frac{K_c I_3}{\sqrt{2g_0' c_f}}, \quad D_b = \frac{I_4}{2I_3} \frac{\langle -g_0'' \rangle}{g_0'} c_f = \frac{I_4 S^2(1)}{2\pi I_3} \frac{\langle -g_0'' \rangle}{g_0'} l_0 \qquad (9.64)$$

Eq. (9.63) is the same as that derived by two different arguments by Bažant and Li (1995a, 1997), and verified by extensive experimental evidence (Bažant and Novák 2000b).

Now we may go back to the generalization with arbitrary positive parameter r. It may be written as

$$\sigma_N = \sigma_\infty \left(1 + \frac{rD_b}{D}\right)^{1/r} \qquad (9.65)$$

This formula gives the same first two terms of the power series expansion of $\sigma_N \sqrt{D}$ in terms of θ, and is therefore equally justified from the large-size asymptotic viewpoint. Parameter r must be found by other arguments or empirically. For the modulus of rupture test of concrete beams, Bažant (1998a) found the value $r = 1.44$ as optimal.

Without altering the essential large-size asymptotic properties, one may further generalize formula (9.65) as follows

$$\sigma_N = \sigma_\infty \left(1 + \frac{rD_b}{D_a + D}\right)^{1/r} \qquad (9.66)$$

where D_a is a positive constant (for $r = 1$ such formula was proposed empirically by Rokugo et al. 1995; see also Bažant and Planas 1998). This formula has a finite limit for $D \to 0$. However it is unclear whether this is important in the case of a FPZ attached to smooth body surface, in which case the FPZ is initially very wide. A realistic value of D_a is so small that D_a makes a significant difference only for bodies with less than realistic sizes compared to the size of material inhomogeneities.

For Case 2, unlike Case 1, the size effect at very large sizes (concrete beam depth over about 5 m) becomes dominated by strength randomness and the asymptotic behavior must approach the Weibull-type weakest link model. Such a statistical generalization of (9.63) and (9.64) was proposed by Bažant and Novák (2000b) and is beyond the scope of the present analysis.

9.4.3. Cases 1 and 2 for Standard Cohesive Crack Model or First Three Terms of Asymptotic Expansion

The foregoing analysis of scaling has been based on the hypothesis that the K-profile, $q(\rho)$, is a fixed material property. Let us now abandon this hypothesis and consider the standard cohesive crack model in which the softening stress-displacement $\sigma(w)$ law is fixed, as in (9.2).

Now one must take into account the second-order near-tip stress and displacement fields characterized by functions $b_1(\xi)$ and $c_1(\xi)$ in (9.7) and (9.9), in which $\xi = \alpha_1 + 2\theta\rho$ and $\alpha = \alpha_1 + 2\theta\omega$. Then it is necessary to consider that the K-profile and the FPZ length are now variables, labeled as $\tilde{q}(\rho)$ and $2\tilde{c}_f$, depending of $1/D$;

$$\tilde{q}(\rho) = q(\rho) + q_1(\rho)\theta + q_2(\rho)\theta^2 + \dots \quad \tilde{c}_f = c_f + c_{f_1}\theta + c_{f_2}\theta^2 + \dots \quad (9.67)$$

The same procedure as that which led to equations (9.20)—(9.30) furnishes equations in which the terms with the zero-th and first powers of θ may be separated. Upon separating them, one obtains for $q_1(\rho)$ an integral equation of the first kind that is similar to (9.20) except that it involves additional terms with functions $q(\rho)$ and $b_1[\xi(\omega)]$ and $c_1[\xi(\omega)]$. For c_{f_1} one obtains a similar equation as (9.32) with (9.30) except that $b_1[\xi(\omega)]$ and $c_1[\xi(\omega)]$ are again involved. This means that the second-order asymptotic profile $q_1(\rho)$ and the second-order FPZ size c_{f_1} depend on the structure shape, albeit not through the LEFM function $k(\alpha)$.

The same procedure as before may now be followed to analyze the scaling for cases 1 and 2 on the basis of the nominal strength expression (9.25).

For Case 1 failure, one gets again the same classical size effect law (9.40) with the same expression for the asymptotic nominal strength σ_0, but the expression for the transitional size D_0 is now slightly different—it depends not only on $k(\alpha)$ but also on $b_1[\xi(\omega)]$ and $c_1[\xi(\omega)]$. This means that the shape dependence of D_0 is not entirely characterized by the dimensionless SIF function $k(\alpha)$ (or energy release function $g(\alpha)$).

Previous tests of notched specimens of different shapes, however, have not so far indicated any systematic disagreement with the expression for D_0 in (9.41). It could be that the effect of $b_1[\xi(\omega)]$ and $c_1[\xi(\omega)]$ is so weak that it is masked by inevitable experimental scatter and is practically unimportant.

It may further be shown that simultaneous consideration of nonzero ρ_0 does not change the foregoing observations.

For Case 2 failure, analogous conclusions are reached. In particular, the law (9.65) describing the size effect at fracture initiation retains the same form.

The geometry dependences of the size effect coefficients that follow for Cases 1 and 2 from a fixed K-profile are identical to those that follow from: (1) equivalent LEFM, (2) R-curve model, (3) two-parameter fracture model characterized by K_c and critical δ_{CTOD} (proposed for metals by Wells and Cottrell and for concrete by Jenq and Shah), and (4) approximate analysis of energy release from stress relief zones (Bažant and Planas 1998). But for the standard cohesive crack model with a fixed stress-displacement curve there is a difference in the second asymptotic term with $1/D$.

A complete large-size asymptotic analysis of Case 1 for the standard cohesive crack model has already been presented by Planas and Elices (1992, 1993). The present results agree with their conclusions.

Furthermore, one may wonder whether better results could be obtained by taking the first three (rather than two) nonzero terms of the asymptotic expansions. This, of course, would lead to a more accurate representation of the size effect for sufficiently large D. However, the resulting expressions then diverge for $D \to 0$ and the possibility of asymptotic matching is lost.

9.4.4. Case 3. Negative-Positive Geometry Transition $(g_0 > 0, g'_0 = 0, g''_0 > 0)$

When the initial fracture geometry is negative, i.e., $k'(\alpha) < 0$, the crack grows stably under increasing load (an example is a panel with a small centric crack loaded on the crack faces). Failure under load control will occur only when the fracture geometry changes from negative to positive (Fig. 9.7), which is what we now consider. So, this case is characterized by

$$k_0 > 0, \qquad k'_0 = 0, \qquad k''_0 > 0 \qquad (9.68)$$

a) Scaling for Nonstandard Model with a Fixed K-Profile

The problem is again much simpler for a fixed K-profile. The location of the FPZ is determined by the maximum load condition, which reads $d\sigma_N/d\alpha_1 = 0$ at constant θ. Denoting $f(\alpha) = q[\rho(\alpha)]/k(\alpha)$, which is a smooth function, we

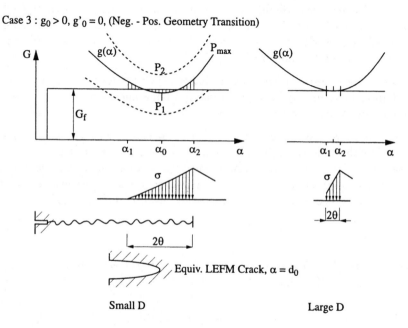

Figure 9.7: Lines of dimensionless energy release function $g(\alpha)$ for increasing values of constant load P_1, P, P_2 for small (left) and large (right) sizes D, along with stress profiles in FPZ for three basic types of failure: Case 3—negative–positive geometry transition

obtain from equation (9.25) the condition

$$\frac{d}{d\alpha_1}\int_{\alpha_1}^{\alpha_1+2\theta} f(\alpha)d\alpha = 0 \qquad (9.69)$$

which indicates that $f(\alpha_1 + 2\theta) - f(\alpha_1) = 0$ or

$$f[\alpha_0 + (2\theta + \alpha_1 - \alpha_0)] - f[\alpha_0 + (\alpha_1 - \alpha_0)] = 0 \qquad (9.70)$$

Expanding now both functions into Taylor series with respect to point α_0 (the crack length at failure when $D \to \infty$, i.e. asymptotic LEFM crack), one concludes that

$$\alpha_0 - \alpha_1 = \theta + O(\theta^3) \approx \theta \qquad (9.71)$$

which means that, with an error of the third order in θ, the center of the FPZ is located at the tip of the asymptotic LEFM crack (Fig. 9.7). In other words, the locations of the FPZ center for structures of different sizes are geometrically similar, with an error of the order of $1/D^3$.

If we truncated the Taylor series expansion (9.36) after the second, linear term in θ, the result would be the pure LEFM size effect $\sigma_N \propto D^{-1/2}$ because $k'(\alpha_0) = 0$. Therefore we must truncate it after the third, quadratic term, and we have

$$\frac{1}{k(\alpha)} = \frac{1}{k_0} - \frac{k_0''}{2k_0^2}(\alpha - \alpha_0)^2 + \ldots \qquad (9.72)$$

where $\alpha = \alpha_1 + 2\theta\rho$. Substituting this into (9.25), and expressing, as before, $1/(\sigma_N^2 D)$, we obtain, after rearrangement,

$$\frac{1}{\sigma_N^2 D} = \frac{k_0^2}{K_c^2}\left(1 - \frac{2k_0'' I_5}{k_0 I_1}\theta^2 + \ldots\right)^{-2} \qquad (9.73)$$

in which (since $\rho_0 = \rho(\alpha_0) = \frac{1}{2}$)

$$I_5 = \int_0^1 (\rho - \tfrac{1}{2})^2 q(\rho) d\rho \qquad (9.74)$$

Eq. (9.73) is accurate only up to the first two terms of the expansion in θ. By Taylor series expansion in may be checked that the following formula with any r has the same terms of Taylor series expansion up to θ^2 and is therefore as equally justified as (9.73);

$$\sigma_N = \frac{K_c}{\sqrt{D}}\left(1 + \frac{4rk_0'' I_5}{k_0}\theta^2\right)^{-1/2r} \qquad (9.75)$$

Consider first the value $r = 1$, which seems appropriate because it yields a formula that is in the middle range close to the well tested original size effect law;

$$\sigma_N = \sigma_0 \left(\frac{D_1}{D} + \frac{D}{D_0} \right)^{-1/2} \tag{9.76}$$

in which σ_0, D_0 and D_1 are certain positive constants.

We will now take again the viewpoint of asymptotic matching. So far we considered only the asymptotic properties for $D \to \infty$. To get a formula approximately applicable through the entire size range, $\lim \sigma_N$ for $D \to 0$ must be finite. However, the limit of (9.76) is zero. Therefore, with the goal of proper asymptotic matching, we modify (9.76) in the following manner;

$$\sigma_N = \sigma_0 \left(\frac{D_1}{D + D_1} + \frac{D}{D_0} \right)^{-1/2} \tag{9.77}$$

This modification does not change the first two terms of the large size asymptotic expansion of $\sigma_N \sqrt{D}$ in powers of D^{-1}. Indeed, as may be checked, the asymptotic expansions of both expressions are:

$$\sigma_N \sqrt{D}/\sigma_0 \sqrt{D_0} = 1 - \frac{1}{2} D_0 D_1 D^{-2} + O(D^{-4}) \tag{9.78}$$

To determine the expressions for D_0 and D_1, we make the analogous replacement of D with $D + D_1$ in (9.75) with $r = 1$;

$$\sigma_N = \frac{K_c}{k_0} \left(D + \frac{4 k_0'' I_5}{k_0} \frac{c_f^2}{D + D_1} \right)^{-1/2} \tag{9.79}$$

By matching this expression to (9.77), one finds

$$D_1 = 4 k_0 k_0'' I_5 \left(\frac{c_f \sigma_0}{K_c} \right)^2, \quad D_0 = \left(\frac{K_c}{k_0 \sigma_0} \right)^2 \tag{9.80}$$

The value of the small-size nominal strength σ_0 cannot be determined from the present asymptotic theory. It can be determined by using plastic analysis to solve the cohesive crack model for the case that $D \ll c_f$. In that case the cohesive stress along the entire crack path is nearly uniform, as if the material were plastic, and such analysis proves that σ_0 must be finite.

One restriction, though, must be imposed. The scaling law (9.77) could exhibit a nonmonotonic dependence on D, with a reverse size effect (increasing σ_N) for small D. But according to diverse other evidence (Bažant and Planas

1998), the scaling law (9.77) must monotonically decrease with increasing D. It may be easily checked that this is satisfied if and only if

$$D_1 \geq D_0 \tag{9.81}$$

When $D_1 = D_0$, the slope $d\sigma_N/dD = 0$ at $D \to 0$, and when $D_1 > D_0$, this slope is negative.

Consider now whether the r-value may differ from 1. If it does, (9.77) is replaced by

$$\sigma_N = \sigma_0 \sqrt{\frac{D+D_1}{D_1}} \left(1 + \frac{rD(D+D_1)}{D_0 D_1}\right)^{-1/2r} \tag{9.82}$$

This causes no change in the first two terms of the power series expansion of $\sigma_N\sqrt{D}$ in terms of θ, and the same expressions for D_0 and D_1 apply. This formula is nothing more than another way to write equation (9.75).

Negative r values give imaginary σ_N when D is less than a certain value, and are therefore unacceptable. On the other hand, σ_N for $D \to 0$ is σ_0, i.e., finite, for any positive r value. Furthermore, one can check that (9.77) is a monotonic function of D if and only if $r \leq 2$ and $D_1 \geq D_0$. Therefore, unlike Case 1, all r values such that $0 < r \leq 2$ are, in principle, acceptable (Fig. 9.6).

Eq. (9.77) is a new size effect formula representing the scaling law at fixed K-profile for failures at negative-positive geometry transition. This result cannot be obtained with the approach used in previous studies, in which the cohesive crack was assumed to be approximately equivalent to an LEFM crack whose tip is located at the middle of the FPZ, i.e., at distance c_f from the end of the stress-free crack (point α_0 in Fig. 9.7, Case 3). That approach obviously works only when the fracture geometry at failure is positive ($g'_0 > 0$), as in notched fracture specimens or in the modulus of rupture test. But here the LEFM scaling, $\sigma_N \propto 1/\sqrt{D}$, would, incorrectly, ensue because $g'_0 = 0$.

The problem of size effect in Case 3 has been one motivation for this study. Previously, the size effects for Cases 1 and 2 have been analytically derived from the equivalent LEFM approach, from the R-curve model, and from J-integral expansion. For Case 3, however, these methods fail to give any deviation from the LEFM size effect because the equivalent crack tip has the same relative coordinate α_0 for all D, and because the crack tip is on the R curve located beyond the initial rising portion.

b) Scaling for Standard Model with a Fixed Stress-Displacement Law

Consider now Case 3 for the standard cohesive crack model with a fixed stress-displacement law. As described below (9.67), separation of the terms with the zero-th and first powers of θ provides for the second-order K-profile $q_1\rho$ an integral equation involving parameters of the second-order near-tip stress and displacement fields $b_1[\xi(\omega)]$ and $c_1[\xi(\omega)]$, and for the second-order process zone size c_{f_1} an equation also involving these second-order fields. The consequence is that, instead of (9.75), one gets an equation of the form:

$$\sigma_N = e_0 \left(1 + e_1 r\theta + e_2 r\theta^2\right)^{-1/2r} \qquad (9.83)$$

in which e_0, e_1 and e_2 depend on structural geometry through functions $b_1[\xi(\omega)]$ and $c_1[\xi(\omega)]$ rather than merely through the SIF function $k(\alpha)$. Due to the presence in (9.83) of a term linear in θ, one gets now a size effect law of the form:

$$\sigma_N = \sigma_0 \left(1 + \frac{D}{D_0}\right)^{-1/2} \qquad (9.84)$$

which is the same as the form of the classical size effect law (9.40) for Case 1; but D_0 now depends on the second-order near-tip stress and displacement fields $b_1[\xi(\omega)]$ and $c_1[\xi(\omega)]$, which gives a different geometry dependence than that in Case 1.

It may be noted that the geometry dependence of the Case 3 scaling law is, for standard cohesive crack model, considerably more complicated than for the nonstandard model with a fixed K-profile. On the other hand, the result for the standard cohesive crack model is simpler in the sense that the scaling law for Case 3 has the same form as for Case 1.

9.5. Small-Size Asymptotics of Cohesive Crack Model

The small size asymptotic behavior of the cohesive crack model has been derived in Sec. 2.12 directly on the basis of the boundary value problem formulation. It is interesting to derive it now on the basis of the traditional compliance (Green's function) formulation, which may be written as

$$\varphi[\sigma(x)] = -b \int_{-x_1}^{x_2} C(x, x')\, \sigma(x')\, dx' + C_P(x)\, P \qquad (9.85)$$

$$b \int_{x_1}^{x_2} \kappa(x)\, \sigma(x)\, dx + P\kappa_P = 0 \qquad (9.86)$$

where $w = \varphi(\sigma) = $ softening stress-displacement law defining the cohesive crack model; $x \in (x_1, x_2)$ is the FPZ, x_2 is the cohesive crack tip; $C_P(x)$ or $\kappa(x)$ and $C(x, x')$ or κ_P are the crack-face displacement (compliance) at x' and the SIF at x_2 caused by a pair of unit normal forces applied to crack faces at x or by unit applied load P. The second equation means that the total SIF, K_I^{tot}, caused by applied load P and crack-bridging stresses $\sigma(x)$, must vanish (which is the condition of smooth closure at tip). Eq. (9.85) is the condition of compatibility of elastic deformation of the body with the given softening stress-displacement law. $C(x, x')$ represents a Green's function.

Introduce now the dimensionless variables

$$\xi = x/D, \quad \xi_1 = x_1/D, \quad \xi_2 = x_2/D \tag{9.87}$$
$$\bar{\sigma}(\xi) = \sigma(x)/f_t, \quad \bar{w}(\xi) = w(x)/w_f \tag{9.88}$$
$$\bar{C}(\xi, \xi') = E'b\, C(x, x'), \quad \bar{C}_P(\xi) = E'b\, C_P(x) \tag{9.89}$$
$$\bar{\kappa}(\xi) = b\sqrt{D}\, \kappa(x), \quad \bar{\kappa}_P = b\sqrt{D}\, \kappa_P \tag{9.90}$$

where f_t = tensile strength = stress at which the cohesive crack begins to open; w_f = crack opening w when the cohesive (crack-bridging) stress σ is reduced to 0, i.e. at full break (Fig. 9.1); x_1 is the beginning of the crack at either the notch tip or the body surface, and x_2 is the cohesive crack tip. The softening stress-displacement law may be written as $\bar{w} = \bar{\varphi}(\bar{\sigma})$ which is a dimensionless function such that $\bar{\varphi}(0) = 1$ and $\bar{\varphi}(1) = 0$. Setting $P = bD\sigma_N$ and substituting the dimensionless variables, one may transform (9.85) and (9.86) to the form:

$$E'b\, w_f\, \bar{\varphi}[\bar{\sigma}(\xi)] = -f_t D \int_{\xi_1}^{\xi_2} \bar{C}(\xi, \xi')\, \bar{\sigma}(\xi')\, d\xi' + \bar{C}_P(\xi)\, \sigma_N\, D \tag{9.91}$$

$$\bar{\kappa}_P \sigma_N + f_t \int_{\xi_1}^{\xi_2} \bar{\kappa}(\xi)\, \bar{\sigma}(\xi)\, d\xi = 0 \tag{9.92}$$

Consider now that $D \to 0$. The first of these two equations requires that $\bar{\varphi}[\bar{\sigma}(\xi)] = 0$ for all ξ, i.e. $w(x) = 0$. So the crack just begins to open, i.e. the corresponding cohesive stress is equal to the tensile strength f_t all along the crack, as if the FPZ were perfectly plastic (or as if a thin strip of a rigid-perfectly plastic material were glued into a slit cut in place of the crack). σ_N is determined by the smooth closure condition (9.92), and because $\bar{\sigma}(0) = 1$ this condition yields

$$\sigma_N = -\frac{f_t}{\bar{\kappa}_P} \int_{\xi_1}^{\xi_2} \bar{\kappa}(\xi)\, d\xi \tag{9.93}$$

This expression depends on ξ_2, and ξ_2 must be found so as to maximize σ_N (note in this regard that $\kappa(\xi)$ has a square-root singularity at $\bar{\xi}_2$).

The foregoing analysis proves that the size effect of the cohesive crack model begins at $D = 0$ with a finite value $\sigma_N = \sigma_0$. A much more realistic model for $D \to 0$ is nonlocal continuous damage mechanics. Although analytical asymptotic solutions are unavailable, numerical solutions have confirmed that σ_0 is finite.

9.6. Nonlocal LEFM—A Simple Approach to Cohesive Fracture and Its Scaling

A simpler analysis of cohesive fracture can be based on the following *hypothesis*: the FPZ moves ahead when the average $\bar{\mathcal{G}}$ of the energy release rates $\mathcal{G}$ for the crack tips located along the FPZ is equal to the fracture energy G_f. The average is defined as

$$\bar{\mathcal{G}}(\alpha_1) = \frac{1}{2\theta} \int_{\alpha_1}^{\alpha_1+2\theta} w[\rho(\alpha)]\mathcal{G}(\alpha)d\alpha = \int_0^1 w(\rho)\mathcal{G}[\alpha(\rho)]d\rho \qquad (9.94)$$

where $w(\rho)$ is the chosen weight function, normalized so that $\int_0^1 w(\rho)d\rho = 1$. For the sake of simplicity, it will be assumed from now on that $w(\rho) = 1$. $\mathcal{G}$ may be expressed in terms of the dimensionless energy release rate function $g(\alpha) = [k(\alpha)]^2$;

$$\mathcal{G} = (\sigma_N^2/E')Dg(\alpha) \qquad (9.95)$$

The present hypothesis is intuitive. Its only justification is that, as will be seen, it yields for all the three cases essentially the same scaling laws, as well as the same kind geometry effect through function $k(\alpha)$, as does the smeared tip method.

This is not completely true for the method of equivalent LEFM with an effective sharp crack tip in the middle of the fracture process zone (Bažant and Kazemi 1991b; Bažant and Planas 1998), nor for the asymptotic analysis based on an R-curve (Bažant 1997a; Bažant and Planas 1998), which both yield no size effect for Case 3, contrary to the smeared-tip method.

Case 1: $k_0 > 0$ and $k'_0 > 0$; this is the case with FPZ attached to the tip of the notch (or stress-free crack). The dimensionless notch length is again denoted as $a_1/D = \alpha_1$, and for geometrically similar structures of different sizes, $\alpha_1 = $ const. We again approximate $g(\alpha)$ by the first two terms of its Taylor series expansion about α_1, i.e., set

$$g(\alpha) \approx g_0 + g'_0(\alpha - \alpha_1) \qquad (9.96)$$

Substituting this and (9.95) into (9.94), one can easily integrate, with the result:

$$\bar{\mathcal{G}} = \frac{\sigma_N^2}{2E'}\frac{D^2}{c_f}\left(g_1\frac{2c_f}{D} + \frac{g'_1}{2}\frac{4c_f^2}{D^2}\right) \qquad (9.97)$$

Setting $\bar{\mathcal{G}} = G_f$ and solving for σ_N, one gets the classical energetic size effect law (9.40):

$$\sigma_N = \sqrt{\frac{E'G_f}{g'_1c_f + g_1D}} = \sigma_0\left(1 + \frac{D}{D_0}\right)^{-1/2} \qquad (9.98)$$

in which the expressions for σ_0 and D_0 are equivalent to those obtained in (9.41) by the smeared tip method except for factors depending on integrals I_1 and I_3 (i.e., on the SIF density profile). In particular, these expressions imply the same geometry dependence of σ_0 and D_0.

Case 2: $g_1 = 0$ and $g_1' > 0$; this is the case of failures at fracture initiation, as in the modulus of rupture test. The first two nonzero terms of the Taylor series expansion are:

$$g(\alpha) \approx g_1'(\alpha - \alpha_1) + \tfrac{1}{2}g_1''(\alpha - \alpha_1)^2 \qquad (9.99)$$

Substituting this into (9.95) and then (9.94), one obtains

$$\bar{\mathcal{G}} = \frac{\sigma_N^2}{2E'}\frac{D^2}{c_f}\left(\frac{g_1'}{2}\frac{4c_f^2}{D^2} + \frac{g_1''}{6}\frac{8c_f^3}{D^3}\right) \qquad (9.100)$$

Setting this equal to G_f and solving for σ_N, one gets the same formula as (9.63) obtained under the hypothesis of a fixed SIF profile:

$$\sigma_N = \sqrt{E'G_f}\left(g_1'c_f - \frac{2\langle -g_1''\rangle}{3}\frac{c_f^2}{D}\right)^{-1/2} = \sigma_\infty\left(1 - \frac{2D_b}{D}\right)^{-1/2} \qquad (9.101)$$

where D_b is a constant expressed similarly as in (9.64) for the smeared-tip approach except for a difference in the coefficients involving the integrals I_3 and I_4, which express the effect of the SIF distribution profile. For the same reason as before, g_1'' has been replaced by $\langle -g_1''\rangle$ because typically g_1'' is negative and if it is positive there is no size effect.

To achieve proper asymptotic matching, which requires a monotonic size effect law and a finite σ_N for $D \to 0$, the last formula is replaced by the formula

$$\sigma_N = \sigma_\infty\left(1 + \frac{rD_b}{D}\right)^{1/r} \qquad (9.102)$$

which gives, for any positive r value, the same first two terms of the power series expansion of $\sigma_N\sqrt{D}$ in terms of θ; D_b is expressed similarly as in (9.64).

Case 3: $g_0 > 0, g_0' = 0, g_0'' > 0$; this is the case of failure at negative-positive geometry transition, after the FPZ has moved away from notch tip and a large traction-free crack has formed. Unlike Cases 1 and 2, the location α_1 of the beginning of FPZ is for this case unknown and must be determined from the condition of maximum load (or stability limit), that is, $d\sigma_N/d\alpha_1 = 0$. According to (9.95) with (9.94) and the condition $\bar{\mathcal{G}} = G_f$,

$$\sigma_N^{-2} = \frac{D}{2E'G_fc_f}\int_{\alpha_1}^{\alpha_1+2\theta} g(\alpha)d\alpha \qquad (9.103)$$

The maximum load condition can be rewritten as $d\sigma_N^{-2}/d\alpha_1 = 0$, and upon differentiating the integral it is reduced to

$$g(\alpha_1 + 2\theta) - g(\alpha_1) = 0 \qquad (9.104)$$

This condition has the same form as (9.70), and its analysis is similar. Writing the Taylor series expansion about the crack tip location α_0 for $D \to \infty$, i.e.

$$g(\alpha_1 + 2\theta) = g(\alpha_0) + \tfrac{1}{2}g_0''(\alpha - \alpha_0)^2 + \ldots \qquad (9.105)$$

for $\alpha = \alpha_1$ and $\alpha = \alpha + 2\theta$, and substituting these expansions into (9.104), one concludes that

$$\alpha_0 - \alpha_1 = \theta + O(\theta^3) \approx \theta \qquad (9.106)$$

So, with only a third order error, the LEFM crack tip is at the middle of FPZ, same as deduced from the smeared tip method. Substituting this and (9.104) into (9.94), one can easily integrate, with the result:

$$\bar{\mathcal{G}} = \frac{\sigma_N^2}{2E'} \frac{D^2}{c_f} \left(g_0 \frac{2c_f}{D} + \frac{g_0''}{6} \frac{16 c_f^3}{D^3} \right) \qquad (9.107)$$

Setting $\bar{\mathcal{G}} = G_f$ and solving for σ_N, one gets the size effect formula

$$\sigma_N = \sqrt{E'G_f} \left(g_0 D + \frac{g_0'' c_f^2}{6 D} \right)^{-1/2} \qquad (9.108)$$

Now we note that this is equivalent to (9.73), and a similar transformation needs to be carried out to obtain a monotonic size effect law and achieve asymptotic matching. The result is

$$\sigma_N = \sigma_0 \left(\frac{D_1}{D+D_1} + \frac{D}{D_0} \right)^{-1/2} \qquad (D_1 \geq D_0) \qquad (9.109)$$

This is the same as equation (9.77) obtained under the hypothesis of a fixed SIF profile except that the expressions for D_0 and D_1 have different coefficients as they do not involve the integrals I_1 and I_5.

9.7. Broad-Range Size Effect Law and Its Dirichlet Series Expansion

The simple size effect law in (9.40) for the basic Case 1 is normally adequate for a size range up to about 1:20, which suffices for most structural engineering applications. The generalized size effect law (9.43) has been shown to give excellent approximation of the numerical results obtained by Hillerborg with the cohesive crack model for the size range of 1:250 (Bažant 1985b). However, only the first term of the large-size power series expansion of that law in $1/D$ is correct. The following broad-range size effect law for Case 1, which is a generalization of formula (9.46), is capable of approximating the large-size asymptotic behavior up to order $n+2$ in $1/D$;

$$\sigma_N^2 = E' \sum_{n=1}^{N} \Gamma_n \left(1 - e^{-D_n/D}\right) \tag{9.110}$$

which has the asymptotic expansion:

$$\sigma_N^2 = \frac{E'}{D}\left(S_1 - \frac{S_2}{D} + \frac{S_3}{D^2} - \ldots\right) \tag{9.111}$$

with the notations:

$$S_1 = \sum_n \Gamma_n D_n, \quad S_2 = \frac{1}{2!}\sum_n \Gamma_n D_n^2, \quad S_3 = \frac{1}{3!}\sum_n \Gamma_n D_n^3, \quad \ldots \tag{9.112}$$

Here $D_1, D_2, \ldots D_N, \Gamma_1, \Gamma_2, \ldots \Gamma_N$ are constants; $D_1, D_2, \ldots D_N$ are positive and may be assumed to form an increasing sequence. It is evident that this formula also preserves the finiteness of $\lim \sigma_0$ for $D \to 0$.

From the foregoing expansion one can further obtain:

$$\frac{E'}{\sigma_N^2 D} = A_0 + \frac{A_1}{D} + \frac{A_2}{D^2} + \ldots \tag{9.113}$$

in which

$$A_1 = \frac{1}{S_1}, \quad A_2 = \frac{S_2}{S_1^2}, \quad A_3 = \frac{S_2^2}{S_1^3} - \frac{S_3}{S_1^2}, \quad \ldots \tag{9.114}$$

Case 1 is obviously characterized by

$$A_0 > 0, \quad A_1 > 0 \quad \text{(Case 1)} \tag{9.115}$$

Formula (9.110) has been deliberately structured so that the expression for $\sigma_N^2 D/E'$ be the well-known Dirichlet series (called also the Prony series). This series is the real counterpart of the Fourier series and is known to be generally effective for modeling broad-range decay, for example the creep or relaxation functions of linear viscoelasticity where the use of this series is well understood. It is advantageous to transplant this knowledge to our problem.

Progressively increasing the number of terms in the broad-range size effect law (9.110), one obtains the 'cascading' bi-logarithmic size effect plot shown in

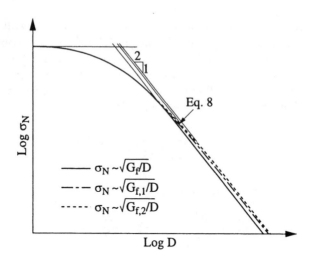

Figure 9.8: Concept of cascading size effect for a very broad size range

Fig. 9.8, in which the individual terms of (9.110) terminate with progressively shifted asymptotes of slope $-1/2$. Each n-th asymptote is associated with a progressively higher fracture energy $G_{f,n}$. Noting that $\sigma_N^2 = E'\mathcal{G}/g(\alpha_0)D$, equating this to (9.110), setting $\mathcal{G} = G_{f,N}$, and renaming n as m and N as n, we get

$$G_{f,n} = g^2(\alpha_0) \sum_{m=1}^{n} \Gamma_m D_m \qquad (9.116)$$

$G_{f,n}$ is the fracture energy associated with the n-th asymptote of slope $-1/2$ in Fig. 9.8, and the equation of this asymptote is

$$\sigma_N = \sqrt{\frac{E'G_{f,n}}{g(\alpha_0)D}} \qquad (9.117)$$

Using the cohesive crack model to obtain very accurate σ_N values for a very broad size range such as $1:10^6$, and fitting these values with formula (9.110), one could thus obtain a set of $G_{f,n}$ values. This set may be regarded as a fracture energy spectrum of the cohesive crack model defined by a certain softening stress-displacement curve.

The first fracture energy, $G_f = G_{f,1}$, corresponds to the area under the initial tangent (Fig. 9.1) of the softening stress-displacement curve (Bažant and Planas 1998, Bažant and Li 1997). This is known to suffice for predicting with the cohesive crack model the maximum loads of structures for a size range of about 1:20 (which covers most practical cases in structural engineering, and certainly the range of laboratory testing).

If the size range is extended much farther, say up to 1:200, then more of the area of the softening curve of the cohesive crack model, except for a remote tail of the curve, must be expected to matter for the maximum load prediction. This area may then be associated with the fracture energy $G_{f,2}$.

If the size range is extended still much farther and many terms are included, the last asymptote should correspond to a fracture energy that coincides with with the fracture energy G_F determined by the work-of-fracture method from the area under the complete load-displacement curve of a notched beam, including its long tail, as introduced for concrete by Hillerborg. It is well known that for concrete G_F is about 2.5 times larger than the fracture energy G_f determined by the size effect method using size ranges between 1:4 and 1:10.

The Dirichlet series in Eq. (9.110) resembles the retardation spectrum in linear viscoelasticity. In analogy to what is known from the theory of retardation spectrum, one may expect the problem of identification of D_n from a given size effect curve to be ill-conditioned. This means that the D_n values would need to be chosen, within certain constraints, just as the discrete retardation times of a viscoelastic material with a broad retardation spectrum must be chosen, within certain constraints. Thus the discrete fracture energy spectrum, defined as the plot of $G_{f,1}, G_{f,2}, ...G_{f,N}$ versus the D_n values, will be nonunique, as it will depend on the choice of $D_1, D_2, ...D_N$ in the foregoing formulae.

However, similar to the theory of a continuous retardation spectrum in linear viscoelasticity, one could consider infinitely closely spaced D_n values and thus generalize (9.110) to a smooth continuous spectrum of fracture energies (which converts formula (9.110) to Laplace transform). This spectrum would then be unique.

With sufficiently many terms, the Dirichlet series can represent any decaying phenomena as closely as desired. Realizing that all the spring-dashpot models of linear viscoelasticity are equivalent to the Kelvin chain corresponding to the Dirichlet series, one must conclude that it would make little sense to look for other formulae for the broad-range size effect. Although infinitely other formulae exist, they must all be equivalent to (9.110).

Formula (9.63) or (9.65) for Case 2 may be generalized in terms of the Dirichlet series by introducing the expansion:

$$\sigma_N = \sigma_\infty + \sum_{n=1}^{N} S_n \left(1 - e^{-H_n/D}\right) \qquad \text{(Case 2)} \qquad (9.118)$$

where H_n is an increasing sequence of positive constants.

This formula may also be written in the form (9.110) in which Γ_n are such

that
$$A_0 = 0, \quad A_1/A_2 < 0 \qquad \text{(Case 2)} \qquad (9.119)$$

Formula (9.77) for Case 3, too, may be generalized in terms of Dirichlet series and can be written in the form of (9.111) provided that

$$A_1 > 0, \quad A_2 = 0, \quad A_3 < 0 \qquad \text{(Case 3)} \qquad (9.120)$$

Should the size effect laws (9.40), (9.63) and (9.77) for size range up to about 1:20 be discarded and replaced by the first term of the Dirichlet series (9.110) or (9.118)? They should not, because the transition of a single exponential term between the small and large size asymptotic behaviors is too abrupt and the size range would have to be further narrowed. At least two terms of the series would be needed.

The Dirichlet series expansion in (9.110) is analogous to the retardation spectrum of Kelvin chain in viscoelasticity. There exists another Dirichlet series expansion analogous to the relaxation spectrum of Maxwell chain in viscoelasticity:

$$\frac{1}{\sigma_N^2 D} = \frac{1}{\sigma_0^2 D_0} \sum_{n=1}^{N} B_n e^{-D_n/D} \qquad (9.121)$$

$$= \frac{1}{\sigma_0^2 D_0} \left(S_0 - \frac{S_1}{D} + \frac{S_2}{D^2} - \frac{S_3}{D^3} + \ldots \right) \qquad (9.122)$$

where $B_1, B_2, \ldots$ are constants and

$$S_0 = \sum_n B_n, \quad S_1 = \sum_n B_n D_n, \qquad (9.123)$$

$$S_2 = \frac{1}{2!} \sum_n B_n D_n^2, \quad S_3 = \frac{1}{3!} \sum_n B_n D_n^3, \quad \ldots \qquad (9.124)$$

The first two terms of this expansion, if both non-zero, yield again the classical size effect law (9.40) (Case 1).

While (9.110) appears suited for up-scale extrapolations (e.g., from laboratory specimens of concrete, rock or ice to dam failure, mountain slide or Arctic ice sheet break-up), (9.121) appears suited for down-scale extrapolations (e.g., from laboratory fracture specimens of silicon to microelectronic or MEMS components which might behave in a non-brittle manner). A different and in some respects more convenient form of the broad-range size effect law is given in Bažant (1999b, 2001a) and discussed in detail in Bažant (2001d).

9.8. Size Effect Law Anchored in Both Small and Large Size Asymptotics

In consequence of sections 9.4, 9.5 and 2.12, the quasibrittle size effect must have the following small and large size asymptotics:

For $D \to 0$:
$$\sigma_N \propto 1 - \frac{D}{D_s} - \ldots \tag{9.125}$$

For $D \to \infty$:

- Case 1:
$$\sigma_N \propto \frac{1}{\sqrt{D}}\left(1 - \frac{D_0}{2D} + \ldots\right) \tag{9.126}$$

- Case 2:
$$\sigma_N \propto 1 + \frac{D_b}{D} + \ldots \tag{9.127}$$

- Case 3:
$$\sigma_N \propto \frac{1}{\sqrt{D}}\left(1 - \frac{D_a{}^2}{D^2} + \ldots\right) \tag{9.128}$$

Here D_0, D_a, D_b, D_s = constants, $D_a = D_0 D_1/2$, $D_s = rD_b$, and $\propto$ is the proportionality sign. Note that cases 1 and 3 are verified by the following expansions:

$$\left(1 + \frac{D}{D_0}\right)^{-1/2} = \sqrt{\frac{D_0}{D}}\left(1 + \frac{D_0}{D}\right)^{-1/2}$$
$$= \sqrt{\frac{D_0}{D}}\left(1 - \frac{D_0}{2D} + \ldots\right) \tag{9.129}$$

$$\left(\frac{D_1}{D + D_1} + \frac{D}{D_0}\right)^{-1/2} = \sqrt{\frac{D_0}{D}}\left(1 + \frac{D_0 D_1}{D^2(1 + D_1/D)}\right)^{-1/2}$$
$$= \sqrt{\frac{D_0}{D}}\left[1 + \frac{D_0 D_1}{D^2}\left(1 - \frac{D_1}{D} + \ldots\right)\right]^{-1/2}$$
$$= \sqrt{\frac{D_0}{D}}\left(1 - \frac{D_0 D_1}{2D^2} + \ldots\right) \tag{9.130}$$

From now on, consider exclusively Case 1. As shown in Sec. 2.12, the first two terms of not only the large-size, but also the small-size, asymptotic expansions can be numerically predicted from the cohesive crack model using elastic analysis only (e.g., elastic finite element analysis). To match these terms, an asymptotic matching formula must contain at least four free parameters. One such formula is:

$$\sigma_N = \left(p_0 + p_1 D + \frac{m_1 D^2}{m_0 - D} \right)^{-1/2} \quad (9.131)$$

where

$$m_0 = \frac{p_0 - s_0}{m_1}, \quad m_1 = p_1 - s_1, \quad p_0 = \frac{1}{\sigma_N^{0\,2}}, \quad p_1 = -\frac{2\sigma_N^{0\,\prime}}{\sigma_N^{0\,3}} \quad (9.132)$$

$$s_0 = \frac{1}{\sigma_0^2} = \frac{g_0' c_f}{E'G_f}, \quad s_1 = \frac{g_0}{E'G_f} \quad (9.133)$$

To verify that Eq. (9.131) has the correct asymptotics, note that, for small enough D:

$$\sigma_N \approx (p_0 + p_1 D)^{-1/2} = [\sigma_N^{0\,-2}(1 - 2\sigma_N^{0\,\prime} D/\sigma_N^0)]^{-1/2}$$
$$\approx \sigma_N^0 (1 + \sigma_N^{0\,\prime} D/\sigma_N^0) = \sigma_N^0 + \sigma_N^{0\,\prime} D \quad (9.134)$$

and for large enough D:

$$\sigma_N^{-2} = p_0 + p_1 D - (p_1 - s_1) D \left(1 - \frac{p_0 - s_0}{(p_1 - s_1)D} \right)^{-1}$$
$$\approx p_0 + p_1 D - (p_1 - s_1) D \left(1 + \frac{p_0 - s_0}{(p_1 - s_1)D} \right)$$
$$= p_0 + p_1 D - p_1 D + s_1 D - p_0 + s_0$$
$$= s_0 + s_1 D = \frac{g_0' c_f + g_0 D}{E'G_f} \quad (9.135)$$

Eq. (9.131) is applicable only if the asymptotes $\sigma_N^{-2} = p_0 + p_1 D$ and $\sigma_N^{-2} = s_0 + s_1 D$ intersect at positive D. This gives for the validity of (9.131) the condition $(p_1 - s_1)(p_0 - s_0) < 0$, which seems to be satisfied for the realistic situations.

Note that formula (9.131) captures not only the LEFM-type shape dependence [through $g(\alpha_0)$ and $g'(\alpha_0)$], which dominates for large sizes, but also the plasticity-like shape dependence [through σ_N^0 and $\sigma_N^{0\,\prime}$], which dominates for small sizes. There is a gradual transition from near-fracture to near-plasticity shape dependence as the size is diminished.

The broad-range size effect law discussed in section 9.7 has also enough parameters to match both the small and large size asymptotics, but is not well suited for that purpose because it has no data fitting flexibility for D less than about $0.3D_0$. That law extends the size effect to sizes orders of magnitude larger than D_0, while Eq. (9.131) extends the size effect to sizes orders of magnitude smaller than D_0.

An ongoing study by Q. Yu at Northwestern University has verified that Eq. (9.131) can match the computed size effect curves of the cohesive crack model for various fracture specimen geometries so closely that a visual distinction in a graph is impossible. Since the maximum load for cohesive crack model at $D \to 0$ depends only on the tensile strength f'_t, it seems to be possible to calibrate the size effect law, (and thus identify the values of G_f and c_f) merely by measuring the load capacity of notched and unnotched specimens of the same size and shape. The tensioned prism with a one-sided notch seems to be particularly attractive as a test specimen. The reason is that the size effect computed for this prism from the cohesive crack model happens to conform practically exactly to the classical size effect law (i.e., the relations $s_0 = p_0$ and $s_1 = p_1$ happen to hold), which means that the zero-size strength limit of the size effect law agrees accurately with the measured direct tensile strength.

Eq. (9.131) offers the tantalizing prospect of being able to dispense with nonlinear structural analysis according to the cohesive crack model (or the crack band model) whenever the crack path is known in advance. It should suffice to use a linear finite element code to determine the small and large size asymptotic properties for the given structure geometry, and then 'interpolate' for any size according to (9.131).

9.9. Recapitulation

1. As an alternative to the standard form of the cohesive crack model characterized by a unique stress-displacement relation, a nonstandard form characterized by a fixed density profile of the stress-intensity factor (SIF) in the sense of the smeared-tip superposition method may be used. This new form facilitates analytical solutions.

2. Asymptotically for large enough structures, both forms are equivalent. The stress-displacement law can be obtained from a given profile of SIF density by solving an integral equation of the first kind with a weakly singular kernel, and vice versa.

3. It is shown that, for the nonstandard form of the cohesive crack model, the laws for the size effect, including their dependence on structure geometry, are essentially the same as for the standard form. This provides justification for the proposed nonstandard form.

4. The asymptotic large-size scaling of the nonstandard model for structures with a notch or preexisting stress-free crack is essentially the same as that

established for the standard model by Planas and Elices (1992, 1993), but it is easier to establish.

5. Matching the large size asymptotic properties to the small size asymptotic properties derived from a compliance formulation leads to approximate size effect formulae for the entire size range.

6. In the nonstandard cohesive crack model, three cases of failure, leading to three different laws for the size effect, must be distinguished, depending on the location of the fracture process zone (FPZ) at maximum load:

   - *Case 1.* When the FPZ is attached to the tip of a notch or pre-existing stress-free crack, one has the classical deterministic size effect law proposed by Bažant (1983, 1984a), representing in the bi-logarithmic plot a smooth transition from a horizontal asymptote to an asymptote of downward slope $-1/2$.
   - *Case 2.* When the FPZ is attached to a smooth body surface (as in the modulus of rupture test of unnotched beams), the size effect plot represents a transition from a downward inclined asymptote to a horizontal one, with a finite large-size asymptotic value.
   - *Case 3.* When the FPZ is detached from a notch or body surface—a failure type that can happen only when a negative fracture geometry is changing to positive—the size effect is similar as in Case 1 but exhibits a less gradual transition from the horizontal asymptote the asymptote of slope $-1/2$.

   The laws for Cases 1 and 2 coincide with those derived previously in several other ways—by simplified energy release analysis (Bažant 1984a), equivalent LEFM (Bažant and Kazemi 1991) and J-integral expansion (Bažant 1998a), and verified by tests and numerical simulations. With the nonstandard model, the law for Case 3 is different; but it has the same large-size and small-size asymptotes, its only difference from the classical law for Case 1 being a more abrupt transition between the asymptotes.

7. The scaling laws obtained for the nonstandard model display only small differences compared to the standard model: (1) for Case 1, the transitional size depends on structure geometry not only through the SIF of LEFM but also through the factor of the second-order near-tip stress field, and (2) for Case 3 one gets a size effect law of the same form as for Case 1 but with a transitional size whose shape dependence is given not by the SIF of LEFM but by the aforementioned factor.

8. In Case 3, the crack length a_0 at maximum load is *a priori* unknown. In previous works it has been tacitly assumed that the a_0-values in geometrically similar structures of different sizes D are geometrically similar.

This assumption is here proven correct, except for an error third-order small in $1/D$.

9. The size effect laws for Cases 1, 2 and 3, based on a fixed K-profile, capture also the effect of structure geometry (shape) on the size effect law.

10. As the simplest approach, it is further proposed to introduce a model called the nonlocal LEFM, in which the energy release rate of the structure is assumed to be the average over the FPZ of the LEFM energy release rates. The nonlocal LEFM is shown to yield the same asymptotic scaling properties and shape effects as the cohesive crack model.

11. To characterize the size effect for a broad size range exceeding about 1:20, an asymptotic Dirichlet series expansion of the size effect law may be used. The terms of the series may be associated with progressively increasing fracture energy values associated with larger and larger scales.

Note: Since the cohesive crack model can be seen as the localization limit of the nonlocal continuum damage mechanics and the crack band model, the same three size effect laws should approximately apply for these models.

Chapter 10

Size Effect at Continuum Limit on Approach to Atomic Lattice Scale

The recent emphasis of continuum mechanics studies on the transition from continuum to atomic lattice models implies interesting questions of scaling. Although the subject is not pertinent to quasibrittle and heterogeneous materials, the main focus of this book, a brief discussion (based on Bažant 2001b) will be presented in this section because of methodological similarities in scaling, because of similarity of the objective (which is the strength or load capacity of structures), and because of the importance of the subject in nano-technology.

10.1. Scaling of Dislocation Based Strain-Gradient Plasticity

Building on the initial ideas of Toupin (1962) and Mindlin (1965), an impressive series of progressively refined studies extended to microscale the theory of metal plasticity (Fleck and Hutchinson 1993, 1997, Hutchinson 1997; Gao and Huang 2000). Careful physical arguments based on the theory of dislocations led Gao et al. (1999a,b) and Huang et al. (2000) to derive the following constitutive relation:

$$\sigma_{ik} = K\delta_{ik}\epsilon_{nn} + \frac{2\sigma}{3\epsilon}\epsilon'_{ik} \tag{10.1}$$

$$\tau_{ijk} = l_\epsilon^2 \left(\frac{K}{6}\eta^H_{ijk} + \sigma\Phi_{ijk} + \frac{\sigma_Y^2}{\sigma}\Psi_{ijk} \right) \tag{10.2}$$

where

$$\Phi_{ijk} = \frac{1}{\epsilon}\left(\Lambda_{ijk} - \Pi_{ijk}\right), \quad \Psi_{ijk} = f(\epsilon)f'(\epsilon)\Pi_{ijk}, \tag{10.3}$$

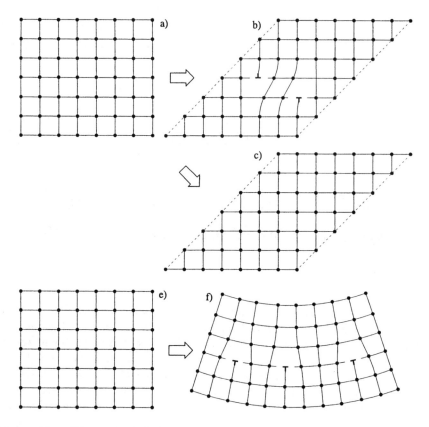

Figure 10.1: Illustration of the difference between (b) statistically stored dislocations and (e) geometrically necessary dislocations; (a,d) show the initial states of square lattices with 56 and 63 atoms, respectively, and (c) shows that for a homogeneous deformation no dislocations are necessary

and

$$\epsilon = \sqrt{\tfrac{2}{3}\epsilon'_{ij}\epsilon'_{ij}}, \quad \eta = \tfrac{1}{2}\sqrt{\eta_{ijk}\eta_{ijk}} \tag{10.4}$$

Here K = elastic bulk modulus; $\epsilon'_{ik} = \epsilon_{ik} - \tfrac{1}{3}\delta_{ik}\epsilon_{nn}$ = deviatoric strains, $\epsilon_{ik} = \tfrac{1}{2}(u_{i,k} + u_{k,i})$ = strains; ϵ, η = 2nd and 3rd order tensors of components $\epsilon_{ij}, \eta_{ijk}$; $\eta_{ijk} = u_{k,ij}$ = displacement curvature (or twist), reflecting the effect of geometrically necessary dislocations (Fig. 10.1e) [the strain gradient is $\epsilon_{ij,k} = \tfrac{1}{2}(\eta_{jki} + \eta_{ikj})$]; η^H_{ijk} = volumetric (hydrostatic) part of η_{ijk}; τ_{ijk} = third-order stresses work-conjugate to η_{ijk} (analogous to Cosserat's couple stresses). While Gao et al. (1999a,b) characterize the plastic constitutive properties by the semi-empirical relation $\sigma = \sigma_Y\sqrt{f^2(\epsilon) + l\eta}$, we will find it interesting to consider a

more general relation:
$$\sigma = \sigma_Y[f^q(\epsilon) + (l\eta)^p]^{1/q} \quad (10.5)$$

with positive exponents p and q (the case $p=1, q=2$ corresponds to Gao et al.); σ_Y = yield stress, σ, ϵ = stress and strain intensities; η = effective strain gradient proportional to the density of geometrically stored dislocations (i.e., to lattice curvature or twist); $f(\epsilon)$ = classical plastic hardening function, reflecting the effect of statistically stored dislocations; $l = \lambda_0 b$; $l_\epsilon = \lambda_\epsilon b$ = size of the so-called 'mesoscale cell', which is the material length characterizing the transition from standard to gradient plasticity and represents the minimum volume on which the macroscopic deformation contributions of the geometrically necessary dislocations (Fig. 10.1e) may be smoothed by a continuum; b = magnitude of Burger's vector of edge dislocation (e.g., 0.255 nm for copper); $\lambda_0, \lambda_\epsilon$ = positive dimensionless material characteristics expressed in terms of Taylor factor, Nye factor and the ratio of elastic shear modulus to the dislocation reference stress ($\lambda_\epsilon \approx 20000$ for copper, and is generally expected to be of the order of 10^4 or 10^5); and (Gao et al. 1999a,b)

$$\Lambda_{ijk} = \tfrac{1}{72}[\eta_{ijk} + \eta_{kji} + \eta_{kij} - \tfrac{1}{4}(\delta_{ik}\eta_{ppj} + \delta_{jk}\eta_{ppi})], \quad (10.6)$$
$$\Pi_{ijk} = [\epsilon_{ik}\eta_{jmn} + \epsilon_{jk}\eta_{imn} - \tfrac{1}{4}(\delta_{ik}\epsilon_{jp} + \delta_{jk}\epsilon_{ip})\eta_{pmn}]\epsilon_{mn}/54\epsilon^2 \quad (10.7)$$
$$\eta^H_{ijk} = \tfrac{1}{4}(\delta_{ik}\eta_{jpp} + \delta_{jk}\eta_{ipp}) \quad (10.8)$$

According to the principle of virtual work, the field equations of equilibrium are
$$\sigma_{ik,i} - \tau_{ijk,ij} + f_k = 0 \quad (10.9)$$

Scaling:

Similarly to (1.8) and (1.9) in Section 1.7, we now again introduce the dimensionless variables (labeled by an overbar):

$$\bar{x}_i = x_i/D, \quad \bar{u}_i = u_i/D, \quad \bar{\epsilon}_{ij} = \epsilon_{ij},$$
$$\bar{\eta}_{ijk} = \eta_{ijk}D, \quad \bar{\epsilon} = \epsilon, \quad \bar{\eta} = \eta D \quad (10.10)$$
$$\bar{\sigma}_{ik} = \sigma_{ik}/\sigma_Y, \quad \bar{\tau}_{ijk} = \tau_{ijk}/(\sigma_Y l_\epsilon),$$
$$\bar{\sigma} = \sigma/\sigma_Y, \quad \bar{f}_k = f_k D/\sigma_N \quad (10.11)$$

While the derivatives with respect to x_i are denoted by subscript i preceded by a comma, the derivatives with respect to dimensionless coordinates will be denoted (similar to Sec. 1.7) as $\partial_i = \partial/\partial \bar{x}_i$. To transform the field equations into dimensionless coordinates, we first note that, since η^H_{ijk}, Λ_{ijk} and Π_{ijk} are defined by Gao et al. (2000a,b) as homogeneous functions of degree 1 of tensors η and ϵ, they transform as $\eta_{ijk}^H = \bar{\eta}_{ijk}^H/D$, $\Lambda_{ijk} = \bar{\Lambda}_{ijk}/D$, $\Pi_{ijk} = \bar{\Pi}_{ijk}/D$, and so $\Phi_{ijk} = \bar{\Phi}_{ijk}/D$, $\Psi_{ijk} = \bar{\Psi}_{ijk}/D$; $\bar{\Lambda}_{ijk}, \bar{\Pi}_{ijk}, \bar{\Phi}_{ijk}, \bar{\Psi}_{ijk}$ and $\bar{\eta}_{ijk}^H$ are given by the same expressions as (10.3), (10.6)–(10.8) except that all the arguments are replaced by the dimensionless ones. We now substitute (10.1) and (10.2) into (10.9) and thus, using (10.10) and noting that $\partial/\partial x_i = (1/D)\partial_i$, we obtain the following dimensionless stresses and third-order stresses

$$\bar{\sigma}_{ik} = \frac{K}{\sigma_Y}\delta_{ik}\bar{\epsilon}_{nn} + \frac{\bar{\sigma}}{\bar{\epsilon}}\bar{\epsilon}'_{ik}, \qquad \bar{\sigma} = \left[f^q(\bar{\epsilon}) + \left(\frac{\lambda_1 l_\epsilon}{D}\bar{\eta}\right)^p\right]^{1/q} \tag{10.12}$$

$$\bar{\tau}_{ijk} = \frac{\lambda_\epsilon}{D}\left(\frac{K}{6\sigma_Y}\bar{\eta}_{ijk}^H + \bar{\sigma}\bar{\Phi}_{ijk} + \frac{1}{\bar{\sigma}}\bar{\Psi}_{ijk}\right) \tag{10.13}$$

where we set $\lambda_1 = \lambda_0/\lambda_\epsilon$. The field equations of equilibrium transform as

$$\partial_i\bar{\sigma}_{ik} - \frac{l_\epsilon}{D}\partial_i\partial_j\bar{\tau}_{ijk} + \frac{\sigma_N}{\sigma_Y}\bar{f}_k = 0 \tag{10.14}$$

To avoid struggling with the formulation of the boundary conditions, consider first that they are homogeneous, i.e., the applied surface tractions and applied couple stresses vanish at all parts of the boundary where the displacements are not fixed as 0. All the loading characterized by σ_N is applied as body forces f_k, and σ_N is considered as the parameter of these forces, varying proportionally. Then the transformed boundary conditions are also homogeneous. For $D/l_\epsilon \to \infty$, $\bar{\tau}_{ijk}$ vanishes and all the equations reduce, as required, to the standard field equations of equilibrium on the macroscale.

For $D/l_\epsilon \to 0$, on the other hand, one has $\bar{\sigma} \approx (\bar{\eta}\lambda_1 l_\epsilon/D)^{p/q}$. After substituting (10.12) and (10.13) into (10.14), we obtain the differential equations of equilibrium in the form:

$$\partial_i\left[\frac{K}{\sigma_Y}\delta_{ik}\bar{\epsilon}_{nn} + \frac{1}{\bar{\epsilon}}\left(\frac{\lambda_1 l_\epsilon\bar{\eta}}{D}\right)^{p/q}\bar{\epsilon}'_{ik}\right]$$
$$-\left(\frac{l_\epsilon}{D}\right)^2\partial_i\partial_j\left[\frac{K}{6\sigma_Y}\bar{\eta}_{ijk}^H + \left(\frac{\lambda_1 l_\epsilon\bar{\eta}}{D}\right)^{p/q}\bar{\Phi}_{ijk} + \left(\frac{D}{\lambda_1 l_\epsilon\bar{\eta}}\right)^{p/q}\bar{\Psi}_{ijk}\right]$$
$$= -\frac{\sigma_N}{\sigma_Y}\bar{f}_k \tag{10.15}$$

Now we multiply this equation by $(D/l_\epsilon)^{2+p/q}$ and take the limit of the left-hand size for $D \to 0$. This leads to the following asymptotic form of the field equations:

$$\partial_i\partial_j\left(\bar{\eta}^{-\frac{p}{q}}\bar{\Phi}_{ijk}\right) = \chi\bar{f}_k, \qquad \text{with} \qquad \chi = \lambda_1^{-\frac{p}{q}}\frac{\sigma_N}{\sigma_Y}\left(\frac{D}{l_\epsilon}\right)^{2+\frac{p}{q}} \tag{10.16}$$

Since D is absent from the above field equation (and from the boundary conditions, too, because they are homogeneous), the dimensionless displacement field as well as parameter χ must be size independent. Thus we obtain the following small-size asymptotic scaling law for the dislocation based gradient plasticity:

$$\sigma_N = \sigma_Y \, \chi \lambda_1^{p/q} \left(\frac{l_\epsilon}{D}\right)^{2+\frac{p}{q}} \tag{10.17}$$

(Bažant 2001b) where the exponent $2 + p/q > 2$. According to Gao et al.'s theory, $p/q = 1/2$, and so $\sigma_N \propto D^{-5/2}$ (an exception is the case of pure bending, for which $\sigma_N \propto D^{-3/2}$ because $\Phi_{ijk} = 0$).

As for the loading by applied surface tractions and applied couple stresses, one may consider them replaced by body forces f_k acting within a surface layer of a very small thickness δ initially proportional to D. In that case the preceding analysis applies, and the limit process $\delta \to 0$ proves very simply that (10.17) must also be valid for such loading.

The asymptotic size effect given by (10.17) is curiously strong. It is much stronger than that for similar LEFM cracks on the macroscale, which is $\sigma_N \propto D^{-1/2}$.

Definition of corresponding nominal stresses:

When the structure is not at maximum load but is hardening, one must decide which are the σ_N values that are comparable and should be described by the scaling law (10.17). In the small-size asymptotic field equation (10.16), if the value of $(\sigma_N/\sigma_Y)(D/l_\epsilon)^{2+p/q}$ is given, then parameter χ is a constant, and (if the problem is physically well posed) the partial differential equation (10.17) with homogeneous boundary conditions must have one solution $\bar{u}_k$, with the corresponding $\bar{\epsilon}_{ik}, \bar{\eta}_{kij}$. Hence, the dimensionless deformation field is the same for all sizes D. It follows that the σ_N values to which the scaling law (10.17) applies are those are those corresponding to the same norm of the relative (dimensionless) displacement, $\| \bar{u}_k \|$. The norm may for example be defined as the angle of twist of a cylinder, θ; or the maximum relative displacement $\bar{u}_{max}$ in the body; or the maximum strain in the body; or the relative depth of indentation $\bar{h} = h/D$; or the relative displacement at any homologous points in the body.

The asymptotic field equation for $D/l_\epsilon \to \infty$ is $\partial_i \bar{\sigma}_{ik} + \bar{f}_k \bar{\sigma}_N/\sigma_Y = 0$, in which $\bar{\sigma}_{ik} = (K/\sigma_Y)\delta_{ik}\bar{\epsilon}_{nn} + (\bar{\sigma}/\bar{\epsilon})\bar{\epsilon}'_{ik}$ and $\bar{\sigma} = f(\bar{\epsilon})$. Elimination of the stresses yields the field equation

$$\frac{K}{\sigma_Y}\delta_{ik}\partial_i\bar{\epsilon}_{nn}\partial_i\left(\frac{f(\bar{\epsilon})}{\bar{\epsilon}}\bar{\epsilon}'_{ik}\right) + \frac{\sigma_N}{\sigma_Y}\bar{f}_k = 0 \qquad (10.18)$$

Now again, if the ratio σ_N/σ_Y is given, then (for a problem properly posed physically) this partial differential equation with homogeneous boundary conditions must have one solution $\bar{\epsilon}_{ik}$, corresponding to one field $\bar{u}_k$. It follows that the comparable σ_N values for different sizes are those corresponding to the same norm of relative displacement, $\|\bar{u}_k\|$.

Strange small-scale asymptotic properties of existing theory:

By virtue of the fact that the hardening function $f(\epsilon)$ disappears from the from the field equation when it is reduced to its asymptotic form (10.16), it turns out that, for the theory of Gao et al. (1999), it is easy to determine the load-deflection curve when the displacement distribution (or relative displacement profile) remains constant during the loading process (this is for example typical of the pure torsion test of a long circular fiber, in which, by arguments of symmetry, the tangential displacements must vary linearly along every radius). For such loading, all the dimensionless displacements $\bar{u}_k$ at all the points in a structure of arbitrary but fixed geometry increase in proportion to a parameter w such that $\bar{u}_k = w\hat{u}_k$ where $\hat{u}_k$ is not only independent of D but also invariable during the proportional loading process.

Noting that η and Φ_{ijk} are homogeneous functions of degree 1 of both η and ϵ, we may write for such deformation behavior $\bar{\eta} = w\hat{\eta}$, $\bar{\Phi}_{ijk} = w\hat{\Phi}_{ijk}$ where $\hat{\eta}$ and $\hat{\Phi}_{ijk}$ are functions of dimensionless coordinates that do not change during the loading process at any small enough size D. Therefore, the asymptotic field equation (10.16) may be rewritten as

$$\partial_i\partial_j\left(\hat{\eta}^{\frac{p}{q}}\hat{\Phi}_{ijk}\right) = \chi\hat{f}_k, \quad \text{with} \quad \hat{f}_k = w^{-\left(1+\frac{p}{q}\right)}\bar{f}_k \qquad (10.19)$$

It follows from the field equation that if the relative displacement distribution (profile) $\hat{u}_k$ is constant during the loading process, as in torsion of a cylinder, then the distribution $\hat{f}_k$ must be constant as well.

At the small-size limit we also have $\bar{f}_k = f_k \sigma_N^0/\sigma_Y$, and using (10.19) we see that

$$f_k = w^{3/2}(\hat{f}_k \sigma_N^0 D) \qquad (10.20)$$

Since the expression in parenthesis is constant during loading, we must conclude that, for the small-size limit and for any loading with a constant relative displacement profile (as in torsion of a cylinder), the load-deflection curve is a power curve of exponent $1 + (p/q) > 1$; for Gao et al.'s theory, the exponent is 3/2 (Fig. 10.2).

Now it is hard to escape noticing that such a behavior is strange. The tangential stiffness of the structure is at infinitely small deflections zero, which is physically hard to accept, and then it increases with increasing deflection. An increase of tangential stiffness with increasing displacement is seen in locking materials (such as rubber or cellular materials), but would be queer as a property attributed to metals, even at the small-size continuum limit.

We included arbitrary positive exponents p and q in the definition of effective stress in order to check whether a change of p and q could remedy this problem. We see that the exponent of the power-law load-deflection curve can be made as close to 1 as desired but cannot be exactly 1. So a complete remedy cannot be achieved by modifying Gao et al. (1999a,b) semi-empirical definition of stress intensity σ.

Thus it seems that elimination of the strange asymptotic behavior of Gao et al.'s (1999) theory might perhaps necessitate some fundamental improvement of the theory. The D-values for which the small-size asymptotic behavior is approached are, of course, below the range of the theory. However, it is always preferable that all the asymptotic properties of a theory be reasonable.

Approximate asymptotic-matching formula for transitional size effect:

The theory of Gao et al. (1999a,b) characterizes the deviation from the classical plasticity as the structure size D becomes too small. But this formulation has only one-sided asymptotic support (on the scale of $\log D$). Ideally, one should seek a theory with a two-sided asymptotic support, having also realistic properties for the small-scale limit. Smooth formulae with such two-sided asymptotic support are generally called the asymptotic matching. Such formulae are applicable over the entire size range and have the potential of being more accurate than formulae with one-sided asymptotic support.

Even though the small-size asymptotic behavior of Gao et al.'s (1999a,b) theory seems questionable, we may use it to illustrate the construction of a simple asymptotic matching formula. Based on the established asymptotic properties, the broad-range transitional scaling law having both the classical macroscale plasticity and the gradient plasticity on the microscale as its asymptotes should be approximately describable by a smooth function approaching (10.17) for $D/l_\epsilon \to 0$ and $\sigma_N = $ const. for $D/l_\epsilon \to \infty$. This can be achieved by several simple formulae, and one of them is

$$\sigma_N = \sigma_0 \left[1 + \left(\frac{D_0}{D}\right)^{2s/r}\right]^{r/2}, \quad s = 2 + \frac{p}{q}, \tag{10.21}$$

$$\sigma_0 = \alpha_0 \sigma_Y \chi \lambda_1^{p/q}, \quad D_0 = \alpha_0^{-1/s} l_\epsilon \tag{10.22}$$

where α_0 and r are dimensionless constants which need to be determined either experimentally or by a numerical solution of the boundary value problem of gradient plasticity for an intermediate size D. In the plot of $\log \sigma_N$ versus $\log D$, the transitional size D_0 represents the intersection of the straight large-size and small-size asymptotes. A similar approach can be used to construct an asymptotic matching formula once an improved theory with a more realistic small-size asymptotic behavior is formulated.

Tests of Micro-Torsion and Micro-Hardness:

One case for which an explicit formula in terms of an integral has been obtained is the circular fiber of radius D subjected to torque T; see Eq. 35 in Huang et al. (1999). After transformation to dimensionless coordinates, that formula (for $p = 1$ and $q = 1/2$) reads:

$$\sigma_N = \frac{T}{D^3} = \sigma_Y \frac{2\pi \bar{\kappa}}{3} \int_0^1 \left\{ \frac{\bar{\sigma}}{\bar{\epsilon}} \left(\rho^2 + \frac{l_\epsilon^2}{12 D^2}\right) + \frac{l_\epsilon^2 f(\bar{\epsilon}) f'(\bar{\epsilon})}{12 D^2 \bar{\sigma}} \right\} \rho d\rho \tag{10.23}$$

where $\bar{\kappa} = \bar{\eta} = \kappa D$ = dimensionless specific angle of twist, κ = actual specific angle of twist (rotation angle per unit length of fiber). By taking the limit of $\sigma_N D^{5/2}$ for $D \to 0$, with σ_N given by the foregoing expression, one may be readily check that the small-size asymptotic form of this formula is

$$\sigma_N = \sigma_Y \left(\bar{\kappa}^{3/2} l_\epsilon^2 \sqrt{\lambda_1 l_\epsilon} \frac{\pi}{18} \int_0^1 \frac{\rho}{\bar{\epsilon}} d\rho\right) D^{-5/2} \tag{10.24}$$

This verifies our previous result (10.17).

By optimal fitting of this formula to several numerical values of (10.21) one could obtain parameters D_0, σ_0 and r appropriate for the case of torsion. Gao et al. (1999b, Fig. 6) compared this formula to torsional tests of fibers of diameters ranging from 12 μm to 170 μm. They achieved good agreement (except that the predicted stress-deformation curve for the smallest size was rising at about double the slope of the data, which might be an indication of a transition to the locking behavior that characterizes the small scale asymptotic behavior according to (10.17)).

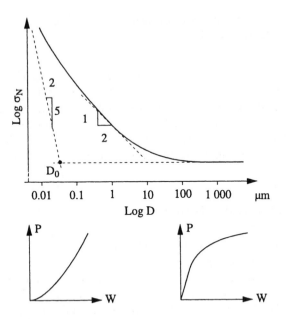

Figure 10.2: (a) Size effect of the existing dislocation-based theory of strain gradient plasticity, and (b) strange load-deflection curve for the small-size limit implied by the existing dislocation-based theory of gradient plasticity for bodies deforming with a constant relative displacement profile

Gao et al. (1999b) have further shown that the test results for Rockwell micro-hardness tests of copper can be well approximated as $\sigma_N = H_0\sqrt{1 + h^*/D}$ (Fig. 10.2a) where H_0 and h^* are constants and σ_N now stands for the hardness (stress average over the indentation area, denoted by Gao as H) and D is taken as the depth of penetration of the diamond cone (denoted by Gao as h). This test has the advantage that the situations at different depth of penetration of the cone are self-similar.

For small D, the foregoing formula has the asymptotic behavior $\sigma_N \approx \sqrt{h^*/D}$, which apparently contradicts our results in (10.17). A closer look, however, suggests that there need not be any contradiction. The test data used were of a very limited size range, ranging from 0.15 μm to 6 μm. This testing range is quite narrow, given that the transition from the large size to the small size asymptotic behavior might be spread over a much broader range of D, perhaps from from 0.005 μm to 100 μm. In the plot of $\log \sigma_N$ versus $\log D$, the asymptotic matching formula has a slope gradually decreasing from $-5/2$ (if $p/q = 1/2$) at $D = 0.005$ μm to 0 at $D = 100$ μm. Within the aforementioned range of the micro-hardness tests of copper, the curve of the formula is almost straight and has the slope of about $-1/2$ (Fig. 10.2a).

10.2. Scaling of Original Phenomenological Theory of Strain-Gradient Plasticity

The development of strain-gradient plasticity for micrometer scale was pioneered by Fleck and Hutchinson (1993). Their first version, called the couple-stress theory (CS theory), was later improved under the name of *stretch and rotation gradients theory* (SG theory) (Fleck and Hutchinson 1997). In this section, which closely follows the exposition in Bažant and Guo (2002a), the scaling of this, by now classical, theory, will be examined.

Fleck and Hutchinson's Formulation

In CS and SG theories, the strain energy density W is assumed to depend on the strain gradient tensor η of components $\eta_{ijk} \equiv u_{k,ij}$ as well as the linearized strain tensor ϵ of components $\epsilon_{ij} \equiv \frac{1}{2}(u_{i,j} + u_{j,i})$ (attention is here restricted to small strains). This assumption comes from the classical work of Toupin (1962) and Mindlin (1965), confined to elastic behavior, and is expressed as

$$W = \frac{1}{2}\lambda \epsilon_{ii}\epsilon_{jj} + \mu \epsilon_{ij}\epsilon_{ij} + a_1 \eta_{ijj}\eta_{ikk} + a_2 \eta_{iik}\eta_{kjj}$$
$$+ a_3 \eta_{iik}\eta_{jjk} + a_4 \eta_{ijk}\eta_{ijk} + a_5 \eta_{ijk}\eta_{kji} \quad (10.25)$$

where λ and μ are the normal Lamé constants and a_n are additional elastic stiffness constants of the material. As in the classical theory, Cauchy stress σ_{ij} is defined as $\partial W/\partial \epsilon_{ij}$, and is work-conjugate to ϵ_{ij}. Furthermore, a higher-order stress tensor τ, work-conjugate to the strain gradient tensor η, is defined as $\tau_{ijk} = \partial W/\partial \eta_{ijk}$. If W is defined by (10.25), the constitutive relation is of course linear. So (10.25) is appropriate to linear isotropic elastic materials only. To extend it to general nonlinear elastic materials, a new variable, an invariant named combined strain quantity, $\mathcal{E}$, is introduced by Fleck and Hutchinson (1997); it is defined as a function of both the strain tensor and the strain gradient tensor, while the strain energy density W is assumed, for general nonlinear elastic material, to be a nonlinear function of $\mathcal{E}$.

To define $\mathcal{E}$, Fleck and Hutchinson (1997) decompose the strain gradient tensor η into its hydrostatic part η^H and deviatoric part η';

$$\eta^H_{ijk} \equiv \frac{1}{4}(\delta_{ik}\eta_{jpp} + \delta_{jk}\eta_{ipp}); \quad \eta' = \eta - \eta^H \quad (10.26)$$

To simplify the problem, only incompressible materials are considered in the modeling of metals, in which case $\epsilon'_{ij} = \epsilon_{ij}$ and $\eta^H_{ijk} = 0$ (which implies deviatoric strain gradient $\eta'_{ijk} = \eta_{ijk}$). Furthermore, Fleck and Hutchinson (1997) introduce the orthogonal decomposition

$$\boldsymbol{\eta}' = \boldsymbol{\eta}'^{(1)} + \boldsymbol{\eta}'^{(2)} + \boldsymbol{\eta}'^{(3)} \tag{10.27}$$

in which the three tensors are defined in the component form as

$$\eta_{ijk}'^{(1)} = \eta_{ijk}'^{S} - \frac{1}{5}\left(\delta_{ij}\eta_{kpp}'^{S} + \delta_{jk}\eta_{ipp}'^{S} + \delta_{ki}\eta_{jpp}'^{S}\right) \tag{10.28}$$

$$\eta_{ijk}'^{(2)} = \frac{1}{6}\left(e_{ikp}e_{jlm}\eta_{lpm}' + e_{jkp}e_{ilm}\eta_{lpm}' + 2\eta_{ijk}' - \eta_{jki}' - \eta_{kij}'\right) \tag{10.29}$$

$$\eta_{ijk}'^{(3)} = \frac{1}{6}\left(-e_{ikp}e_{jlm}\eta_{lpm}' - e_{jkp}e_{ilm}\eta_{lpm}' + 2\eta_{ijk}' - \eta_{jki}' - \eta_{kij}'\right)$$
$$+ \frac{1}{5}\left(\delta_{ij}\eta_{kpp}'^{S} + \delta_{jk}\eta_{ipp}'^{S} + \delta_{ki}\eta_{jpp}'^{S}\right) \tag{10.30}$$

Here $\boldsymbol{\eta}'^{S}$ is a fully symmetric tensor defined as

$$\eta_{ijk}'^{S} = \frac{1}{3}\left(\eta_{ijk}' + \eta_{jki}' + \eta_{kij}'\right) \tag{10.31}$$

Using the foregoing three tensors $\boldsymbol{\eta}'^{(i)}$, Fleck and Hutchinson (1997) define the combined strain quantity $\mathcal{E}$ as

$$\mathcal{E}^2 = \frac{2}{3}\epsilon_{ij}'\epsilon_{ij}' + \ell_1^2 \eta_{ijk}'^{(1)}\eta_{ijk}'^{(1)} + \ell_2^2 \eta_{ijk}'^{(2)}\eta_{ijk}'^{(2)} + \ell_3^2 \eta_{ijk}'^{(3)}\eta_{ijk}'^{(3)} \tag{10.32}$$

where ℓ_i are three length constants which are given different values in the CS and SG theories (which is the only major difference between these two theories):

$$\text{For CS:} \quad \ell_1 = 0, \quad \ell_2 = \frac{1}{2}\ell_{CS}, \quad \ell_3 = \sqrt{\frac{5}{24}}\ell_{CS} \tag{10.33}$$

$$\text{For SG:} \quad \ell_1 = \ell_{CS}, \quad \ell_2 = \frac{1}{2}\ell_{CS}, \quad \ell_3 = \sqrt{\frac{5}{24}}\ell_{CS} \tag{10.34}$$

Here ℓ_{CS} is called the material characteristic length.

Based on the combined strain quantity $\mathcal{E}$ as defined, the strain energy density W can be defined as a function of $\mathcal{E}$ instead of ϵ. Then Cauchy stress tensor $\boldsymbol{\sigma}$ and the higher-order stress tensor $\boldsymbol{\tau}$ (couple stress tensor) can be expressed as:

$$\sigma_{ik} = \frac{\partial W}{\partial \epsilon_{ik}} = \frac{\mathrm{d}W}{\mathrm{d}\mathcal{E}}\frac{\partial \mathcal{E}}{\partial \epsilon_{ik}} \tag{10.35}$$

$$\tau_{ijk} = \frac{\partial W}{\partial \eta_{ijk}} = \frac{\mathrm{d}W}{\mathrm{d}\mathcal{E}}\frac{\partial \mathcal{E}}{\partial \eta_{ijk}} \tag{10.36}$$

Using (10.32) and the condition of incompressibility, one has

$$\frac{\partial \mathcal{E}}{\partial \epsilon_{ik}} = \frac{2\epsilon_{ik}}{3\mathcal{E}} \tag{10.37}$$

$$\frac{\partial \mathcal{E}}{\partial \eta_{ijk}} = \frac{1}{\mathcal{E}}\left(\ell_1^2 \eta_{lmn}^{\prime(1)} \frac{\partial \eta_{lmn}^{\prime(1)}}{\partial \eta_{ijk}} + \ell_2^2 \eta_{lmn}^{\prime(2)} \frac{\partial \eta_{lmn}^{\prime(2)}}{\partial \eta_{ijk}} + \ell_3^2 \eta_{lmn}^{\prime(3)} \frac{\partial \eta_{lmn}^{\prime(3)}}{\partial \eta_{ijk}}\right)$$

$$= \frac{\ell_{CS}^2 C_{ijkmnl}\eta_{mnl}}{\mathcal{E}} \tag{10.38}$$

where C_{ijkmnl} is a six-dimensional constant dimensionless tensor which could be determined from (10.28), (10.29), (10.30) and (10.38). Obviously, the CS and SG theories will be characterized by different tensors C, although, for each of them, tensor C is constant, that is, independent of ϵ, η and ℓ_{CS}.

For the sake of simplicity, the following power law is assumed for the strain energy density W (Fleck and Hutchinson 1997):

$$W = \frac{n}{n+1}\Sigma_0 \mathcal{E}_0 \left(\frac{\mathcal{E}}{\mathcal{E}_0}\right)^{(n+1)/n} \tag{10.39}$$

where Σ_0, $\mathcal{E}_0$ and the strain hardening exponent n are taken to be material constants (and, for hardening materials, $n \geq 1$; typically $n \approx 2$ to 5). Thus (10.35) and (10.36) yield the constitutive relations

$$\sigma_{ik} = \frac{2}{3}\Sigma_0 \left(\frac{1}{\mathcal{E}_0}\right)^{1/n} \mathcal{E}^{(1-n)/n}\epsilon_{ik} \tag{10.40}$$

$$\tau_{ijk} = \Sigma_0 \left(\frac{1}{\mathcal{E}_0}\right)^{1/n} \ell_{CS}^2 \mathcal{E}^{(1-n)/n} C_{ijklmn}\eta_{lmn} \tag{10.41}$$

The principle of virtual work yields the following field equations of equilibrium (Fleck and Hutchinson 1997):

$$\sigma_{ik,i} - \tau_{ijk,ij} + f_k = 0 \tag{10.42}$$

Dimensionless Variables

To analyze scaling, conversion to dimensionless variables (labeled by an overbar) is needed. Among many possible sets of such variables, the following will be convenient:

$$\bar{x}_i = x_i/D, \quad \bar{u}_i = u_i/D, \quad \bar{\epsilon}_{ij} = \epsilon_{ij},$$

$$\bar{\eta}_{ijk} = \eta_{ijk} D, \quad \bar{f}_k = f_k D/\sigma_N \tag{10.43}$$

$$\bar{\tau}_{ijk} = \tau_{ijk}/(\Sigma_0 \ell_{CS}), \quad \bar{\sigma}_{ik} = \sigma_{ik}/\Sigma_0, \quad \bar{\mathcal{E}} = \mathcal{E} \tag{10.44}$$

$$\bar{\eta}'^{(l)}_{ijk} = \eta'^{(l)}_{ijk} D \quad \text{where} \quad l = 1, 2, 3 \tag{10.45}$$

Here D is the characteristic length of the structure and σ_N is the nominal strength. Using these dimensionless variables, the constitutive law of the SGP theory can be rewritten as:

$$\bar{\sigma}_{ik} = \frac{2}{3}\left(\frac{1}{\mathcal{E}_0}\right)^{1/n} \bar{\mathcal{E}}^{(1-n)/n} \bar{\epsilon}_{ik} \tag{10.46}$$

$$\bar{\tau}_{ijk} = \left(\frac{1}{\mathcal{E}_0}\right)^{1/n} \frac{\ell_{CS}}{D} C_{ijklmn} \bar{\mathcal{E}}^{(1-n)/n} \bar{\eta}_{lmn} \tag{10.47}$$

The field equations of equilibrium transform as

$$\partial_i \bar{\sigma}_{ik} - \frac{\ell_{CS}}{D} \partial_i \partial_j \bar{\tau}_{ijk} + \frac{\sigma_N}{\Sigma_0} \bar{f}_k = 0 \tag{10.48}$$

where $\partial_i = \partial/\partial \bar{x}_i$ = derivatives with respect to the dimensionless coordinates. After substituting (10.46) and (10.47) into (10.48), we obtain the differential equations of equilibrium in the form:

$$\frac{2}{3}\left(\frac{1}{\mathcal{E}_0}\right)^{1/n} \partial_i \left(\bar{\mathcal{E}}^{(1-n)/n} \bar{\epsilon}_{ik}\right)$$

$$- \left(\frac{\ell_{CS}}{D}\right)^2 \left(\frac{1}{\mathcal{E}_0}\right)^{1/n} \partial_i \partial_j \left(C_{ijklmn} \bar{\mathcal{E}}^{(1-n)/n} \bar{\eta}_{lmn}\right) = -\frac{\sigma_N}{\Sigma_0} \bar{f}_k \tag{10.49}$$

To avoid struggling with the formulation of the boundary conditions, consider first that they are homogeneous, i.e., the applied surface tractions and applied couple stresses vanish at all parts of the boundary where the displacements are not fixed as 0. All the loading characterized by nominal stress σ_N is applied as body forces f_k whose distributions are assumed to be geometrically similar; σ_N is considered as the parameter of these forces, all of which vary proportionally to σ_N. Then the transformed boundary conditions are also homogeneous. In terms of the dimensionless coordinates, the boundaries of geometrically similar structures of different sizes are identical.

When the structure is not at maximum load but is hardening, one must decide which are the σ_N values that are comparable. What is meaningful is to compare structures of different sizes for the same dimensionless displacement field $\bar{u}_k$. Thus, the comparable structures will have same $\bar{\epsilon}_{ik}$ and $\bar{\eta}_{ijk}$.

Scaling and Size Effect

The problem of scaling and size effect can now be fully discussed. The limit $D/\ell_{CS} \to \infty$ is simple because the dimensionless third-order stresses $\bar{\tau}_{ijk}$ vanish and all the equations reduce to the standard field equations of equilibrium on the macroscale. The combined strain quantity $\mathcal{E}$ reduces to the classical effective strain, and (10.39) becomes the normal strain energy density function.

The opposite asymptotic behavior for $D/\ell_{CS} \to 0$ is a little more complex. From (10.32) we know that when $D/\ell_{CS} \to 0$,

$$\bar{\mathcal{E}} = \sqrt{\frac{2}{3}\epsilon'_{ij}\epsilon'_{ij} + \frac{1}{D^2}\left(\ell_1^2 \bar{\eta}'^{(1)}_{ijk}\bar{\eta}'^{(1)}_{ijk} + \ell_2^2 \bar{\eta}'^{(2)}_{ijk}\bar{\eta}'^{(2)}_{ijk} + \ell_3^2 \bar{\eta}'^{(3)}_{ijk}\bar{\eta}'^{(3)}_{ijk}\right)} \propto D^{-1} \quad (10.50)$$

It is useful to define another dimensionless variable as follows:

$$\bar{H} = \frac{1}{\ell_{CS}}\sqrt{\ell_1^2 \bar{\eta}'^{(1)}_{ijk}\bar{\eta}'^{(1)}_{ijk} + \ell_2^2 \bar{\eta}'^{(2)}_{ijk}\bar{\eta}'^{(2)}_{ijk} + \ell_3^2 \bar{\eta}'^{(3)}_{ijk}\bar{\eta}'^{(3)}_{ijk}} \quad (10.51)$$

Obviously $\bar{H}$ is independent of size D, and we have

$$\bar{\mathcal{E}} \approx \frac{\ell_{CS}}{D}\bar{H} \quad \text{when} \quad D/\ell_{CS} \to 0 \quad (10.52)$$

After substituting (10.52) into (10.49), we obtain the differential equations of equilibrium in the form:

$$\frac{2}{3}\left(\frac{\ell_{CS}}{D}\right)^{(1-n)/n}\left(\frac{1}{\mathcal{E}_0}\right)^{1/n}\partial_i\left(\bar{H}^{(1-n)/n}\bar{\epsilon}_{ik}\right)$$
$$-\left(\frac{\ell_{CS}}{D}\right)^{(1+n)/n}\left(\frac{1}{\mathcal{E}_0}\right)^{1/n}\partial_i\partial_j\left(C_{ijklmn}\bar{H}^{(1-n)/n}\bar{\eta}_{lmn}\right)$$
$$= -\frac{\sigma_N}{\Sigma_0}\bar{f}_k \quad (10.53)$$

Now we multiply this equation by $(D/\ell_{CS})^{(n+1)/n}$ and take the limit of the left-hand side for $D \to 0$. This leads to the following asymptotic form of the field equations:

$$\partial_i\partial_j\left(C_{ijklmn}\bar{H}^{(1-n)/n}\bar{\eta}_{lmn}\right) = \chi \bar{f}_k, \quad \text{with} \quad \chi = \bar{\mathcal{E}}_0^{\frac{1}{n}}\frac{\sigma_N}{\Sigma_0}\left(\frac{D}{\ell_{CS}}\right)^{\frac{n+1}{n}} \quad (10.54)$$

Since D is absent from the foregoing field equation (and from the boundary conditions, too, because they are homogeneous), the dimensionless displacement field as well as the parameter χ must be size independent. Thus we obtain the following small-size asymptotic scaling law for Fleck and Hutchinson's theories of gradient plasticity:

$$\sigma_N = \Sigma_0 \, \chi \bar{\mathcal{E}}_0^{-1/n} \left(\frac{\ell_{CS}}{D} \right)^{(n+1)/n} \qquad (10.55)$$

or

$$\sigma_N \propto D^{-(n+1)/n} \qquad (10.56)$$

For hardening materials, we have $1 < (n+1)/n \leq 2$.

Since the surface loads may be regarded as the limit case of body forces applied within a very thin surface layer, the same scaling law must also apply when the load is applied at the boundaries.

Although the result (10.56) applies only to the special case of strain energy density function (10.39), the same analytical technique can be used for general strain energy functions.

Equation (10.56) indicates that the asymptotic behavior on the microscale depends on the hardening relation on the macroscale since the macro-strain-hardening exponent n is involved. Moreover, the asymptotic behavior depends only on that exponent. Generally, the present technique can be used for any strain energy function defined in terms of the strain and strain gradient tensors, even if no combined strain quantity were defined.

For example, a similar technique can also be used for the strain energy density function (10.25) defined for linear isotropic elastic material for which the combined strain quantity is not used. The constitutive relation in that case is:

$$\sigma_{ik} = \frac{\partial W}{\partial \epsilon_{ik}} = \lambda \delta_{ik} \epsilon_{ll} + 2\mu \epsilon_{ik} = (2\mu \delta_{il} \delta_{km} + \lambda \delta_{ik} \delta_{lm}) \, \epsilon_{lm}$$

$$= D_{iklm} \epsilon_{lm}$$

$$\tau_{ijk} = \frac{\partial W}{\partial \eta_{ijk}} = [2a_1 \delta_{il} \delta_{jk} \delta_{lm} + a_2 (\delta_{in} \delta_{jk} \delta_{lm} + \delta_{ij} \delta_{kl} \delta_{mn}) \qquad (10.57)$$

$$+ 2a_3 \delta_{ij} \delta_{lm} \delta_{kn} + 2a_4 \delta_{il} \delta_{jm} \delta_{kn} + 2a_5 \delta_{in} \delta_{jm} \delta_{kl}] \eta_{lmn}$$

$$= C'_{ijklmn} \eta_{lmn} \qquad (10.58)$$

where the dimension of tensor D_{iklm} is that of a stress, and the dimension of tensor C'_{ijklmn} is that of a force. The dimensionless variables defined in (10.43) can again be adopted. In terms of the dimensionless variables, the differential equations of equilibrium are:

$$\frac{1}{D} \partial_i (D_{iklm} \bar{\epsilon}_{lm}) - \frac{1}{D^3} \partial_i \partial_j (C'_{ijklmn} \bar{\eta}_{lmn}) + \frac{1}{D} \bar{f}_k \sigma_N = 0 \qquad (10.59)$$

For sufficiently large D, the term with $\bar{\eta}$ will vanish and (10.59) will become the classical differential equation of equilibrium. For the opposite case, for which

D is sufficiently small, the term with $\bar{\epsilon}$ will vanish, and we get the following asymptotic behavior:

$$\sigma_N \propto D^{-2} \qquad (10.60)$$

Of course, (10.60) is a special case of (10.56) because, for a linear material, the strain hardening exponent $n = 1$.

Examples

It is instructive to verify the scaling law in (10.56) for the basic types of experiments. One important test is that of micro-torsion. It was initially the size effect in this test (Fleck et al. 1997) which motivated the development of gradient plasticity.

An effective stress measure Σ may be defined as the work conjugate to $\mathcal{E}$:

$$\Sigma = \frac{\mathrm{d}W(\mathcal{E})}{\mathrm{d}\mathcal{E}} \qquad (10.61)$$

A simple power law relationship between Σ and $\mathcal{E}$ may be adopted;

$$\Sigma = \Sigma_0 \mathcal{E}^N \qquad (10.62)$$

Compared with (10.39), one sees that $N = 1/n$. For the analysis of size effect, the radius of the wire, D, may be chosen as the characteristic dimension (size). The deformation is characterized by the twist per unit length, κ.

For geometrically similar structures of different sizes, we compare the nominal stresses corresponding to the same dimensionless twist $\bar{\kappa} = \kappa D$. The nominal stress σ_N may be defined as T/D^3, where T is the torque. If the CS theory is used, one has

$$\sigma_N = \frac{T}{D^3} = \frac{6\pi}{N+3}\Sigma_0 \bar{\kappa}^N \left\{ \left[\frac{1}{3} + \left(\frac{\ell_{CS}}{D}\right)^2\right]^{(N+3)/2} - \left(\frac{\ell_{CS}}{D}\right)^{N+3} \right\} \qquad (10.63)$$

For $\ell_{CS}/D \to \infty$,

$$\left[\frac{1}{3} + \left(\frac{\ell_{CS}}{D}\right)^2\right]^{(N+3)/2} - \left(\frac{\ell_{CS}}{D}\right)^{N+3} \approx \frac{N+3}{2}\left(\frac{D}{3\ell_{CS}}\right)^2 \left(\frac{\ell_{CS}}{D}\right)^{N+3} \qquad (10.64)$$

from which

$$\sigma_N \propto D^{-N-1} = D^{-(n+1)/n} \qquad (10.65)$$

Small-Size Asymptotic Load-Deflection Response

For some special cases such as the pure torsion of a long thin wire or the bending of a slender beam, the displacement distribution can be figured out by the arguments of symmetry and the relative displacement profile remains constant during the loading process. For such problems, the asymptotic load-deflection curve for a very small size D can be determined very easily (Bažant 2001, 2002).

For such loading, all the dimensionless displacements $\bar{u}_k$ at all the points in a structure of a fixed geometry increase in proportion to one parameter, w, such that $\bar{u}_k = w\hat{u}_k$ where $\hat{u}_k$ is not only independent of D but also invariable during the proportional loading process. Parameter w may be defined as the displacement norm, $w = \| \bar{u}_k \|$.

From (10.31), (10.28), (10.29), (10.30) and (10.32), we know that

$$\bar{\mathcal{E}} = w\hat{\mathcal{E}} \tag{10.66}$$

where $\hat{\mathcal{E}}$ is a function of dimensionless coordinates that does not change during the loading process when size D is very small. Substituting this into the dimensionless constitutive law (10.46) and (10.47), we have

$$\bar{\sigma}_{ik} = w^{1/n}\hat{\sigma}_{ik}, \quad \bar{\tau}_{ijk} = w^{1/n}\hat{\tau}_{ijk} \tag{10.67}$$

where $\hat{\sigma}_{ik}$ and $\hat{\tau}_{ijk}$ are constant during the loading process if the size D is small enough. Thus the load f_k can now be expressed as a function of w as well as the size D. Since the first two terms on the left-hand side of (10.48) are proportional, respectively, to the functions

$$w^{1/n}D^{(n-1)/n}, \quad w^{1/n}D^{-(n+1)/n} \tag{10.68}$$

one reaches the conclusion that

$$\bar{f}_k \propto w^{1/n} \tag{10.69}$$

For hardening materials, we have $1/n \leq 1$. So the load deflection curve begins with a vertical tangent, similar to the MSG theory. We conclude that the small-size asymptotic load-deflection response is similar to the stress–strain relation for the macroscale, except that the initial elastic response gets wiped out when $D \to 0$.

10.3. Scaling of Strain-Gradient Generalization of Incremental Plasticity

Finally, a brief look at the theory of Acharya and Bassani (2000) and Bassani (2001) is appropriate. These authors developed a simple gradient theory which differs significantly from the previous four theories. It is a generalization of the classical incremental theory of macro-scale plasticity, rather than the deformation (total strain) theory. In contrast to the previous four theories, in which the strain gradient tensor η is defined as a third-order tensor representing the gradient of total strain, the lattice incompatibility is measured by a second-order tensor defined by the following contraction of the gradient of plastic strain ϵ_{ij}^p:

$$\alpha_{ij} = e_{jkl}\epsilon_{il,k}^p \tag{10.70}$$

where e_{jkl} is the alternating symbol. The plastic hardening is assumed to be governed by the invariant:

$$\alpha = \sqrt{2\alpha_{ij}\alpha_{ji}} \tag{10.71}$$

Then the basic equations of the classical J_2 flow theory are modified as follows:

$$\tau = \sqrt{\frac{\sigma'_{ij}\sigma'_{ij}}{2}} = \tau_{cr}, \quad \dot{\tau} = \dot{\tau}_{cr} = h(\gamma^p, \alpha)\dot{\gamma}^p \tag{10.72}$$

$$\dot{\epsilon}_{ij}^p = \left(\frac{\dot{\gamma}^p}{2\tau}\right)\sigma'_{ij}, \quad \dot{\sigma}_{ij} = C_{ijkl}(\dot{\epsilon}_{kl} - \dot{\epsilon}_{kl}^p), \quad \gamma^p = \sqrt{\frac{2}{3}\epsilon_{ij}^p\epsilon_{ij}^p} \tag{10.73}$$

The variables used in the above are almost the same as in the classical J_2 flow theory except that the instantaneous hardening-rate function h depends not only on plastic strain invariant γ^p but also on α. The following hardening function $h(\gamma^p, \alpha)$ is used by Bassani (2001) for numerical simulation of the micro-torsion test:

$$h(\gamma^p, \alpha) = h_0\left(\frac{\gamma^p}{\gamma_0} + 1\right)^{N-1}\left[1 + \frac{l^2(\alpha/\gamma_0)^2}{1 + c(\gamma^p/\gamma_0)^2}\right]^{1/2} \tag{10.74}$$

where l is a material length introduced for dimensionality reasons, and h_0, γ_0, c and N are further material constants (all positive).

Although a full analysis of scaling of this theory is beyond the scope of this paper, some simple observations can be made. From the scale transformations $\bar{u}_i = u_i/D$, $\bar{\epsilon}_{il,k} = \epsilon_{il,k}D$, it follows that $\alpha_{ij} = \bar{\alpha}_{ij}/D$, where the overbars again denote the dimensionless variables. Since $\bar{\gamma}^p = \gamma^p$, the plastic hardening modulus defined by (10.74) scales for $D \to 0$ as

$$h(\gamma^p, \alpha) = h_0\left(\frac{\gamma^p}{\gamma_0} + 1\right)^{N-1}[1 + c(\gamma^p/\gamma_0)^2]^{-1/2}\frac{\bar{\alpha}}{\gamma_0}\frac{l}{D} \propto D^{-1} \tag{10.75}$$

This means that, at the same strain level, the slope of the plastic hardening curve increases as D^{-1} when $D \to 0$. When the plastic strain becomes much larger than the elastic strain, and when the strain distributions and history are similar, then of course the nominal stress σ_N must also scale asymptotically as D^{-1}. This is again a curiously strong asymptotic size effect, not much less strong than that found for the MSG and CS theories. Even though this excessive size effect is approached only outside the range of applicability of the theory, one must expect that it would impair the representation of test data in the middle range.

The excessive asymptotic size effect could be avoided by redefining the plastic hardening modulus in (10.74) as follows:

$$h(\gamma^p, \alpha) = h_0 \left(\frac{\gamma^p}{\gamma_0} + 1\right)^{N-1} \left[1 + \frac{l\alpha/\gamma_0}{1 + c(\gamma^p/\gamma_0)}\right]^{1/2} \qquad (10.76)$$

With this revision, which should be checked against test data, the asymptotic scaling would become

$$h(\gamma^p, \alpha) \propto D^{-1/2} \quad \text{when } D \to 0 \qquad (10.77)$$

which seems more reasonable and similar to Bažant's (2000) proposal for revision of MSG theory, as well as to the TNT theory.

10.4. Closing Observations

The small-size asymptotic scaling laws for the nominal stress σ_N in all the existing theories are power laws, but there are wide disparities among them.

For Gao et al.'s (1999) MSG theory, in general, $\sigma_N \propto D^{-5/2}$, which is an unreasonably strong size effect. For two special cases of this theory, $\sigma_N \propto D^{-2}$ and $D^{-3/2}$, which is also very strong.

For the classical Fleck and Hutchinson (1963, 1997) CS and SG theories, $\sigma_N \propto D^{-(n+1)/n}$ where n is the exponent of the strain hardening law on the macroscale, and typically $\sigma_N \propto D^{-1.3}$, which is also quite strong. For the TNT theory, as well as for the modification of the MSG theory proposed by Bažant (2002, 2000), $\sigma_N \propto D^{-1/2}$, which seems reasonable.

The plastic hardening modulus in the theory of Acharya and Bassani (2000) scales asymptotically as D^{-1}, which also seems excessive. However, a simple modification can achieve the scaling of $D^{-1/2}$.

Although the small-size asymptotic behavior is closely approached only at sizes much smaller than the range of applicability of the strain-gradient theories of plasticity (which is about 0.1 μm–100 μm), the knowledge of this behavior is useful for developing asymptotic matching approximations for the realistic middle range.

Knowledge of both the large-size and small-size makes it possible to obtain good analytical approximations for the middle range of interest, agreeing with experimental as well as numerical results (see Bažant and Guo 2002a). The availability of such formulae means that the stress analysis for the middle range, which is much more difficult than the asymptotic analysis, can in principle be avoided. Such an approach is, of course, possible only if the small-size asymptotic behavior is realistic. The detrimental consequence of unrealistic small-size asymptotic properties is that the possibility of asymptotic matching approximations is lost.

Chapter 11

Future Perspectives

To close on a philosophical note, consider the gradual expansion of human knowledge (Fig. 11.1). What is known may be imagined to form a circle. The unknown is what lies outside. What can be discovered at any given stage of history is only what is in contact with the circle. Questions about what lies farther into the future, not in contact with the circle, cannot even be raised.

In our field, the problem of strength of elastic frames was not even posed before Hooke. It started to be tackled in the middle of the 19th century and has been for the most part solved around 1960.

One of the most formidable problems in physics and mathematics has been that of turbulent flow. It has occupied the best minds for over a century and, as experts say, complete understanding is not yet in sight. The problem of scaling in quasibrittle materials is a part of damage mechanics, in which serious

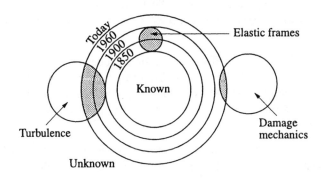

Figure 11.1: Damage mechanics in the perspective of the expansion of human knowledge

research started around 1960. Although much has been learned, it appears that damage mechanics is a formidable problem whose difficulty may be of the same dimension as turbulence. It will take a long time to resolve completely.

For the immediate future—and only such a view is possible now, the following is a sample of research directions that may be identified as necessary and potentially profitable:

1. Micromechanical basis of softening damage.

2. Physically justified nonlocal model (based on the interactions of cracks and inclusions).

3. Scaling of brittle compression fracture and shear fracture.

4. Scaling of fracture at interfaces (bond rupture).

5. Rate and load duration effects on scaling, and size effects in long-time fracture or fatigue.

6. Softening damage and scaling for large strains.

7. Size effect on ductility of softening structures, and on their energy absorption capability.

8. Acquisition of size effect test data for all kinds of quasibrittle materials, many of them high-tech materials (see the Introduction), and data for real structures of various types.

9. Statistical characteristics of the size effect due to energy release and stress redistribution during fracture.

10. Scaling problems in geophysics, e.g., earthquake prediction or ocean ice dynamics.

11. Downsize extrapolation of size effect into a range of reduced brittleness, which is of interest for miniature electronic components and micromechanical devices.

12. Incorporation of size effect into design procedures and code recommendations for concrete structures, geotechnical structures, fiber composites (e.g. for aircrafts and ships), nuclear power plants, ocean oil platforms, mining and drilling technology (especially rockburst and borehole breakout), etc.

13. Incorporation of extreme value statistics and the scaling of loads giving a specified extremely low probability of failure, such as 10^{-7} (Bažant 2001e).

Addendum

Most of the manuscript for the first printing was completed in 2000. Since then research in strength scaling has been progressing rapidly; recent advances are briefly described in this Addendum, and some additional references are added at the end of the bibliography.

A1. Size Effect Derivation by a New Method for Asymptotic Matching

Summary of asymptotic properties of cohesive crack model

A simpler, more general and more fundamental derivation of the scaling laws can be obtained by combining dimensional analysis with a new method of asymptotic matching of the known asymptotics of the cohesive crack model (Bažant 2004a). Asymptotic matching is a broad range of diverse techniques used for a long time in fluid mechanics (Barenblatt 1996, 2003), none of which, however, seems suited for fracture mechanics of solids.

The basic hypothesis is that the cohesive crack model represents a good enough compromise between reality and simplicity. The fracture process zone (FPZ) at the crack front is modeled as a fictitious line crack transmitting cohesive (crack-bridging) stresses $\sigma = f(w)$; w = crack opening (separation), $f(w)$ = monotonically decreasing (softening) function characterizing the material (Fig. A.1 bottom left), with $f(0) = f_t$ = tensile strength of material. The initial tangent of $f(w)$ is assumed to have a finite negative slope $f'(0)$, which is proportional to fracture energy G_f. In Chapter 2 and 9 it was shown that, depending on the first two non-zero terms of the asymptotic expansions of σ_N in terms of powers of D and of D^{-1}, there exist three and only three types of size effect (Bažant 2001):

$$\text{For } D \to 0: \quad \sigma_N = b_0 - c_0 D + \cdots \quad \text{(all types)} \quad (A1.1)$$

$$\text{For } D \to \infty: \quad \sigma_N = b_1 + c_1 D^{-1} + \cdots \quad \text{(type 1)} \quad (A1.2)$$

$$\sigma_N = D^{-1/2}[b_2 - c_2 D^{-1} + \cdots] \quad \text{(type 2)} \quad (A1.3)$$

$$\sigma_N = D^{-1/2}[b_3 - c_3 D^{-2} + \cdots] \quad \text{(type 3)} \quad (A1.4)$$

where $b_0, c_0, \ldots, c_3$ are positive constants determined by structure geometry. Type 1 (Fig. A.1 left) occurs if the geometry is such that P_{max} is reached at crack initiation from an FPZ attached to a smooth surface, i.e., as soon as the FPZ is fully formed. Type 2 (Fig. A.1 right) occurs if there is a large notch or preexisting stress-free (fatigued) crack and if the geometry is positive, i.e., such

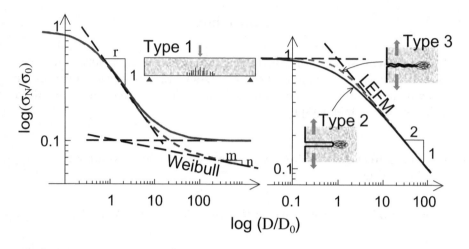

Figure A.1: Size effects of types 1, 2, and 3, and their asymptotes

that P_{max} occurs while the FPZ is still attached to the tip. Type 3 occurs if a large crack can grow stably prior to P_{max} (negative geometry). The size effect types 1 and 2 are very different, but types 2 and 3 are quite similar and hardly distinguishable in fracture testing (note that, in Chapter 9, type 1 was called Case 2, type 2 Case 1, and type 3 Case 3).

Asymptotic matching law for the case of large notch or crack

From Buckingham's Π-theorem of dimensional analysis (Barenblatt 1979), two special size effects have been proven in section 1.7:

(a) If the failure depends on f_t (dimension N/m²) but not G_f, then there is no size effect, i.e., σ_N = constant (this is so for all elasto-plastic failures).

(b) If the failure depends on G_f (dimension J/m²) but not on f_t, then there is a size effect of the type $\sigma_N D^{-1/2}$ (this is so for LEFM, if the cracks or notches are geometrically similar).

Nothing more can be deduced from dimensional analysis alone.

We can, however, deduce more information upon considering the physical meaning of the material characteristic length governing the FPZ size, $l_0 = EG_f/f_t^2$, introduced by Irwin (1958). The length and width of the FPZ for $D \to \infty$ are ηl_0 and $\eta' l_0$, in which η and η' are constants of the order of 1, depending on microstructure characteristics (note that the parameter $l_e = G_f/E$, which is also a length, is irrelevant for failure and controls only the ratio of deformation to stress intensity factor K_I in LEFM). Depending on the ratio D/l_0, two asymptotic cases may be distinguished:

(a) If $D/l_0 \to 0$, the body is much smaller than a fully developed FPZ. Then a derivative of potential energy with respect to the crack length makes no sense, and so G_f cannot matter. Therefore, the case of no size effect, $\sigma_N = $ constant, is the small-size asymptote.

(b) If $D/l_0 \to \infty$, the FPZ becomes a point in dimensionless coordinates $\xi = x/D$ and the stress field approaches the LEFM singularity. So, f_t cannot matter. Therefore, the LEFM scaling, $\sigma_N \propto D^{-1/2}$, is the large-size asymptote of quasibrittle failure (which is represented by a straight line of slope $-1/2$ in the plot of $\log \sigma_N$ versus $\log D$).

The size effect curve for the intermediate sizes may be expected to be a gradual transition between these two asymptotes. The approximate form of this transition can be deduced upon noting how the asymptotes are approached, i.e., by exploiting the higher-order asymptotic terms in (A1.1) and (A1.3). By transformation to dimensionless variables in the formulation of the boundary value problem with cohesive crack, one concludes that (if the ratios of structural dimensions characterizing the structure geometry are fixed) σ_N depends on only three parameters, f_t, D and K_c, where, for plane stress states, $K_c = (EG_f)^{1/2} = $ mode I fracture toughness and $E = $ Young's modulus (for plane strain states, E needs to be replaced by $E' = E/(1-\nu^2)$ where $\nu = $ Poisson ratio).

Therefore, we have now four governing parameters, σ_N, D, f_t and K_c. Since they involve 2 independent physical dimensions (length and force), the Π-theorem (Barenblatt 1979) implies that there can be only $4-2$, i.e., 2 independent dimensionless parameters, Π_1 and Π_2. So, the equation governing failure may be written as $F(\Pi_1, \Pi_2) = 0$ where function F may be assumed to be sufficiently smooth. Although many diverse choices of Π_i ($i = 1, 2$) are possible, the *key idea* was to make a choice for which, *in each asymptotic case, all Π_i vanish except one* (Bažant 2004a). If consideration is limited to dimensionless monomials, this can be most generally achieved by choosing:

$$\Pi_1 = (\sigma_N/f_t)^p (D/l_0)^u, \quad \Pi_2 = (\sigma_N/f_t)^q (D/l_0)^v \tag{A1.5}$$

where p, q, u, v are four unknown real constants.

If we let $\Pi_1 = 0$ correspond to $D \to 0$, then $F(0, \Pi_2) = 0$. This implies that $\Pi_2 = $ constant or $\sigma_N{}^q D^v = $ constant for $D \to 0$, which must be the case of no size effect; hence $v = 0$.

If we let $\Pi_2 = 0$ correspond to $D \to \infty$, then $F(\Pi_1, 0) = 0$, which implies that $\Pi_1 = $ constant or $\sigma_N{}^p D^u = $ constant, or $\sigma_N \propto D^{-u/p}$ for $D \to \infty$. This must be the LEFM scaling; hence $u/p = 1/2$ or $u = p/2$.

To find p and q, we expand function F into Taylor series about the state $\Pi_1 = \Pi_2 = 0$, and truncate this series after the linear terms; i.e.,

$$F(\Pi_1, \Pi_2) \approx F_0 + F_1 \Pi_1 + F_2 \Pi_2 = 0 \quad (A1.6)$$

$$\text{or} \quad F_1(\sigma_N \sqrt{D}/f_t \sqrt{l_0})^p + F_2(\sigma_N/f_t)^q = -F_0 \quad (A1.7)$$

where $F_1 = [\partial F/\partial \Pi_1]_0$ and $F_2 = [\partial F/\partial \Pi_2]_0$ (evaluated at $\Pi_1 = \Pi_2 = 0$) and $F_0 = F(0,0)$ ($F_0, F_1, F_2 \neq 0$). Now we need to compare the last equation with the asymptotic expansion in (A1.3). But this cannot be done in general because, for general p/q, the last equation cannot be solved explicitly for σ_N. But it can be solved for D, which gives

$$D = l_0 f_t{}^2 (-F_0/F_1)^{2/p} \sigma_N{}^{-2} \left[1 + (F_2/F_0 f_t^q)\sigma_N{}^q\right]^{2/p} \quad (A1.8)$$

which may be compared to the inverse expansion of the large-size asymptotic expansion in (A1.3). This inverse expansion has the form

$$D = B_2 \sigma_N{}^{-2}[1 - C_2 \sigma_N{}^2 + \cdots] \quad \text{for } \sigma_N \to 0 \quad (A1.9)$$

where B_2 and C_2 are positive constants. Evidently, matching of the first two terms of this expansion requires that $p = q = 2$. Then Eq. (A1.7) can be solved for σ_N. This yields, and verifies, the size effect law, Eq. (2.8), i.e.,

$$\sigma_N = \sigma_0 (1 + D/D_0)^{-1/2} \quad (A1.10)$$

in which

$$Bf_t = f_t(-F_0 F_2)^{-1/2}, \quad D_0 = l_0 F_2/F_1 \quad (A1.11)$$

For $D \to 0$, Eq. (2.8) has the approximation:

$$\sigma_N \approx Bf'_t(1 - D/2D_0) \quad (A1.12)$$

which verifies that the form of the second term of small-size expansion in (A1.1) can be matched, too. So, Eq. (A1.10) is the simplest formula for type 2 size effect that can match the form of four asymptotic terms, two for $D \to 0$ and two for $D \to \infty$.

If, however, not only the form of the second-order terms but also the values of all four coefficients b_0, c_0, b_2, c_2 are given, a slightly more general formula with four adjustable parameters is needed. This formula was recently developed and examined in Bažant and Yu (2002), with the purpose of a identifying fracture energy G_f from the maximum load of notched and unnotched test specimens of relatively small sizes.

More complex formulas of the same asymptotic accuracy, but more flexible in data fitting, can be obtained by replacing (A1.5) with dimensionless polynomials (or other monotonic functions of Π_1 and Π_2). Such formulas can make a significant difference only if the size effect needs to be modeled for a size range exceeding about $1\frac{1}{2}$ order of magnitude of D. Their merit is that they can capture a spectrum of fracture energies (Bažant 2002f), each of which is associated with a different order of magnitude of D (see section 9.7).

The type 3 (Case 3) size effect formula (Fig. A.1 right) can be derived similarly.

Asymptotic matching law for failures at crack initiation (type 1 scaling)

The deterministic (type 1) size effect for failures at macrocrack initiation from a smooth surface necessitates a somewhat different approach. As explained in section 2.8 (Fig. 2.13), the source of this size effect is the stress redistribution due to a boundary layer of cracking (or formation of the FPZ at the surface), which is associated with energy release. The deterministic type 1 size effect is also called the strain gradient effect, but this is not quite accurate because, aside from the local strain gradient at the surface, the geometry of the whole structure also has some (albeit small) influence on the strain redistribution and energy release. This is clear from the alternative approach used in section 2.7, in which the type 1 failure has been treated as a limit case of finite crack for $a_0 \to 0$ (Bažant 1997, 1998), with the energy release rate approaching zero.

For general dimensional analysis, we need a slightly different procedure than for type 2. Let us consider the **deterministic** type 1 first ($m \to \infty$). Because, according to (A1.1) and (A1.2), both asymptotes in the plot of $\log \sigma_N$ versus $\log D$ must be horizontal, the size effect curve of σ_N versus $\log D$ must have an inflexion point. Thus it is convenient to postulate the existence of what is called an *intermediate asymptote* (Barenblatt 1979, 2003), which consists of some unknown power law.

Because the intermediate asymptote separates the large-size and small-size asymptotics, we must first match the large-size asymptotic terms alone (which was the approach taken in sections 2.7 and 9.4.2). We choose the large-size asymptote to correspond to $\Pi_1 = 0$, i.e., to $F(0, \Pi_2) = 0$. This requires that $\Pi_2 =$ constant for $D \to \infty$, and so $v = 0$ in (A1.5). We again truncate the Taylor series expansion of function $F(\Pi_1, \Pi_2)$ after the linear terms, i.e., use the linear approximation (A1.6). This equation cannot, in general, be solved for σ_N, but it can for D, which yields:

$$l_0/D = (-F_0/F_1)^{-1/u}[(f_t/\sigma_N)^p + (F_2/F_0)(f_t/\sigma_N)^{p-q}]^{-1/u} \quad (A1.13)$$

The large-size expansion (A1.2) may be generalized as $\sigma_N = (b_1 + rc_1 D^{-1} + \cdots)^{1/r}$ (r = arbitrary constant $\neq 0$). Although we cannot match the asymptotic expansion (A1.2) directly, we can match the inverse expansion, which can be shown to be

$$1/D = (-b_1 + \sigma_N{}^r + \cdots)/rc_1 \quad (A1.14)$$

Matching of (A1.13) obviously requires that $u = -1$ and $p = q = -r$ in (A1.5). So

$$\Pi_1 = (f_t/\sigma_N)^r l_0/D, \quad \Pi_2 = (f_t/\sigma_N)^r \quad (A1.15)$$

With these particular expressions, Eq. (A1.6) can be solved for σ_N. This yields, and thus also verifies, the type 1 size effect law

$$\sigma_N = f_r^0 (1 + r\lambda\ell/D)^{1/r} \quad \text{(type 1, } m \to \infty\text{)} \quad (A1.16)$$

(equivalent to Eq. 2.39), in which

$$\lambda = F_1/rF_2, \quad f_r^0 = f_t(-F_2/F_0)^{1/r} \quad (A1.17)$$

Second, consider the **statistical** generalization, i.e., Weibull modulus $m < \infty$. Instead of the horizontal asymptote (A1.1), the mean large-size asymptotic size effect (as explained in section 3.1, Eq. 3.6) is now given by the Weibull power law

$$\text{for } D \to \infty: \quad \sigma_N = k_0 D^{-n/m} + \cdots \quad \text{(type 1)} \quad (A1.18)$$

where m = Weibull modulus, k_0 = constant, and n = number of dimensions in which the structure is scaled ($n = 1, 2, 3$); typically, $m = 10$ to 40 and $n/m \ll 1$. In the plot of $\log \sigma_N$ versus $\log D$, this power law represents an inclined straight line asymptote of slope $-n/m$. The Weibull power law size effect (A1.18) must apply for failures at crack initiation if the FPZ is negligible compared with D.

We again choose the large size asymptote to correspond to $\Pi_1 = 0$ and Π_2 = constant. To match (A1.18), it is necessary that $\sigma_N = C_1 D^{-n/m} \propto f_t(D/l_0)^{-v/q}$, and so, $v/q = n/m$. If we should not loose the deterministic limit already figured out, we must keep $p = q = -r$ and $u = -1$. With these values, Eq. (A1.6) can now be solved for σ_N. This leads to the *mean* type 1 energetic-statistical size effect:

$$\sigma_N = f_r^0 \left[(\lambda l_0/D)^{rn/m} + r\kappa(\lambda l_0/D) \right]^{1/r} \quad (A1.19)$$

equivalent to Eq. 3.9; here $f_r^0 = \lambda^{-n/m}(-F_2/F_0)^{1/r}$, $\kappa = \lambda^{rn/m-1}F_1/F_2 r$, and r is a parameter of the order of 1, sensitive to structure geometry. For large D, the first term dominates, and so $\sigma_N \propto D^{-n/m}$, which agrees with the asymptotic requirement in (A1.18). Furthermore, for $m \to \infty$, the asymptotic scaling of Eq. (A1.2) is matched because Eq. (A1.19) takes the form $\sigma_N = s_0(1 + \kappa r \lambda l_0/D)^{1/r} \approx s_0(1 + \kappa \lambda l_0/D)$.

Eq. (A1.19), however, is separated by the intermediate asymptote from the small-size asymptotics in (A1.1), and does not match that asymptotics. It must, therefore, be further modified, but without affecting the first two large-size asymptotic terms. To bridge the small-size and intermediate asymptotes, we could engage in similar arguments as we did for bridging the large-size and intermediate asymptotes. Suffice to say, the complete law for the mean size effect of type 1 (Fig. A.1 left), matching the small-size asymptotics in (A1.1), can simply be obtained by replacing $\lambda l_0/D$ in Eq. (A1.19) with θ, as follows:

$$\sigma_N = f_r^0 \left(\theta^{rn/m} + rs\kappa\theta \right)^{1/rs}, \quad \theta = (1 + D/s\eta l_0)^{-s} \lambda/\eta \qquad (A1.20)$$

where η = positive constant of the order of 1, and s may be taken as 1.

Note that the intermediate asymptote is defined by infinite separation of the characteristic lengths involved, ηl_0 and λl_0. If we let $\eta l_0 \ll D \ll \lambda l_0$, i.e., if we consider the limiting process $\lambda/\eta \to \infty$, Eq. (A1.20) converges to an intermediate asymptote, given by the power law

$$\sigma_N = f_r^0 (l_a/D)^{1/r} \qquad (A1.21)$$

in which $l_a = s\eta(rs\kappa\lambda/\eta)^{1/r}$ = constant, and becomes infinitely close to the small-size asymptote of (A1.19), as well as to the large size asymptote of $\sigma_N = f_r^0 \theta^{n/m}$ (given by Eq. 3.6), which passes through the aforementioned inflexion point. The physical reason for the existence of an intermediate asymptote is that normally the size of averaging domain in nonlocal Weibull theory (roughly equal to FPZ width) is much smaller than the FPZ length.

Eq. (A1.20) is supported by finite element simulations with nonlocal Weibull theory, as well as test data from ten different labs (all combined in one dimensionless plot in Fig. 3.6). However, test data for concrete and composites show that the D values for which the difference between (A1.19) and (A1.20) is significant are less than the material inhomogeneities, which means that (A1.19) should mostly suffice in practice.

For bending of laminates or unreinforced concrete beams, the Weibull statistical component in (A1.19) or (A1.20) is usually insignificant for normal sizes (Bažant Zhou, Nov'ak and Daniel 2004). It becomes significant only for bending of very large structures, such as arch dams. For bending of unreinforced concrete beams or plates, for which the size effect is generally of type 1, the Weibull statistical component normally becomes significant only for cross sections at least several meters thick (as typical of arch dams).

Universal Size Effect Law

The size effect for crack initiation from a smooth surface ($a_0 = 0$) has been shown to be very different from the size effect for large notches or large stress-free (fatigued) cracks at maximum load (for which a_0/D is not too small). As far as the mean nominal strength of structure, σ_N, is concerned, the former is always energetic (i.e. purely deterministic), while the latter is purely energetic only for small enough sizes and becomes statistical for large enough sizes. It is of interest to find a universal size effect law that includes both of these size effects and spans the transition between them. A formula for this purpose was proposed in Bažant and Li (1996) (also Bažant and Chen 1997); see Eq. (2.52) and Fig. 2.14. However, that formula is not smooth and does not include the statistical (Weibull) part for crack initiation failures.

A better formula (Bažant and Yu 2004) is shown in Fig. A.2—on the left (Fig. A.2a) without, and on the right (Fig. A.2b) with, the statistical (Weibull)

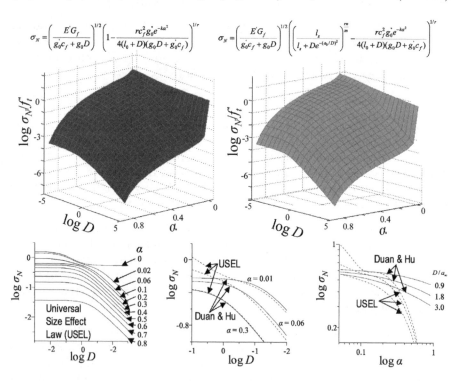

Figure A.2: Universal size effect law (improvement of Fig. 2.14); top left—without, and top right—with Weibull statistics; bottom—profiles obtained from the universal size effect law, compared to Duan and Hu's (2003) approximation part (E' = effective Young's modulus for plane stress or plane strain), f'_t = local tensile strength of material; g_0, g'_0, g''_0 are values of dimensionless energy release function $g(a)$ and its derivatives at $\alpha_0 = a_0/D$; $l_0 = E'G_f/f'_t{}^2$ = Irwin's characteristic length (corresponding to the initial fracture energy G_f); $g(a) = [k(a)]^2$, $k(a)$ is the dimensionless stress intensity factor; m = Weibull modulus of concrete (about 24), n = number of dimensions for scaling; r, k = empirical positive constants; and c_f = constant (the ratio c_f/l_0 depends on the softening curve shape, and $c_f \approx l_0/2$ for triangular softening). The formulas in Figs. F1a,b were derived by asymptotic matching of 7 cases: the small-size and large-size asymptotic behaviors (first two terms of expansion for each), of the large-notch and vanishing-notch behaviors, and of Weibull size effect.

A2. Size Effect of Finite-Angle Notches

In elastic bodies, a sharp notch of a finite angle (Fig. A.3) causes stress singularity $\sigma \approx r^{\lambda-1}$ that is weaker than the crack singularity (i.e., $\lambda > 0.5$) and is given by Williams' (1952) formulas (a)–(e) shown in Fig. A.3, in which r, ϕ = polar coordinates and $\sigma = \sigma_{rr}, \sigma_{\phi\phi}, \sigma_{r\phi}$ = near-tip stresses. If the structure has a positive geometry, it will fail as soon as a FPZ of a certain characteristic length $2c_f$ is fully formed at the notch tip. In the limit of $D \to \infty$, the structure will fail as soon as a crack can start propagating from the notch tip, in which case the critical energy release rate must equal G_f. Experiments show that the load (or nominal stress σ_N) at which this occurs increases with angle γ. In previous studies (e.g., Carpinteri 1987, Dunn et al. 1997a,b), some arguments in terms of a non-standard 'stress intensity factor' K_γ corresponding to singularity exponent $1-\lambda < 0.5$ were used to propose that the nominal stress $\sigma_N \propto D^{\lambda-1}$.

A notch of finite angle cannot propagate, and a stress singularity of a power other than $1/2$ cannot supply a finite energy flux. So, a realistic approach requires considering that a cohesive crack must propagate from the notch tip (Fig. A.3 left; Bažant and Yu 2004). Circular bodies with notches of various angles 2γ (and ligament dimension D, Fig. A.3a) were simulated by finite elements with a mesh progressively refined as $r \to 0$ (the first and second rings of elements spanned from 0 to $r = D/6000$ and then to $D/3000$). The circular boundary was loaded by normal and tangential surface tractions equal to stresses σ_{rr} and $\sigma_{r\phi}$ taken from Williams symmetric (mode I) solution; P = load parameter representing the resultant of these tractions; and $\sigma_N = P/bD$ = nominal stress ($b = 1$). First, ligament D was considered to be so large that the length of the FPZ, l_c, was less than $0.01D$. In that case, the angular distribution of stresses along each circle with $r \geq 0.1D$ ought to match Williams functions $f_{rr}, f_{r\phi}, f_{\phi\phi}$. Indeed, the numerical results cannot be visually

250   Scaling of Structural Strength

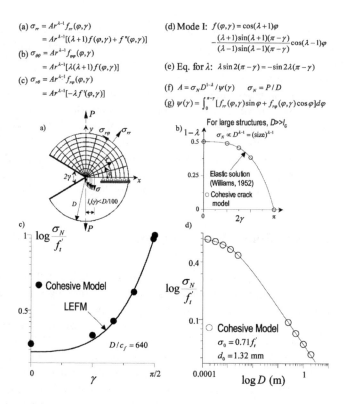

Figure A.3: (a) Angle-notched circular specimen considered for analysis. (b) Numerically computed singularity exponent compared to Williams' elastic solution. (c) Computed variation of nominal strength with notch angle. (d) Curve of analytical size effect law for angular notches compared to the results of cohesive crack model

distinguished from these functions. The logarithmic plots of the calculated stress versus r for any fixed ϕ (and any γ) were straight lines, and their slopes agreed with the exponent $\lambda - 1$ required by Williams solution; see Fig. A.3 (right). This serves to confirm the correctness of the cohesive finite element simulation.

From Eqs. (b), (f), and (g) of Williams' LEFM solution in Fig. A.3,

$$\sigma_N = (r/D)^{1-\lambda} \sigma_{\phi\phi} \psi(\gamma) / f_{\phi\phi}(0, \gamma) \quad (8) \qquad (A2.1)$$

According to the equivalent LEFM approximation of cohesive fracture, $\sigma_{\phi\phi}$ for $r = c_f$ (the middle of FPZ), should be approximately equal to material tensile strength f'_t. This condition yields

$$\sigma_N = f'_t (D/H_r)^{\lambda(\gamma)-1}, \quad H_r = c_f [f_{\phi\phi}(0, \gamma) / \psi(\gamma)] \quad (9) \qquad (A2.2)$$

which is for $D \geq 250l_0$ indistinguishable graphically. Eq. (A1.19), however, violates the small-size asymptotics in (A1.1), and must therefore be further modified, but without affecting the first two large-size asymptotic terms. To bridge the small-size and intermediate asymptotes, we could engage in similar arguments as we did for bridging the large-size and intermediate asymptotes. Suffice to say, the complete law for the mean size effect of type 1 (Fig. A.1d), matching the small-size asymptotics in (A1.1), is obtained by replacing $\lambda \ell/D$ in Eq. (A1.19) with θ:

$$\sigma_N = f_r^0 \left(\theta^{rn/m} + r\kappa\theta \right)^{1/r}, \quad \theta = (1 + D/s\eta\ell)^{-s}\lambda/\eta \qquad (A2.3)$$

where η = positive constant of the order of 1 and s may be taken as 1; $\lambda(\gamma)$ is the λ value for angle γ. To check this equation, geometrically similar scaled circular bodies of different ligament dimensions D (Fig. A.3) were analyzed by finite elements for various angles γ using the same linear softening stress-separation diagram of cohesive crack. The numerically obtained values of $\log \sigma_N$ for various fixed D/c_f are plotted in Fig. A.3c as a function of angle γ. We see that this size effect curve matches perfectly the curve of Eq. (A2.2) for $D/c_f > 500$, confirming that the equivalent LEFM approximation obtained for $r = c_f$ is good enough.

A general approximate formula for the size effect of notches of any angle, applicable to any size D, may be written as follows:

$$\sigma_N = \sigma_0 \left(1 + \frac{H_0}{H_\gamma} \frac{D}{D_0} \right)^{\lambda(\gamma)-1} \qquad (A2.4)$$

where $H_0 = h_\gamma$-value for $\gamma = 0$, and D_0 is given in terms of $g(\alpha)$ and is the same as for a crack ($\gamma = 0$). Eq. (A2.4), which is of course valid only for large enough notches penetrating through the boundary layer of concrete, has been derived by asymptotic matching of the following asymptotic conditions:

1. for $\gamma \to 0$, the classical size effect law $\sigma_N = \sigma_0(1 + D/D_0)^{-1/2}$ must be recovered;
2. for $D/l_0 \to 0$, there must be no size effect;
3. for $D/l_0 \to \infty$, Eqs. A2.4 and A2.2 must coincide;
4. for $\gamma = \pi/2$ (flat surface), the formula must give no size effect for $D \to \infty$.

In reality, there is of course a size effect in the last case, but it requires a further generalization of Eq. (A2.4) (which will be presented separately). Therefore, Eqs. (A2.4) and (A2.2) can be applied only when the notch is deeper than the boundary layer, which is at least one aggregate size. Complete generality will require amalgamating Eq. (A2.4) with the universal size effect law in Fig. A.2.

The plot of $\log \sigma_N$ versus $\log D$ for $\gamma = \pi/3$ according to Eq. (F28) is compared to the finite element results for notched circular bodies with cohesive cracks in Fig. A.3d. The agreement is seen to be excellent.

A3. Size Effect on Flexural Strength of Fiber-Composite Laminates

The failure of fiber-polymer laminates is in general practice still treated according to the strength theory or plastic limit analysis, which exhibits no size effect, and all the size effects are considered as purely statistical. A recent systematic analysis by Bažant, Zhou and Daniel (2003), based on the flexural strength tests of Jackson (1992), Johnson et al. (2000), Wisnom (1991), Wisnom and Atkinson (1997) and others, confirms that such practice in unrealistic. The problem is that if the available size effect data on flexural strength of fiber-polymer laminates are considered alone, they can be fitted well enough by Weibull statistical size effect theory. However, to conclude that this theory applies is deceptive for two reasons: (1) the size range of the existing data is quite limited, and the energetic theory for size effect at crack initiation due to a boundary layer can fit these data equally well; and (2) checks of the coefficient of variation of flexural strength, ω_W, have been omitted. These checks are essential.

Weibull statistical theory cannot be correct if the ω_W-value does not agree with the value of Weibull modulus m, and if this agreement is not obtained for every size. The ω_W values must satisfy the well-known formula of Weibull theory, Eq. (3.3), i.e., $\omega_W = [(\Gamma(1 + 2/m))/(\Gamma^2(1 + 1/m) - 1]^{-1/2}$. Even though the available size effect data for laminate flexure did not include measurements of ω_W, it is easy to ascertain that the foregoing equation is not satisfied. For example, Jackson et al.'s (1992) tests included four different layups—unidirectional, quasi-isotropic, cross-ply and angle ply. To fit these size effect data, the optimum m values vary widely from case to case, ranging from $m = 5$ to $m = 30$. According to the foregoing relation,

$$\omega_W = 22.8\% \text{ for } m = 5, \qquad \omega_W = 4.18\% \text{ for } m = 30 \qquad (A3.1)$$

for angle-ply and unidirectional laminates, respectively. Although the coefficient of variation of flexural strength has not been reported for the test data considered, such huge differences in scatter go against all experience, and are evidently implausible. The angle-ply laminate, due to its greater ductility

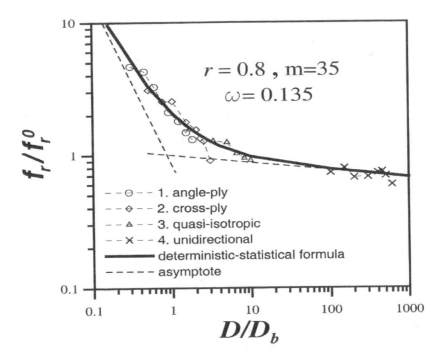

Figure A.4: Type 1 size effect plot for Jackson's data from flexural tests of laminates of four different layups, in dimensionless coordinates

(lower brittleness), might be expected to give a somewhat smaller scatter (larger m) than the unidirectional, cross-ply or quasi-isotropic laminate, but the opposite is systematically noted in pure Weibull fitting of the data. Therefore, the classical hypothesis of a purely statistical size effect in flexural strength of laminates is untenable.

On the other hand, as shown by Bažant, Zhou and Daniel (2003), the data of Jackson can be fitted consistently by the energetic-statistical theory, with the same value of m for all the data; see Fig. A.4 in which the data for various layups are plotted in dimensionless coordinates, after optimal fitting for each individual layup. However, it must further be noted that these data can be fitted about equally well by a purely energetic theory. From this it follows that the Weibull-type statistical part of size effect becomes significant only for laminate thicknesses well beyond the normal usage (this observation is similar to that made for flexural strength of concrete, for which the statistical part of size effect becomes important only for concrete thickness found, for example, in arch dams).

A further implication is that fracture mechanics, rather than some strength criterion (or material failure criterion expressed in terms of stresses and strains), needs to be used for evaluating the strength of laminates. The fracture analysis of laminates must take into account the quasibrittle (or cohesive) nature of fracture.

A4. Size Effect in Fracture of Closed-Cell Polymeric Foam

Very light closed-cell polymeric (vinyl) foam is a favored material for the core of sandwich structures for ships. In classical works (e.g., Gibson and Ashby 1997, Shipsha et al. 2000, Gdoutos et al. 2001), the failure of this material in tension or compression has usually been described as ductile, involving a yield plateau followed by locking, and failure criteria expressed in terms of stresses have been used. Such failure can exhibit no size effect.

Such ductile response, however, does not take place when high tensile stress concentrations exist, induced for example by notches in laboratory specimens, or various structural holes or accidental damage of real structures. In such a case, the failure of the foam may be brittle. This was first revealed by the notched specimen tests of Zenkert and Bäcklund (1989) in which the fracture toughness of foam was measured, by the tests of holed panels tests by Fleck and co-workers in Cambridge in (2001), and by the finite element studies of foam based on microplane model for foam by Brocca et al. (2001). Consequently, the foam, when notched or damaged, must be expected to exhibit size effect, and this is what is confirmed by a recent study (Bažant, Zhou, Novák and Daniel 2004), in which the size effect in Divinycell H100 foam, both the case of failure with large cracks or notches (type 1) and failure at crack initiation (type 2).

In this study, geometrically similar single edge-notched prismatic specimens of various sizes were tested under tension. The results were shown to agree with Bažant's size effect law. Fitting this law to the test results furnished the values of the fracture energy G_f of the foam as well as the characteristic size c_f of the fracture process zone. Fracture analysis based on these properties was shown to match the test data of Fleck, Olurin and co-workers (2001) obtained on dissimilar panels with holes of different sizes. These results were also matched using the eigenvalue method for finite element analysis based on the cohesive crack model with properties corresponding to the size effect law. Finally, compressed foam specimens with V-shaped notches (with an angle wide enough to prevent the notch faces from coming into contact) were found to exhibit no size effect. This implies that the cell collapse at the tip of the notch must be essentially a plastic process rather than a softening damage process.

The results demonstrate that the current design practice, in which the tensile failure of foam is generally predicted on the basis of strength criteria or plasticity, is acceptable only for small structural parts. In the case of large structural parts, the size effect must be taken into account, especially if the foam can suffer large fatigue cracks or large damage zones prior to critical loading to failure.

A5. Variation of Cohesive Softening Law Tail in Boundary Layer of Concrete

It has been well established experimentally that the total fracture energy G_F of a heterogeneous material such as concrete, defined as the area under the cohesive softening curve, is not constant but varies during crack propagation across the ligament. The variation of G_F at the beginning of fracture growth, which is described by the R-curve, is known to be only an apparent phenomenon which is perfectly consistent with the cohesive crack model (with a fixed softening stress-separation law), and can in fact be calculated from it. However, the variation during crack propagation through the boundary layer at the end of the ligament is not consistent with the cohesive crack model and implies that the softening curve of this model is not an invariant property in the boundary layer. The analysis of this problem by Bažant and Yu (2004) will now be presented.

The fact that the fracture energy representing the area under the softening curve should decrease to zero at the end of the ligament was pointed out in a paper by Bažant (1996), motivated by the experiments of Hu and Wittmann (1991 and 1992a), and was explained by a decrease of the fracture process zone (FPZ) size, as illustrated on the left of Fig. A.5a reproduced from Bažant's (1996) paper. An experimental verification and detailed justification of this property was provided in the works of Hu (1997, 1998), Hu and Wittmann (1992b, 2000), Duan, Hu and Wittmann (2002, 2003), and Karihaloo, Abdalla, and Imjai (2003). As mentioned by Bažant (1996), as well as Hu and Wittmann, the consequence of these experimental observations is that:

$$\bar{G}(a) = \frac{1}{D - a_0} \int_0^\infty P \, du = \int_{a_0}^D \Gamma(x) \, dx < G_F \qquad (A5.1)$$

where $\bar{G}(a)$ = average fracture energy in the ligament (Fig. A.5b), D = specimen size (Fig. A.5a), a_0 = notch depth, P = load, u = load-point deflection, $\Gamma(x)$ = local fracture energy as a function of coordinate x along the ligament (Fig. A.5a left), $G_F = \int_0^\infty \sigma \, dw = \Gamma(x)$-value at points x remote from the

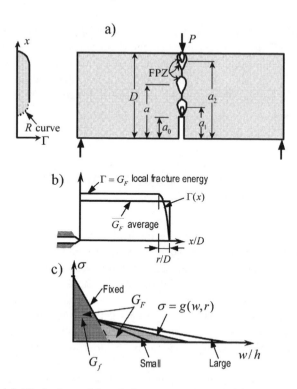

Figure A.5: (a) Variation of local fracture energy $\Gamma(x)$ across the ligament, decreasing in the boundary layer (reproduced from Bažant 1996). (b) Average fracture energy G_F. (c) Required modification of cohesive (or fictitious) crack model boundary (= area under the complete $\sigma(w)$-diagram, Fig. A.5c), σ = cohesive (crack-bridging) stress, and w = crack opening = separation of crack faces.

Is this behavior compatible with the cohesive crack model? To check it, consider a decreasing FPZ attached to the boundary at the end of ligament, Fig. A.5a. Extending to this situation Rice's (1968) approach, we calculate the J-integral along a path touching the crack faces as shown in Fig. A.6a,b:

$$J = \oint \left(\bar{U} - t_i \frac{\partial u_i}{\mathrm{d}x} \mathrm{d}s \right] (2) = 2 \int_{x_{tip}}^{D} \sigma \frac{\mathrm{d}v(x)}{\mathrm{d}x} \mathrm{d}s - [t_2 \Delta w]_{end} \quad (A5.2)$$

$$= \int_0^\infty \sigma(w) \mathrm{d}w - [\sigma \Delta w]_{end} = \text{Area below } \sigma_{end} \quad (A5.3)$$

$$= J_{end}(D) \text{ or } J_{tail}(x) \text{ or } J_{middle}[(x+D)/2] \quad (A5.4)$$

in which s = integration contour length; $x = x_1$, $y = x_2$ are the cartesian coordinates; u_i = displacements; t_i = tractions acting on the contour from the

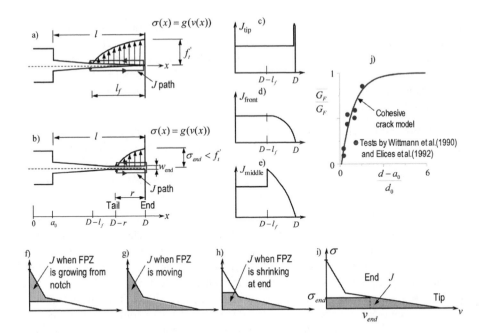

Figure A.6: (a,b) J-integral path. (c,d,e) Ambiguity in J-integral variation. (f,g,h,i) Fracture energies corresponding to J-integral. (j) Test data fitted by modified cohesive crack model with variable tail

outside; U = strain energy density; $v = w/2$; $x = D$ is the end point of the ligament; w_{end} is the opening at the end of ligament (Fig. A.6b). From these equations, we see that the instantaneous flux of energy, J, into the shrinking FPZ attached to the end of ligament (Fig. A.5a) represents the area below the line $\sigma = \sigma_{end}$ in the softening diagram, cross-hatched in Fig. A.6i.

It is, however, a matter of choice with which coordinate x in the FPZ this flux J should be associated. If we associate J with the front, the tail, or the middle (Fig. A.6b) of the FPZ, we get widely different plots of $\Gamma = J_{end}, J_{tail}, J_{middle}$, as shown in Fig. A.6(c,d,e), respectively (the first one terminating with Dirac delta function). This ambiguity means that the boundary layer effect, experimentally documented by Hu and Wittmann, cannot be represented by the standard cohesive crack model, with a fixed stress-separation diagram.

Can the cohesive crack model be adapted for this purpose? It follows from Eq. (A5.4) and Fig. A.6(h,i) (and has been computationally verified) that Wittmann et al.'s (1990) and Elices et al.'s (1992) data can be matched (Fig. A.6j) if the slope of the tail segment of the bilinear stress-separation diagram for concrete is assumed to decrease (Fig. A.5c) in proportion to diminishing distance $r = D - x$ (Fig. A.6b) from the end of ligament. After such an adaptation, the cohesive (or fictitious) crack model has a general applicability, including the boundary layer.

However, the consequence is that the total fracture energy G_F (area under the complete stress-separation curve) is not constant. Noting that, the larger the structure, the smaller is the length fraction of the boundary layer, one must conclude that the diminishing tail slope in Fig. A.5c automatically implies a certain size effect on the apparent G_F, as given by Eq. (A5.4).

It further follows from Fig. A.6(h,i), and has been computationally verified, that the initial tangent of the stress-separation diagram, the area under which represents the initial fracture energy G_f (Bažant 2002f, Bažant and Becq-Giraudon 2002, Bažant, Yu and Zi 2002), can be considered as fixed—in other words, G_f, unlike G_F, is a material constant. Aside from the fact that the maximum loads of specimens and structures are generally controlled by G_f, not G_F, this suggests that the standard fracture test that should be introduced is that which yields not G_F but G_f (the size effect method, as well as the method of Guinea, Planas and Elices, 1994a,b, and the zero brittleness method of Bažant, Yu and Zi 2002, serve this purpose, while the work-of-fracture method does not). This conclusion is not surprising in the light of the abundance of experimental data revealing that G_F is statistically much more variable than G_f (Bažant and Becq-Giraudon 2002, Bažant, Yu and Zi 2002).

Jirásek (2003) showed that Hu and Wittmann's data can be matched by a nonlocal continuum damage model in which the characteristic softening curve is kept fixed. Consequently the nonlocal model is a more general, and thus more fundamental, characterization of fracture than the cohesive (or fictitious) crack model. This finding should be taken into account in fracture testing. It appears that G_F would better be defined by the area under the softening curve of the nonlocal model, multiplied by the characteristic length of material corresponding to the effective width of FPZ, which equals the minimum possible spacing of parallel cracks (Bažant 1985, Bažant and Jirásek 2002), to be distinguished from Irwin's characteristic length $l_0 = EG_f/f'^{t2}$.

A6. Can Fracture Energy Be Measured on One-Size Specimens with Different Notch Lengths?

The fact that specimens of different sizes are needed for the size effect method of measuring G_f is considered by some a disadvantage. For this reason, Bažant and Kazemi (1990), Bažant and Li (1996) and Tang et al. (1996) generalized the size effect method to dissimilar specimens, the dissimilarity being caused by the use of different notch lengths a_0 in specimens of one size. If the random scatter of test data were small (coefficient of variation CoV < 4%), this approach would work. However, for the typical scatter of maximum loads of concrete specimens (CoV = 8%), the range of brittleness numbers attainable by variation of notch length in a specimen of any geometry (about 1:3, Bažant and Li 1996) does not suffice to get a sharp trend in the regression of test data, and thus prevents determination of G_f accurately.

Recently, this problem was considered independently by Duan and Hu (2003). They proposed the semi-empirical formula:

$$\sigma_n = \sigma_0(1 + a/a_\infty^*)^{-1/2} \qquad (A6.1)$$

where $\sigma_0 = f'_t$ for small 3PB specimens; a_∞^* is a certain constant; and σ_n represents the maximum tensile stress in the ligament based on a linear stress distribution over the ligament, $\sigma_n = \sigma_N/A(a_0)$. This alternative formula, intended for specimens of the same size when the notch length a_0 is varied, in effect attempts to replace the profile of the universal size effect law (Fig. A.2) at constant size D, scaled by the ratio $\sigma_n/\sigma_N = 1/A(a_0)$, where $A(a_0)$ depends on specimen geometry; $A(a_0) = (1 - a_0)^2$ for notched three-point bend beams. However, the curve of the proposed formula has, for $a_0 \to 0$, a size independent limit approached, in $\log a_0$ scale, with a horizontal asymptote, while the correct curve, amply justified by tests of modulus of rupture (or flexural strength) of unnotched beams (Bažant and Li 1995, Bažant 1998, 2001, 2002f, Bažant and Novák 2000a,b,c), terminates with a steep slope for $a_0 \to 0$ and has a size dependent limit, as seen in the aforementioned profiles in Fig. A.2a,b, and better in Fig. A.2c. Fig. A.2d,e shows the profiles in D and in a, and it is seen that they cannot be matched well by Duan and Hu's approximation converted from σ_n to σ_N.

The test data on the dependence of σ_n on a_0, which Duan and Hu fitted by their formula, should be fitted, more accurately, with the usual size effect law (Bažant and Kazemi 1990, Bažant and Planas 1998, and Bažant 1997, 2002f) $\sigma_n = \sigma_N/A(a_0)$ in which $\sigma_N = \sqrt{E'G_f}[g'(\alpha_0)c_f + g(\alpha_0)D]^{1/2}$ where D is constant and $\alpha_0 = a_0/D$ is varied (and function $g(\alpha_0)$ is available from handbooks). However, very short or zero notches ($\alpha_0 < 0.15D$) must be excluded, which means that the value of strength f'_t cannot be used with Duan and Hu's approach. To use it, it is necessary to adopt either the approach of Guinea et al. (1994a,b) or the zero-brittleness method (Bažant, Yu and Zi 2002).

A7. Notched–Unnotched Size Effect Method and Standardization of Fracture Testing

Accurate simulations with the cohesive crack model (Guinea et al. 1997, Bažant and Yu 2002) revealed that, at maximum load, the cohesive stresses through the entire process zone in concrete fracture specimens (as well as normal-size structures) remain on the initial straight sloping segment of the stress-separation curve (Fig. 9.1, bottom left, Fig. A.7 top left). This means that (for type 2 fracture), the long tail of the stress-separation curve, and thus also the total fracture energy G_F representing the area under the entire curve, do not matter. The maximum load depends only on the initial fracture energy G_f, representing the area under the initial tangent of this curve. Furthermore, statistical studies of published test data (Bažant and Becq-Giraudon 2002) as well as lattice particle simulations of microstructure (Cusatis et al. 2003a,b) revealed that the scatter of G_f has a much smaller coefficient of variation than the scatter of G_F (about one half).

For these two reasons, it is preferable that a standard fracture test of concrete would yield directly G_f rather than G_F. This means that the size effect method is, in principle, preferable because it provides directly G_f, rather than G_F. However, the need to use in this method specimens of very different sizes may be seen as inconvenient. When G_F is measured directly by the work-of-fracture method, one can roughly estimate $G_f \approx 0.4 G_F$ (Bažant 2002f, Bažant and Becq-Giraudon 2002). But this relation involves a large error. So, together with greater scatter of G_F, this leads to a much greater uncertainty in G_f (and thus maximum load prediction), compared to direct measurement of G_f.

Guinea et al. (1997) found that, if an accurate solution with the cohesive crack model is available for the given specimen geometry, then the initial slope of the stress-separation curve (and thus G_f) can be easily obtained by measuring only the maximum loads of: (1) the maximum load of a small notched specimen (the smallest practical), and (2) an unnotched specimen. No testing method can be simpler and more robust.

However, in view of large random scatter in fracture testing of concrete, statistical evaluation is important. In this regard, the simple method of Guinea et al. (1997) can still be improved—by recasting it in the form of zero-brittleness size effect method (Bažant and Li 1996b), in a way that permits simultaneous statistical regression of the data for notched and unnotched tests. The key is to realize that the maximum load of unnotched specimens is independent of G_f, and thus depends only on material tensile strength f'_t. This is the property of the zero-size limit, $\sigma_N = B f'_t$, of the size effect curve. So it suffices to relate B to the zero-size limit $B_0 f'_t$ of the exact size effect curve of the cohesive crack model for the given geometry, which can be computed once for all for a chosen standardized specimen geometry (Bažant, Yu and Zi 2002).

This exact size effect curve is, for normal sizes, very close to the classical size effect law $\sigma_N = B f'_t (1 + D/D_0)^{-1/2}$ (Eq. 2.8) (and is identical to it for very large sizes). But for extrapolation to zero size, $D \to 0$, a correction is needed. To this end, the size effect curve of cohesive crack model for any fixed specimen geometry may be written in a dimensionless form as

$$(f'_t/\sigma_N)^2 = g(\alpha_0)\xi + B^{-2} - \Delta(\xi), \quad \text{with } \xi = D/l_1, \ l_1 = E'G_f/f'_t \quad (A7.1)$$

where $g(\alpha_0)$ = dimensionless energy release rate for the given geometry, and $\Delta(\xi)$ = deviation from the classical size effect law in Eq. (2.8), which is calculated accurately from the cohesive crack model (and vanishes for large ξ); $\Delta(\xi)$ can be ignored in the usual size effect method (section 2.5) (Bažant, Yu and Zi 2002) (l_1 must not be confused with $l_0 = E'G_F/f'_t$). The statistical regression of test data may proceed as follows:

1. Calculate the mean tensile strength $\bar{f}'_t = \frac{1}{m}\sum_{i=1}^m f'_{ti}$. Then, for each of the notched tests ($j = 1, 2, \ldots, n$), calculate $\eta_j = (\bar{f}'_t/\sigma_{Nj})^2$. After that, from the inverse relation of (A7.1) (determined explicitly once for all in advance), calculate $\xi_j = \psi(\eta_j)$ (the vertical line on the right of each plot on the top of Fig. A.7). Then calculate $\Delta_j = \Delta(\xi_j)$ ($j = 1, 2, \ldots, n$), and compute the notched specimen data for linear regression (Fig. A.7 right):

$$Y_{\nu+j} = (\eta_j + \Delta_j)/(\bar{f}'_t)^2 \quad (j = 1, 2, \ldots, n) \quad (A7.2)$$

(these are the points that would ideally lie on the straight size effect asymptote, if there were no scatter). Set $Y_i = 1/(Bf'_{ti})^2$ ($i = 1, 2, \ldots, \nu$).

2. Consider the standard linear regression relation $Y = AX + C$ of the size effect method (Fig. A.7, top) where $X = D$, and set $X_i = 0$ for $i = 1, 2, \ldots, \nu$ and $X_i = D$ for $i = \nu+1, \nu+2, \ldots, \nu+n$. Then run linear regression of all data points (X_i, Y_i), $i = 1, 2, \ldots, \nu+n$, using the weights $w_i = 1/\nu$ for $i = 1, \ldots, \nu$ and $w_i = 1/n$ for $i = \nu+1, \nu+2, \ldots, N$ ($N = \nu + n$). The well known regression formulas yield the mean slope $\bar{A}$ and mean intercept $\bar{C}$, and also the coefficients of variation of regression slope and intercept, ω_A, ω_C where $\omega_A = [\sum_{i=1}^N (Y_i - Y'_i)^2/(N-2)]^{1/2}/\bar{Y}$, $\bar{Y} = \sum_{i=1}^N (Y_i - Y'_i)/N$ and $Y'_i = AX_i + C$ = values on the regression line. Finally, according to the formulas of the size effect method (section 2.5), estimate the mean fracture energy and its coefficient of variation: $\bar{G}_f \approx g(\alpha_0)/\bar{E}'\bar{A}, \omega_{G_f} \approx \sqrt{\omega_A^2 + \alpha_E \omega_E^2}$.

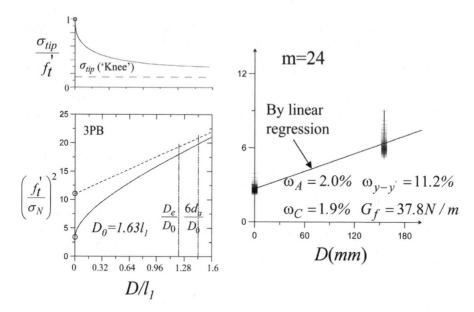

Figure A.7: Left bottom: Regression line based on classical size effect law (Eq. 2.8, dashed line) and deviation $\Delta(\xi)$ from this line for small specimen sizes (solid line), computed from the cohesive crack model (the fist vertical dash-dot line indicates the smallest possible specimen for which $D \approx 6$ aggregate sizes, and the second one the smallest size for the deviation $\Delta(\xi)$ can be ignored). Left top: The smallest cohesive stress in the FPZ, calculated for a standard bilinear softening law with $G_F = 2.5 G_f$ ('knee' = point of slope). Right: Regression of data for notched and unnotched specimens randomly generated according to Weibull distribution with modulus $m = 24$ (after Bažant, Yu and Zi 2002)

A minor weak point of the foregoing procedure is that the initial calculation of ξ_j is based on the mean of the f'_t rather than the random f'_t-data. This may be avoided by a certain kind of iteration (Bažant, Yu and Zi 2002).

A8. Designing Reinforced Concrete Beams against Size Effect

The time has become ripe for adding size effect in a rational form to all the design specifications dealing with brittle failures of concrete structures

$$v_c = \mu \rho^{3/8}\left(1+\frac{d}{a}\right)\sqrt{\frac{f_c'}{1+d/d_0}} \qquad d_0 = \kappa f_c'^{-2/3}$$

$$\kappa = 3800\sqrt{d_a} \text{ if } d_a \text{ is known}, \quad \kappa = 3330 \text{ if not}$$

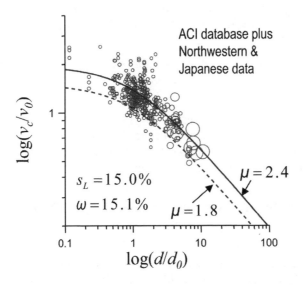

Figure A.8: Improved design formula for shear failure of reinforced concrete beams without stirrups (the circle areas represents the weights, which are taken inversely proportional to the histogram of beam depths in the database)

(shear and torsion of R.C. beams with or without stirrups, slab punching, column failure, bar embedment length, splices, bearing strength, plain concrete flexure, compression failure of prestressed beams, etc.). In ACI Standard 318, so far, only the recently introduced formula for anchor pullout (which is of the LEFM type) includes size effect. By analysis of the latest ACI 445 database with 398 data (Reineck et al. 2003), representing an update of 1984 and 1986 Northwestern University databases (with 296 data), an improved formula for shear failure of reinforced concrete beams without stirrups, having the simplicity desired by concrete designers, has recently been developed by second-order asymptotic matching (Bažant and Yu 2003); see the comparison with database in Fig. A.8, with CoV of the vertical deviations from data points (the solid line is the mean formula, the dashed line is a formula scaled down to achieve additional safety, as practiced in ACI; $v_c = \sigma_N =$ mean shear stress in the cross section at failure, $d =$ beam depth from top face to the reinforcement centroid, $f_c' =$ standard compression strength of concrete).

In the derivation of these formulas, the following three principles were adhered to:

1. Only theoretically justified formulas must be used in data fitting because the size effect (which is of main interest for beam depth d ranging from 1 m to 10 m) requires enormous extrapolation of the ACI 445 database (in which 86% of data pertain to $d \leq 0.6$ m, 99% to $d \leq 1.1$ m and 100% to $d \leq 1.89$ m).

2. The validity of the formula must be assessed by comparing it only with (nearly) geometrically scaled beams of broad enough size range (only 11 such test series exist, among 398).

3. The entire database must be used only for the final calibration of the chosen formula (but not for choosing the best formula, because data for many different concretes and geometries are mixed in the database, and only 2% of the data have a non-negligible size range).

A9. Scaling of Probability Density Distribution (pdf) of Failure Load

In practice, structures must typically be designed for failure probabilities p_f of the order of 10^{-7}. Such a low probability cannot be determined experimentally, especially for large structures. It can be assessed only theoretically. One must deduce the size effect on the pdf of σ_N by a rational argument. This can be done on the basis of the nonlocal Weibull theory expounded in section 3.2.

For $D \to 0$, failure is non-propagating and must occur nearly simultaneously along the entire failure surface. So the pdf of σ_N must be Gaussian (except in far-off tails), as deduced from Daniels' (1945) fiber bundle model. With increasing D, the asymptotic size effect on the mean strength is nil and the coefficient of variation (CoV) asymptotically decreases as $D^{-1/2}$. For $D \to \infty$, the pdf must be Weibull, corresponding to the weakest-link chain model, which implies a size-independent CoV. The gradual transition of pdf from Gaussian to Weibull, exemplified by various load-sharing models (Phoenix and Beyerlein 2000, Mahesh et al. 2002), can be obtained by asymptotic matching (Bažant 2004a, 2004b) and calibrated by a chain-of-bundles model (Fig. A.1f middle) (Bažant and Pang 2004).

A10. Recent Debates on Fractal Theory of Size Effect and Their Impact on Design Codes

An engineering field such as aeronautical can exist without simple design codes because only a few large aircraft are designed annually. But structural engineering, in which many thousands of structures are designed annually, each of them different, cannot get by without codes. During the last few years, the need to introduce the size effect into the design code specifications guarding against brittle failures of concrete structures has finally been widely recognized on all continents.

Members of the code-making committees of course notice that there are currently promulgated two radically different theories of size effect in concrete structures—the energetic-statistical theory expounded here and a purely fractal theory. Therefore, unless the conflict between the proponents of these two theories is resolved, these committees would opt for a purely empirical approach. Yet, an empirical approach is doomed to yield an incorrect formula because a statistically significant set of test results on a sufficiently large number of identical large structures can hardly ever be acquired, because of prohibitive cost. The largest concrete structures, sea ice bodies or rock formations are way beyond the range of failure testing, and small-size data vastly dominate, because of cost. For instance, 86% of the data for shear failure of beams pertain to beam depths less than 0.6 m, and 99% less than 1.1 m, while the main interest is in beam depths from 1 m to 10 m. Consequently, the design of large structures must rely on extrapolation from the range of testing. A theoretical basis for such an extrapolation is, therefore, imperative. But progress on the design code appears impossible unless the fractal-nonfractal conflict is resolved.

Motivated by the need to resolve the conflict, an extensive critical study has recently been undertaken (Bažant and Yavari 2004). It greatly expands the analysis of fractal approach in sections 3.5 and 3.6, and focuses on the recent arguments.

A11. Boundary Layer Size Effect in Thin Metallic Films at Micrometer Scale

Recent pure tension tests of free polycrystal thin metallic films of Au, Al and Cu, conducted by H. Espinosa at Northwestern University revealed a surprising and very strong size effect. The strength of these films increases by a factor of 3 to 5 when the thickness is reduced from 5 μm to 0.5 μm, while for smaller thickness the size effect in pure tension apparently disappears. This size effect can be explained by the existence of a boundary layer of a fixed thickness, located at the face of film that was during deposition in contact with the substrate.

The boundary layer, having a thickness of the order of the crystal grain size of metal, has a significantly higher yield strength than the rest of film. The layer forms during film deposition on the face of the substrate and is influenced by the epitaxial effects of crystal growth on the dislocation density and texture.

The increase of yield strength of the epitaxial boundary layer is considered to be caused partly by a difference in the mean density of statistically stored dislocations due to blockage of dislocation movements at grain boundaries and formation of dislocation pile-ups near grain boundaries, compared to the rest of film, and partly also by prevalent crystal orientation unfavorable to dislocation glide, which are engendered in the granular epitaxial boundary layer by the substrate during film deposition.

Assuming a simple error function decay of the yield strength across the film thickness from the substrate side leads to a good agreement with pure tension tests, and well as other kinds of experimental observations. The characteristic length l_0 governing the boundary layer size effect is the natural grain size of the metal, $l_0 \approx 0.5$ μm. Therefore the boundary layer size effect is expected to vanish for film thickness $D < l_0$ and $D > 10 l_0$, and this is what is observed in tests. The boundary layer size effect does not conflict with Hutchinson and Fleck's strain gradient size effect. That size effect, which has been used to explain micro-indentation tests down to 1 μm, dominates for the range 5–100 μm. In the range from 1 to 5 μm, in which both kinds of size effect overlap, the fit of the micro-indentation and micro-bending test data by the strain gradient theory is not very close.

The epitaxial boundary layer model captures the extra plastic hardening due to statistically stored dislocations, whereas the strain gradient theory explained in section 10.1 captures the effect of geometrically necessary dislocations. A combined theory is found to be necessary in order to fit with the same theory the film test data on both pure tension and bending.

Thin metallic films of thickness less than 5 μm are found to fail in a quasibrittle manner, exhibiting gradual postpeak softening, which can also be explained and fitted by the epitaxial boundary layer model if the damage due to void or microcrack growth is taken into account. The classical Hall-Petch relation for the dependence of yield strength on the natural grain size, as well as Nix's model for the strength of films thinner than the grain size, needs to be included in the formulation to obtain a general model with proper asymptotics, the former for very thick films ($D > 100$ μm) and the latter for extremely thin films ($D < 0.3$ μm). Inclusion of the Hall-Petch formula is important when not only the film thickness but also the grain size is varied. To this end, the Hall-Petch formula is enhanced by a smooth approach to a cutoff yield strength for very small grain size. In the case of films thinner than than the natural grain size, in which the crystal grains are flattened to fit the film thickness, the grain size to be substituted into the Hall-Petch formula is the film thickness, which means that the Hall-Petch effect disappears for very thin films (being replaced by Nix's effect).

A12. Size Effect on Triggering of Dry Snow Slab Avalanches

A size effect model analogous to that expounded in this book has recently been formulated for dry snow slab avalanches in a study by Bažant, Zi and McClung (2003). The ratio of the thickness of snow layer to the length is, in that study, assumed to be small enough for one-dimensional analysis to be applicable. Equivalent LEFM is used to obtain the law of the decrease of the nominal shear strength of the snow layer with an increasing thickness of snow. This law is shown to closely agree with two-dimensional numerical solutions according to the cohesive crack model for mode II fracture with finite residual shear stress. By fitting these solutions for various snow thicknesses D with the structural size effect law, an approximate one-to-one relationship between the material parameters in the size effect law and the cohesive crack model is established.

The failure mode and size effect are found to depend on the pre-existing (initial) shear stress τ_i in a fracture-triggering weak layer that normally exists at the base of snow slab. There exists a critical stress $\tau_i = \tau_{cr}$ and a critical snow depth D_{cr} such that, for $D > D_{cr}$ or $\tau_i < \tau_{cr}$, there is no structural size effect, no peak load, and no postpeak softening. The value of τ_{cr} is approximately 85% of the residual shear stress at the base of a sliding layer.

Because the fracture process zone can be many meters long, it is not easy to perform laboratory experiments simulating avalanches. Direct measurement of the size effect per se is next to impossible because the effects of snow weight and snow aging prevent finding different-sized samples with similar mechanical properties.

The size effect law needs to take into account the dependence of snow properties in the weak layer at base on the snow slab thickness. This is achieved by replacing the constant fracture toughness and the limiting zero-size strength with quantities that increase with snow slab thickness. Fitting of the data from over a hundred field observations of avalanches suggests that this increase is very strong, with the toughness increasing roughly as (snow thickness)$^{1.8}$.

Bibliography

1501–1850

da Vinci L (1500s)—see *The Notebooks of Leonardo da Vinci* (1945), Edward McCurdy, London (p. 546); and *Les Manuscrits de Léonard de Vinci*, transl. in French by C. Ravaisson-Mollien, Institut de France (1881–91), Vol. 3.

Galileo Galilei Linceo (1638), "Discorsi i Demostrazioni Matematiche intorno à due Nuove Scienze", Elsevirii, Leiden; English transl. by T. Weston, London (1730), pp. 178–181.

Mariotte E (1686), *Traité du mouvement des eaux*, posthumously edited by M. de la Hire; Engl. transl. by J.T. Desvaguliers, London (1718), p. 249; also *Mariotte's collected works*, 2nd ed., The Hague (1740).

Young T (1807), *A course of lectures on natural philosophy and the mechanical arts*. London, vol. I, p. 144.

1851–1960

Barenblatt GI (1959), "The formation of equilibrium cracks during brittle fracture. General ideas and hypothesis, axially symmetric cracks." *Prikl. Mat. Mekh.*, **23** (3), 434–444.

Blanks RF and McNamara CC (1935), "Mass concrete tests in large cylinders." *J. of American Concrete Institute*, **31**, 280–303.

Bridgman PW (1922), *Dimensional analysis*, Yale University Press, New Haven.

Buckingham E (1914)," On physically linear systems; illustrations of the use of dimensional equations." *Physical Review* Ser.2, **IV** 4, 345–376.

Buckingham E (1915), "Model experiments and the form of empirical equations", *Trans. ASME*, **37**, 263–296 (also *Phys. Rev.*, **4**, 345–376).

Dugdale DS (1960), "Yielding of steel sheets containing slits." *J. of Mech. and Phys. of Solids*, **8**, 100–108.

Fisher RA and Tippett LHC (1928), "Limiting forms of the frequency distribution of the largest and smallest member of a sample." *Proc., Cambridge Philosophical Society*, **24**, 180–190.

Fréchet M (1927), "Sur la loi de probabilité de l' écart maximum." *Ann. Soc. Polon. Math.* (Cracow), **6**, p. 93.

Freudenthal AM (1956), "Physical and statistical aspects of fatigue." in *Advances in Applied Mechanics*, Vol. 4, Academic Press, 117–157.

Gonnermann HF (1925), "Effect of size and shape of test specimen on compressive strength of concrete." *Proc. ASTM*, **25**, 237–250.

Griffith AA (1921), "The phenomena of rupture and flow in solids." *Phil. Trans.*, **221A**, 179–180.

Hadamard J (1903), *Leçons sur la propagation des ondes*, Chap. VI, Hermann Paris.

Irwin GR (1958), "Fracture". in *Handbuch der Physik*, Vol. **VI**, ed. by W. Flügge, Springer Verlag, Berlin, 551–590.

Mörsch E (1922),"Der Eisenbetonbau — Seine Theorie und Anwendung." (*Reinforced Concrete Construction—Theory and Application*"), Wittwer, Stuttgart, 5th ed., Vol. **1**, Part 1, 1920 and Part 2.

Peirce FT (1926),"Tensile Tests of Cotton Yarns-V. The Weakest Link." *J. Textile Inst.*, **17**, 355–368.

Porter, A.W. (1933). *The method of dimensions*, Methuen.

Prandtl L (1904), "Uber die Flüssigkeitsbewebung bei sehr kleiner Reibung." *Verhandlungen, III. Int. Math.-Kongr.*, Heidelberg, Germany.

Ritter W (1899),"Die Bauweise Hennebique." *Schweizerische Bauzeitung Zürich*, **33**, 59–61.

Sedov LI (1959), *Similarity and dimensional methods in mechanics*, Academic Press, New York.

Talbot AN (1909), "Tests of reinforced concrete beams—resistance to web stresses." *Bulletin* **29**, Univ. of Illinois Engrg. Exp. Station, 85 p.

Taylor GI (1938), "Plastic strain in metals." *J. Inst. Metals*, **62**, 307–324.

Tippett LHC (1925), "On the extreme individuals and the range of samples", *Biometrika*, **17**, p. 364.

von Mises R (1936), "La distribution de la plus grande de n valeurs." *Rev. Math. Union Interbalcanique*, **1**, p. 1.

Weibull W (1939), "The phenomenon of rupture in solids." Proc., Royal Swedish Institute of Engineering Research (Ingenioersvetenskaps Akad. Handl.), **153**, Stockholm, 1–55.

Weibull W (1949), "A statistical representation of fatigue failures in solids." *Proc., Roy. Inst. of Techn.* No. **27**.

Weibull W (1951), "A statistical distribution function of wide applicability." *J. of Applied Mechanics ASME*, Vol. **18**.

Weibull W (1956), "Basic aspects of fatigue." *Proc., Colloquium on Fatigue*, Stockholm, Springer–Verlag.

Williams E (1957), "Some observations of Leonardo, Galileo, Mariotte and others relative to size effect." *Annals of Science* **13**, 23–29.

Withey MO (1907-08), "Tests of plain and reinforced concrete", *Bulletin* of the University of Wisconsin, Engineering Series **4** (1 and 2), 1–66.

1961–1965

Assur A (1963), "Breakup of pack-ice floes." *Ice and Snow: Properties, Processes and Applications*, MIT Press, Cambridge, Mass.

Barenblatt GI (1962), "The mathematical theory of equilibrium cracks in brittle fracture", *Advanced Appl. Mech.*, **7**, 55–129.

Biot MA (1965), *Mechanics of Incremental Deformations*, John Wiley & Sons, New York.

Cottrell AH (1963), *Iron and Steel Institute Special Report*, **69**, p. 281.

Giles RV (1962), *Theory and problems of fluid mechanics and hydraulics*, Mc-Graw Hill, New York (chapter 5).

Hill R (1962), "Acceleration waves in solids." *J. of Mechanics and Physics of Solids*, **10**, 1–16.

Hoek E and Bieniawski ZJ (1965), "Brittle fracture propagation in rock under compression." *Int. J. of Fracture Mech.*, **1**, 137–155.

Kaplan MF (1961),"Crack propagation and the fracture concrete", *ACI J.*, **58**, No. 11.

Kupfer H (1964), 'Erweiterung der Mörsch-schen Fachwerkanalogie mit Hilfe des Prinzips vom Minimum der Formänderungsarbeit (Generelization of

Mörsch's truss analogy using the principle of minimum strain energy)", Comite Euro-International du Beton, *Bulletin d'Information*, No. **40**, Paris, 44–57.

Leonhardt F, and Walther R (1962),"Beiträge zur Behandlung der Schubprobleme in Stahlbetonbau," *Beton- und Stahlbetonbau* (Berlin) **57** (3), 54–64, (6) 141–149. *Rheology and Soil Mechanics* (Proc., IUTAM Symp., Grenoble), ed. by J Kravtchenko and PM Sirieys, Springer Verlag, Berlin, 58–68.

Mandel J (1964), "Conditions de stabilité et postulat de Drucker."

Mindlin RD (1965) "Second gradient of strain and surface tension in linear elasticity." *Int. J. of Solids and Structures*, **1**, 417–438.

Nakayama J (1965), "Direct measurement of fracture energies of brittle heterogeneous material." *J. of the Amer. Ceramic Soc.*, **48** (11).

Rosen BW (1965). "Mechanics of composite strengthening." *Fiber Composite Materials*, Am. Soc. for Metals Seminar, Chapter 3.

Rüsch H, Haugli FR, and Mayer H (1962). "Schubversuch an Stahlbeton-Rechteckbalken mit gleichmässing verteilter Belastung," *Bulletin* No. **145**, Deutscher Ausschuss für Stahlbeton, Berlin, 4–30.

Rüsch H, and Hilsdorf H (1963), *Deformation characteristics of concrete under axial tension*, Vorunterschungen, Munich, Bericht **44**.

Thomas TY (1961), *Plastic flow and fracture in solids*, Academic Press, New York.

Toupin RA (1962), "Elastic materials with couple stresses." *Arch. Mech. and Rational Analysis*, **1**, 285–414.

Wells AA (1961), "Unstable crack propagation in metals-cleavage and fast fracture." *Symp. on Crack Propagation*, Cranfield, Vol.1, 210–230.

Williams ML (1965), in *International Journal of Fracture*, **1**, 292–310.

1966–1970

Bahl NS (1968), "Über den Einfluss der Balkenhöhe auf Schubtragfähigeit von einfeldrigen Stahlbetonbalken mit und ohne Schubbewehrung," *Dissertation*, Universität Stuttgart, 124 p.

Bažant ZP (1967), "L'instabilité d'un milieu continu et la résistance en compression." (Continuum instability and compression strength), *Bulletin RILEM* (Paris) (No. **35**), 99–112.

Bažant ZP (1968), "Effect of folding of reinforcing fibers on the elastic moduli and strength of composite materials." (in Russian), *Mekhanika Polimerov* (Riga), **4**, 314–321.

Evans RH, and Marathe MS (1968), "Microcracking and stress-strain curves for concrete in tension." *Mater. and Struct.*, **1**, pp. 61-64.

Freudenthal AM (1968), "Statistical approach to brittle fracture." Chapter 6 in *Fracture*, Vol. **2**, ed. by H. Liebowitz, Academic Press, 591–619.

Hawkes I and Mellor M (1970). "Uniaxial testing in rock mechanics laboratories." *Eng. Geol.* 4, 177–285.

Kani GNJ (1967), "Basic Facts Concerning Shear Failure." *ACI Journal, Proceedings*, **64**, 128–141.

Knauss WG (1970), *International Journal of Fracture*, **6**, 7–20.

Leicester RH (1969), "The size effect of notches." *Proc. 2nd Australasian Conf. on Mech. of Struct. Mater.*, Melbourne, pp. 4.1–4.20.

Maier G and Zavelani A (1970), "On the behavior of axially compressed metallic beams.", (in Italian), *Costruz. Metall.*, **22**, 282–297.

Paris P and Erdogan F (1967), "A critical analysis of crack propagation laws." *J. of Basic Engng.*, **87**, 528–534.

Paul B (1968), "Macroscopic criteria for plastic flow and brittle fracture." in *Fracture, an Advanced Treatise*, ed. by H. Liebowitz, Vol. **2**, Chapter 4.

Rice, J.R. (1968). "Path independent integral and approximate analysis of strain concentrations by notches and cracks." *ASME J. of Applied Mechanics*, **35**, 379–386.

Rice, J.R. (1968), in *Fracture*, H. Liebowitz, ed., Vol. **2**, Academic Press, 192–311.

Tattersall HG and Tappin G (1966), "The work of fracture and its measurement in metals, ceramics and other materials." *J. of Mater. Sci.*, **1**, 296–301.

Willis, J.R. (1967). "A comparison of fracture criteria of Griffith and Barenblatt." *J. of the Mech. and Phys. of Solids*, **15**, 151–162.

1971

Cundall PA (1971), "A computer model for simulating progressive large scale movements in blocky rock systems." *Proc. Int. Symp. on Rock Fracture*, ISRM, Nancy, France.

Kesler CE, Naus DJ, and Lott JL (1971), "Fracture Mechanics—Its applicability to concrete", *Proc. Int. Conf. on the Mechanical Behavior of Materials*, Kyoto; see also *The Soc. of Mater. Sci.*, **IV**, 1972, 113–124.

1972

Argon AS (1972), "Fracture of composites." *Treatise of Materials Science and Technology*, Academic Press, New York, Vol. **1**, p. 79.

Cotterell B (1972), "Brittle fracture in compression." *Int. J. of Fracture Mech.*, **8**, 195–208.

Taylor HPJ (1972), "Shear strength of large beams." *Proceedings ASCE*, **98** (ST11), 2473–2490.

Walsh PF (1972), "Fracture of plain concrete." *Indian Concrete Journal*, **46**, No. 11.

1973

Cotterell B (1972), "Brittle fracture in compression." *Int. J. of Fracture Mech.*, **8**, 195–208.

Cruse TA (1973), "Tensile strength of notched composites." *J. of Composite Materials*, **7**, 218–228.

Knauss, WC (1973). "On the steady propagation of a crack in a viscoelastic sheet; experiment and analysis." *The Deformation in Fracture of High Polymers*, HH Kausch, Ed., Plenum, New York 501–541.

Nesetova V, and Lajtai EZ (1973), "Fracture from compressive stress concentration around elastic flaws." *Int. J. of Rock Mech. and Mining Sci.*, **10**, 265–284.

Palmer, A.C. and Rice, J.R. (1973), "The growth of slip surfaces on the progressive failure of over-consolidated clay." *Proc. Roy. Soc. Lond. A.*, **332**, 527–548.

1974

Bieniawski ZT (1974), "Estimating the strength of rock materials." *J. of S. Afr. Inst. Min. Metal*, **74**, 312–320.

Knauss WC (1974), "On the steady propagation of a crack in a viscoelastic plastic solid." *J. of Appl. Mech. ASME*, **41**, 234-248.

Smith, E. (1974). "The structure in the vicinity of a crack tip: A general theory based on the cohesive crack model." *Engineering Fracture Mechanics*, **6**, 213-222.

Whitney JM and Nuismer RJ (1974), "Stress fracture criteria for laminated composites containing stress concentrations." *J. of Composite Materials*, **8**, 253-264.

Wnuk MP (1974), "Quasi-static extension of a tensile crack contained in viscoelastic plastic solid." *J. Appl. Mech. ASME*, **41**, 234-248.

Zaitsev JW and Wittmann, FH (1974), "A statistical approach to the study of the mechanical behavior of porous materials under multiaxial state of stress." *Proc. of the 1973 Symp. on Mechanical Behavior on Materials*, Kyoto, Japan, 705 p.

1975

Nielsen MP and Braestrup NW (1975), "Plastic shear strength of reinforced concrete beams." *Techn. Report* No. 3, Bygningsstatiske Meddelesler, Vol. **46**.

Streeter VL and Wylie EB (1975), *Fluid mechanics* (6th ed.), McGraw Hill, New York (chapter 4).

1976

Bažant ZP (1976), "Instability, ductility, and size effect in strain-softening concrete." *J. Engng. Mech. Div., Am. Soc. Civil Engrs .*, **102**, EM2, 331-344; disc. **103**, 357-358, 775-777, **104**, 501-502.

Hillerborg A, M Modéer and Petersson PE (1976), "Analysis of crack formation and crack growth in concrete by means of fracture mechanics and finite elements." *Cement and Concrete Research*, **6**, 773-782.

Thürlimann B (1976), "Shear strength of reinforced and prestressed concrete beams, CEB approach." *Technical Report*, E.T.H. Zürich, 33 p.

Walsh PF (1976), "Crack initiation in plain concrete." *Magazine of Concrete Research*, **28**, 37-41.

1977

Ingraffea AR (1977), "Discrete fracture propagation in rock: Laboratory tests and finite element analysis." *Ph.D. Dissertation*, University of Colorado, Boulder.

Kfouri AP and Rice JR (1977), "Elastic-plastic separation energy rate for crack advance in finite growth steps." *Fracture 1977* (Proc., 4th Int. Conf. on Fracture, ICF4, Waterloo), DMR Taplin, Ed., Univ. of Waterloo, Ontario, Canada, Vol. **1**, 43–59.

Leonhardt F (1977), "Schub bei Stahlbeton und Spannbeton—Grundlagen der neueren Schubbemessung." *Beton- und Stahlbetonbau*, **72** (11), 270-277, (12), 295-392.

Mihashi H and Izumi M (1977), "Stochastic theory for concrete fracture." *Cem. Concr. Res.*, **7**, 411-422.

Zech B and Wittmann FH (1977), "A complex study on the reliability assessment of the containment of a PWR, Part II. Probabilistic approach to describe the behavior of materials." *Trans. 4th Int. Conf. on Structural Mechanics in Reactor Technology*, T.A. Jaeger and B.A. Boley, eds., European Communities, Brussels, Belgium, Vol. **H**, J1/11, 1-14.

1978

Bender MC and Orszag SA (1978), *Advanced mathematical methods for scientists and engineers*, McGraw Hill, New York (Chapters 9–11).

Collins MP (1978), "Towards a rational theory for RC members in shear." *ASCE J. of the Structural Division*, **104**, 396–408.

Daniel IM (1978), "Strain and failure analysis of graphite/epoxy plate with cracks." *Experimental Mechanics*, **8**, 246–252.

Kendall K (1978), "Complexities of compression failure." *Proc. Royal Soc. London.*, A. **361**, 245–263.

Schapery RA (1978), *Int. J. of Fracture*, **14**, 293–309.

Walraven JC (1978), "The influence of depth on the shear strength of lighweight concrete beams without shear reinforcement." *Stevin Laboratory Report* No. 5-78.4, Delft University of Technology, 36 p.

1979

Barenblatt GI (1979), "Similarity, self-similarity and intermediate asymptotics." *Consultants Bureau*, New York.

Bažant ZP and Cedolin L (1979), "Blunt crack band propagation in finite element analysis." *J. of the Engng. Mech. Div., Proc. ASCE*, **105**, 297–315.

Bažant ZP and Estenssoro L F (1979), "Surface singularity and crack propagation." *Int. J. of Solids and Structures*, **15**, 405–426. Addendum Vol. **16**, 479–481.

Cundall PA and Strack ODL (1979), "A discrete numerical model for granular assemblies." *Geotechnique*, London, Vol. **29**, 47–65.

1980

Bažant ZP and Cedolin L (1980), "Fracture mechanics of reinforced concrete." *J. of the Engng. Mech. Div., Proc., ASCE*, **106**, 1257–1306.

Collins MP and Mitchell D (1980), "Shear and torsion design of prestressed and non-prestressed concrete beams." *Journal of the Prestressed Concrete Institute*, **25**, 32–100. Also, Discussion, Vol. **26**, 96–118.

Daniel IM (1980), "Behavior of graphite/epoxy plates with holes under biaxial loading." *Exper. Mechanics*, **20**, 1–8.

Ingraffea AR and Heuzé FE (1980), "Finite element models for rock fracture mechanics." *Intern J. of Numerical and Analytical Methods in Geomechanics*, **4**, 25–43.

Marti P (1980), " Zur plastischen Berechnung von Stahlbeton." *Bericht Nr.* **104,** Institute für Baustatik und Konstruktion, ETH Zürich.

1981

Chana PS (1981), "Some aspects of modelling the behaviour of reinforced concrete under shear loading." *Technical Report* No. **543**, Cement and Concrete Association, Wexham Springs, 22 p.

Daniel IM (1981), "Biaxial testing of graphite/epoxy laminates with cracks." ASTM STP 734, *American Soc. for Testing and Materials*, 109–128.

Fairhurst C and Cornet F (1981), "Rock fracture and fragmentation." Proc., 22nd U.S. *Symp. on Rock Mechanics* (held at MIT, June), 21–46.

Selected Papers by Alfred M. Freudenthal (1981). Am. Soc. of Civil Engrs., New York.

McKinney KR and Rice RW (1981), "Specimen size effects in fracture toughness testing of heterogeneous ceramics by the notch beam test." *Fracture Mechanics Methods for Ceramics, Rocks and Concrete*, Freiman SW and Fuller ER, eds., Am. Soc. for Testing Materials, Philadelphia, (ASTM Special Technical Publication No. 745) 118–126.

Mihashi H and Zaitsev JW (1981), "Statistical nature of crack propagation." Section 4-2 in *Report to RILEM TC 50 - FMC*, ed. Wittmann, F.H.

Petersson PE (1981), "Crack growth and development of fracture zones in plain concrete and similar materials." *Report* TVBM-1006, Div. of Building Materials, Lund Inst. of Tech., Lund, Sweden.

Wittmann FH and Zaitsev Yu V (1981), "Crack propagation and fracture of composite materials such as concrete." *Proc., 5th Int. Conf. on Fracture (ICF5)*, Cannes.

Zaitsev Yu V and Wittmann FH (1981), "Simulation of crack propagation and failure of concrete." *Materials and Structures* (Paris), **14**, 357–365.

1982

Bažant ZP (1982), "Crack band model for fracture of geomaterials." *Proc. 4th Intern. Conf. on Num. Meth. in Geomechanics*, ed. by Eisenstein Z, held in Edmonton, Alberta, Vol. **3**, 1137–1152.

Daniel IM (1982), "Failure mechanisms and fracture of composite laminates with stress concentrations." *VIIth International Conference on Experimental Stress Analysis*, Haifa, Israel, Aug. 23–27, 1–20.

Horii H and Nemat-Nasser H (1982). "Compression-induced nonplanar crack extension with application to splitting, exfoliation and rockburst." *J. of Geophys. Res.*, **87**, 6806–6821.

Kachanov M (1982), "A microcrack model of rock inelasticity—Part I. Frictional sliding on microcracks." *Mechanics of Materials*, **1**, 19–41.

Schapery RA (1982), *Proc., 9th U.S. Nat. Congress of Applied Mechanics*, Am. Soc. of Mech. Engrs. (ASME), 237–245.

1983

Bažant ZP (1983), "Fracture in concrete and reinforced concrete." in *Preprints, Prager Symp. on Mechanics of Geomaterials: Rocks, Concretes, Soils*, ed. by Bažant ZP, Northwestern University Press, 281–316.

Bažant ZP and Cedolin L (1983), "Finite element modeling of crack band propagation." *J. of Structural Engineering*, ASCE, **109**, 69–92.

Bažant ZP and Oh B-H (1983), "Crack band theory for fracture of concrete." *Materials and Structures* (RILEM, Paris), **16**, 155–177.

Beremin FM (1983), "A local criterion for cleavage fracture of a nuclear pressure vessel steel." *Metallurgy Transactions A*, **14**, 2277–2287.

Budianski B (1983), "Micromechanics." *Computers and Structures*, **16**, 3–12.

Hillerborg A (1983), "Examples of practical results achieved by means of the fictitious crack model. " in *Preprints, Prager Symp. on Mechanics of Geomaterials: Rocks, Concretes, Soils*, ed. by Bažant ZP, Northwestern University Press, 611–614.

Mihashi H (1983), "Stochastic theory for fracture of concrete." *Fracture mechanics of concrete*, Wittmann FH ed., Elsevier Science Publishers, B.V., Amsterdam, The Netherlands, 301–339.

1984

Bažant ZP (1984a), "Size effect in blunt fracture: Concrete, rock, metal." *J. of Engng. Mechanics*, ASCE, **110**, 518–535.

Bažant ZP (1984b), "Imbricate continuum and its variational derivation." *J. of Engng. Mech.*, ASCE, **110**, 1693–1712.

Bažant ZP (1984c), "Microplane model for strain controlled inelastic behavior." Chapter 3 in *Mechanics of Engineering Materials* (Proc., Conf. held at U. of Arizona, Tucson, Jan. 1984), Desai CS and Gallagher RH eds., John Wiley, London, 45–59.

Bažant ZP, Belytschko TB, and Chang T-P (1984), "Continuum model for strain softening." *J. of Engng. Mechanics* ASCE, **110**, 1666–1692.

Bažant ZP, Kim JK, and Pfeiffer P (1984), "Determination of nonlinear fracture parameters from size effect tests." Preprints, *NATO Advanced Research Workshop* on "Application of Fracture Mechanics to Cementitiious Composites." Northwestern University Press, IL, Shah SP ed., 143–169.

Bažant ZP and Kim J-K (1984), "Size effect in shear failure of longitudinally reinforced beams." *Am. Concrete Institute Journal*, **81**, 456–468; Disc. & Closure **82**(1985), 579–583.

Bažant ZP and Oh Byung H (1984), "Rock fracture via strain-softening finite elements." *J. of Engng. Mechanics*, ASCE, **110**, 1015–1035.

Carpinteri A (1984), "Scale effects in fracture of plain and reinforced concrete structures." in *Fracture Mechanics of Concrete: Structural Application and Numerical Calculation*, ed. by Sih GC and DiTommaso A, Martinus Nijhoff Publ., The Hague, 95–140.

Mandelbrot BB, Passoja DE, and Paullay A (1984), "Fractal character of fracture surfaces of metals." *Nature*, **308**, 721–722.

Schapery RA (1984), *International Journal of Fracture*, **25**, 195–223.

Steif PS (1984), "Crack extension under compressive loading." *Eng. Frac. Mech.*, **20**, 463–473.

1985

Bažant ZP (1985a), "Fracture mechanics and strain-softening in concrete." Preprints, *U.S.- Japan Seminar on Finite Element Analysis of Reinforced Concrete Structures*, Tokyo, Vol. **1**, pp. 47–69.

Bažant ZP (1985b), "Comment on Hillerborg's size effect law and fictitious crack model." *Dei Poli Anniversary Volume*, Politecnico di Milano, Italy, ed. by L. Cedolin et al., 335–338.

Bažant ZP and Belytschko TB (1985), "Wave propagation in strain-softening bar: Exact solution." *J. of Engng. Mechanics*, ASCE, **111**, 381–389.

Bažant ZP and Kim J-K (1985), "Fracture theory for nonhomogeneous brittle materials with application to ice." *Proc. ASCE Nat. Conf. on Civil Engineering in the Arctic Offshore ARCTIC*, **85**, San Francisco, ed. by Bennett LF, ASCE New York, 917–930.

Daniel IM (1985), "Mixed-mode failure of composite laminates with cracks." *Experimental Mechanics*, **25**, 413–420.

Hasegawa T, Shioya T, and Okada T (1985), "Size effect on splitting tensile strength of concrete." *Proc., 7th Conference of Japan Concrete Institute*, 305–312.

Hillerborg A (1985a), "Theoretical basis of method to determine the fracture energy G_f of concrete." *Materials and Structures*, **18**, 291–296.

Hillerborg A (1985b), "Results of three comparative test series for determining the fracture energy G_f of concrete." *Materials and Structures*, **18**.

Hsu TTC (1985), "Softened truss model theory for shear and torsion." *ACI Structural Journal*, **85**, 624–635.

Iguro M, Shiyoa T, Nojiri Y, and Akiyama H (1985), "Experimental studies on shear strength of large reinforced concrete beams under uniformly distributed load." *Concrete Library International, Japan Soc. of Civil Engrs.* No. 5, 137–154 (translation of 1984 article in Proc. JSCE).

Jenq YS and Shah SP (1985), "A two parameter fracture model for concrete." *Journal of Engineering Mechanics*, **111**, 1227–1241.

Kachanov M (1985), "A simple technique of stress analysis in elastic solids with many cracks." *Int. J. of Fracture*, **28**, R11–R19.

Marti P (1985), " Basic tools of reinforced concrete beam design." *ACI Journal* **82**, 46–56. Discussion, **82**, 933-935.

Reihnardt HW (1985), "Crack softening zone in plain concrete under static loading." *Cement and Concrete Research*, **15**, 42–52.

RILEM Recommendation (1985), "Determination of fracture energy of mortar and concrete by means of three-point bend tests of notched beams." RILEM TC 50–FMC, *Materials and Structures*, **18**.

Zaitsev YV (1985), "Inelastic properties of solids with random cracks." *Mechanics of Composite Materials: Rocks, Concretes, Soils.* Bažant ZP ed., J. Wiley & Sons, Chichester and New York, 89–128.

1986

Ashby MF and Hallam SD (1986), "The failure of brittle solids containing small cracks under compressive stress states." *Acta Metall.*, **34**, 497–510.

Bažant ZP (1986), "Mechanics of distributed cracking." *Appl. Mech. Reviews ASME*, **39**, 675–705.

Bažant ZP and Cao Z (1986a), "Size effect in brittle failure or unreinforced pipes." *Am. Concrete Institute Journal*, **83**, 365–373.

Bažant ZP and Cao Z (1986b), "Size effect in shear failure of prestressed concrete beams." *Am. Concrete Inst. Journal*, **83**, 260–268.

Bažant ZP, Kim J-K, and Pfeiffer PA (1986), "Nonlinear fracture properties from size effect tests." *J. of Structural Engng., ASCE*, **112**, 289–307.

Bažant ZP and Pfeiffer PA (1986), "Shear fracture tests of concrete." *Materials and Structures* (RILEM, Paris), **19**, 111-121.

Belytschko TB, Bažant ZP, Hyun YW, and Chang T-P (1986), "Strain-softening materials and finite element solutions." *Computers and Structures*, **23**, 163-180.

Carpinteri A (1986), Mechanical damage and crack growth in concrete. *Martinus Nijhoff Publ.—Kluwer*, Dordrecht-Boston.

Foote RML, Mai Y-W, and Cottrell B (1986), "Crack growth resistance curves in strain-softening materials." *J. of the Mech. and Phys. of Solids*, **34**, 593-607.

Horii H and Nemat-Nasser H (1986). "Brittle failure in compression, splitting, faulting and brittle-ductile transition." *Phil. Trans. of Royal Soc.* London 319 (1549), 337-374.

Maier G (1986), "On softening flexural behavior in elastic-plastic beams." *Studi i Ricerche* (Politecnico di Milano), **8**, 85-117.

Nallathambi P and Karihaloo BL (1986), "Determination of specimen- size independent fracture toughness of plain concrete." *Mag. of Concrete Res.*, **38**, 67-76.

Planas J and Elices M (1986), "Un nuevo método de análisis de comportamiento de una fisura en Modo I." *Anales de Mecánica de la Fractura*, **3**, 219-227.

Sammis CG and Ashby MF (1986), "The failure of brittle porous solids under compressive stress state." *Acta Metall*, **34**, 511-526.

Shetty DK, Rosenfield AR, and Duckworth WH (1986), "Mixed mode fracture of ceramics in diametrical compression." *J. Am. Ceram. Soc.*, **69**, 437-443.

van Mier JGM (1986), "Multiaxial strain-softening of concrete." *Materials and Structures* (RILEM, Paris) **19**, 179-200.

Vecchio F, Collins MP (1986), "The modified compression field theory for reinforced concrete elements subjected to shear". *ACI Journal*, **83**, 219-231.

Zaitsev Yu V (1986), "Inelastic properties of solids with random cracks." in *Mechanics of Geomaterials*, ed. by Bažant ZP, (Proc., IUTAM Prager Symp., held at Northwestern University, 1983), John Wiley & Sons, 89-128.

1987

Barenblatt GI (1987), *Dimensional analysis.*, Gordon and Breach Sci. Publ., New York.

Bažant ZP (1987a), "Why continuum damage is nonlocal: Justification by quasi-periodic microcrack array." *Mechanics Research Communications*, **14**, 407–419.

Bažant ZP (1987b), "Fracture energy of heterogeneous material and similitude." Preprints, SEM-RILEM *Int. Conf. on Fracture of Concrete and Rock* (held in Houston, Texas, June 1987), ed. by Shah SP and Swartz SE, publ. by SEM (Soc. for Exper. Mech.) 390–402.

Bažant ZP and Cao Z (1987), "Size effect in punching shear failure of slabs." *ACI Structural Journal* (Am. Concrete Inst.), **84**, 44–53.

Bažant ZP and Pfeiffer PA (1987), "Determination of fracture energy from size effect and brittleness number." *ACI Materials Jour.*, **84**, 463–480.

Bažant ZP, and Pijaudier-Cabot G (1987), "Modeling of distributed damage by nonlocal continuum with local strains", *Numerical Methods in Fracture Mech.* (Proc., 4th Int. Conf. held in San Antonio, Texas), ed. by Luxmore AR et al., Pineridge Press, Swansea, U.K., 411–431.

Bažant ZP, Pijaudier-Cabot G, and Pan J-Y (1987a), "Ductility, snapback, size effect and redistribution in softening beams and frames." *ASCE J. of Structural Engng.*, **113**, 2348–2364.

Bažant ZP, Pijaudier-Cabot G, and Pan J-Y (1987b), "Ductility, snapback, size effect and redistribution in softening beams and frames." ASCE *J. of Structural Engng.*, **113**, 2348–2364.

Bažant ZP and Şener S (1987), "Size effect in torsional failure of concrete beams." *J. of Struct. Engng. ASCE*, **113**, 2125–2136.

Bažant ZP and Sun H-H (1987), "Size effect in diagonal shear failure: Influence of aggregate size and stirrups." *ACI Materials Journal*, **84**, 259–272.

Belytschko T, Wang X-J, Bažant ZP, and Hyun T (1987), "Transient solutions for one-dimensional problems with strain-softening." *Trans. ASME, J. of Applied Mechanics*, **54**, 513–516.

Brown SR (1987) "A note on the description of surface roughness using fractal dimension." *Geophysical Res. Letters* **14**, No.11, 1095–1098, and **15**, No. 11 (1987) 286.

de Borst R (1987), *Computer Methods in Applied Mechanics and Engineering*, **62**, 89–110.

Kachanov M (1987), "Elastic solids with many cracks: A simple method of analysis." *Int. J. of Solids and Structures*, **23**, 23–43.

Kemeny JM and Cook NGW (1987), "Crack models for the failure of rock under compression." *Proc., 2nd Int. Conf. on Constitutive Laws for Eng. Mat.* (held in Tucson), ed. by C.S. Desai et al., Elsevier Science Publ., New York, Vol. **2**, 879–887.

Murakami Y (1987), *Stress intensity factors handbook*, Pergamon Press.

Pijaudier-Cabot G and Bažant ZP (1987), "Nonlocal damage theory." *J. of Engng. Mechanics, ASCE*, **113**, 1512–1533.

Schlaich J, Schafer K, and Jannewein M (1987), "Toward a consistent design for structural concrete." *PCI Journal*, **32**, 75–150.

Xie H (1987), "The fractal effect of irregularity of crack branching on the fracture toughness of brittle materials." *International Journal of Fracture*, **41**, 267–274.

Zubelewicz A and Bažant ZP (1987), "Interface modeling of fracture in aggregate composites." *ASCE J. of Engng. Mech.*, **113**, 1619–1630.

1988

Bažant ZP (1988a), "Softening instability: Part I — Localization into a planar band." *J. of Appl. Mech. ASME*, **55**, 517–522.

Bažant ZP (1988b), "Softening instability: Part II — Localization into ellipsoidal regions." *J. of Appl. Mech. ASME*, **55**, 523–529.

Bažant ZP and Lin F-B (1988a), "Nonlocal smeared cracking model for concrete fracture." *J. of Struct. Engng. ASCE*, **114**, 2493–2510.

Bažant ZP and Lin F-B (1988b), "Nonlocal yield limit degradation." *International J. for Numerical Methods in Engineering*, **26**, 1805–1823.

Bažant ZP and Pijaudier-Cabot G (1988), "Nonlocal continuum damage, localization instability and convergence." *ASME J. of Applied Mechanics*, **55**, 287–293.

Bažant ZP and Prat PC (1988), "Measurement of mode III fracture energy of concrete." *Nuclear Engineering and Design*, **106**, 1–8.

Bažant ZP and Şener S (1988), "Size effect in pullout tests." *ACI Materials Journal*, **85**, 347–351.

Bažant ZP, Şener S, and Prat PC (1988), "Size effect tests of torsional failure of plain and reinforced concrete beams." *Materials and Structures* RILEM, Paris, **21**, 425–430.

Hsu TTC (1988), "Softened truss model theory for shear and torsion." *ACI Structural Journal*, **85** (6), 624–635.

Kittl P and Diaz G (1988), "Weibull's fracture statistics, or probabilistic strength of materials: state of the art." *Res Mechanica*, **24**, 99–207.

Mecholsky JJ and TJ Mackin (1988), "Fractal analysis of fracture in ocala chert." *Journal of Materials Science and Letters*, **7**, 1145–1147.

Planas J and Elices M (1988), "Conceptual and experimental problems in the determination of the fracture energy of concrete." *Proc. Int. Workshop on "Fracture Toughness and Fracture Energy, Test Methods for Concrete and Rock*, Tohoku Univ., Sendai, Japan, 203–212.

Pijaudier-Cabot G, Bažant ZP, and Tabbara M (1988), "Comparison of various models for strain-softening." *Engineering Computations*, **5**, 141–150.

Pijaudier-Cabot G and Bažant ZP (1988), "Dynamic stability analysis with nonlocal damage." *Computers and Structures*, **29**, 503–507.

Rots JG (1988), *Computational modeling of concrete structures*. Ph.D. Thesis, Delft University of Technology, Netherlands.

Sanderson TJO (1988), *Ice Mechanics: Risks to Offshore Structures*, Graham and Trotman Limited, London.

1989

Bažant ZP (1989), "Identification of strain-softening constitutive relation from uniaxial tests by series coupling model for localization." *Cement and Concrete Research*, **19** (6), 973–977.

Bažant ZP and Pijaudier-Cabot G (1989), "Measurement of characteristic length of nonlocal continuum." *J. of Eng. Mech. ASCE*, **115**, 755–767.

Cahn R (1989), "Fractal dimension and fracture." *Nature*, **338** (Mar.), 201–202.

Carpinteri A (1989), "Decrease of apparent tensile and bending strength with specimen size: Two different explanations based on fracture mechanics." *Int. J. Solids Struct.*, **25**, 407–429.

Chen CT and J Runt (1989), "Fractal analysis of polystyrene fracture surfaces." *Polymer Communications*, **30**, 334–335.

Droz P and Bažant ZP (1989), "Nonlocal analysis of stable states and stable paths of propagation of damage shear bands." *Cracking and Damage, Strain Localization and Size Effect* (Proc. of France-U.S. Workshop, Cachan, France) Mazars J and Bažant ZP Eds., Elsevier, 183–207.

Elices M and Planas J (1989), "Material Models", Chapter 3 in *Fracture Mechanics of Concrete Structures*, L. Elfgren (Ed.), Chapman & Hall, London, 16–66.

Haimson BC and Herrick CG (1989), "In-situ stress calculation from borehole breakout experimental studies." *Proc., 26th U.S. Symp. on Rock Mech.*, 1207–1218.

Hornbogen E (1989), "Fractals in microstructure of metals." *International Materials Review*, 6, 277–296.

Kittl P and Diaz G (1989), "Some engineering applications of the probablistic strength of materials." *Appl. Mech. Rev.*, 42, 108-112.

Knauss WG (1989), in *Advances in Fracture Research*, 4, 7th Int. Conf. on Fracture, Houston, Texas, 2683–2711.

Marti P (1989), "Size effect in double-punch tests on concrete cylinders." *ACI Materials Journal*, 86, 597–601.

Planas J, Elices,M, and Toribio J (1989), "Approximation of cohesive carck models by R-CTOD curves." *Fracture of Concrete and Rock: Recent Developments*, eds. Shah SP, Swartz SE, and Barr B (Int. Conf. held at Cardiff, U.K.), Elsevier, London, 203-212.

Planas J and Elices M (1989a), "Size-effect in concrete structures: Mathematical approximation and experimental validation." *Cracking and Damage, Strain Localization and Size Effect* (Proc. of France-U.S. Workshop, Cachan, France) Mazars J and Bažant ZP Eds., Elsevier, 462–476.

Planas J and Elices M (1989b), "Conceptual and experimental problems in the determination of the fracture energy of concrete," in *Fracture Toughness and Fracture Energy* (Proc., RILEM Intern. Workshop held in 1988 at Tohoku Univ., Sendai, Japam), Mihashi et al. eds., Balkema, Rotterdam, 165–181.

Shioya T, Iguro M, Nojiri Y, Akiayama H, and Okada T (1989), "Shear strength of large reinforced concrete beams." *Fracture Mechanics: Application to Concrete*, SP-118, American Concrete Institute, Detroit, 25–279.

Xi Y and Bažant ZP (1989), "Sampling analysis of concrete structures for creep and shrinkage with correlated random material parameters." *Probabilistic Engineering Mechanics*, 4, 174–186.

Xie H (1989), "Studies on fractal models of microfractures of marble." *Chinese Science Bulletin*, **34**, 1292-1296.

1990

Bažant ZP (1990a), "Smeared-tip superposition method for nonlinear and time-dependent fracture." *Mechanics Research Communications*, **17**, 343-351.

Bažant ZP (1990b), "Equilibrium path bifurcation due to strain-softening localization in ellipsoidal region." *Journal of Applied Mechanics, ASME*, **57**, 810-814.

Bažant ZP and Kazemi MT (1990), "Determination of fracture energy, process zone length and brittleness number from size effect, with application to rock and concrete." *Int. J. of Fracture*, **44**, 111-131.

Bažant ZP and Kazemi MT (1990), "Size effect in fracture of ceramics and its use to determine fracture energy and effective process zone length." *J. of American Ceramic Society*, **73**, 1841-1853.

Bažant ZP and Ožbolt J (1990), "Nonlocal microplane model for fracture, damage, and size effect in structures." *ASCE J. of Engng. Mech.*, **116**, 2484-2504.

Bažant ZP, Prat PC, and Tabbara MR (1990), "Antiplane shear fracture tests (Mode III)." *ACI Materials Journal*, **87**, 12-19.

Bažant ZP, Tabbara MR, Kazemi MT, and Pijaudier-Cabot G (1990), "Random particle model for fracture of aggregate or fiber composites." *ASCE J. of Engng. Mech.*, **116**, 1686-1705.

Bhat SU (1990), "Modeling of size effect in ice mechanics using fractal concepts." *Journal of Offshore Mechanics and Arctic Engineering*, **112**, 370-376.

Bouchaud E, Lapasset G, and Planes J (1990), "Fractal dimension of fractured surfaces: a universal value?" *Europhysics Letters* **13**, No.1, 73-79.

Chelidze T and Gueguen Y (1990), "Evidence of fractal fracture," *International Journal of Rock Mechanics and Mininig Sciences*, **27**, 223-225.

Gettu R, Bažant ZP, and Karr ME (1990), "Fracture properties and brittleness of high-strength concrete." *ACI Materials Journal*, **87**, 608-618.

Eligehausen R and Ožbolt J (1990), "Numerical Analysis of headed studs embedded in large plain concrete blocks." *Computer Aided Analysis and Design of Concrete Structures* (Proc., Int. Symp. held in Zell am See, Austria). Bićanić N and Mang H eds., Pineridge Press, Swansea, 645-656.

Hermann H and Roux S (1990), *Statistical models for the fracture of disordered media*, North Holland, Amsterdam (chapter 5).

Jishan X and Xixi H (1990), "Size effect on the strength of a concrete member." *Engrg. Fracture Mechanics*, **35**, 687–696.

Kittl P and Diaz G (1990), "Size effect on fracture strength in the probabilistic sterngth of materials." *Reliability Engrg. Sys. Saf.*, **28**, 9-21.

Lemaitre J and Chaboche J-L (1990), *Mechanics of solid materials*, Cambridge University Press, Cambridge.

Needleman A (1990), "An analysis of tensile decohesion along an interface." *J. of the Mech. and Phys. of Solids*, **38**, 289–324.

Peng G and Tian D (1990), "The fractal nature of a fracture surface." *Journal of Physics A: Matematics and General*, **23**, 3257–3261.

RILEM Recommendation (1990), "Size effect method for determining fracture energy and process zone of concrete." *Materials and Structures*, **23**, 461-465.

Sanderson TJO (1990), "Ice mechanics." *Graham and Trotman*, London.

Saouma VC, Barton C, and Gamal-el-Din N (1990), "Fractal characterization of concrete crack surfaces." *Engineering Fracture Mechanics*, **35**, No. 1.

Schulson EM (1990), "The brittle compressive failure of ice." *Acta Metall. Mater.*, **38**, 1963-1976.

Slepyan LI (1990), "Modeling of fracture of sheet ice." *Izvestia AN SSSR, Mekh. Tverd. Tela*, **25**, 151–157

1991

Bažant ZP (1991), "Why continuum damage is nonlocal: Micromechanics arguments." *Journal of Engineering Mechanics ASCE*, **117**, 1070–1087.

Bažant ZP, Editor (1991) Fracture Mechanics of Concrete Structures (Part I) (*Proc., First Int. Conf. on Fracture Mech. of Concrete Structures* (FraMCoS-1), held in Breckenridge, Colorado), Elsevier, London.

Bažant ZP and Cedolin L (1991), *Stability of Structures: Elastic, Inelastic, Fracture and Damage Theories*, Oxford University Press, New York.

Bažant ZP, Gettu R, and Kazemi MT (1991), "Identification of nonlinear fracture properties from size-effect tests and structural analysis based on geometry-dependent R-curves." *International Journal of Rock Mechanics and Mining Sciences*, **28**, 43–51.

Bažant ZP and Kazemi MT (1991a), "Size effect on diagonal shear failure of beams without stirrups." *ACI Structural Journal*, **88**, 268–276.

Bažant ZP and Kazemi MT (1991b), "Size dependence of concrete fracture energy determined by RILEM work-of-fracture method." *International J. of Fracture*, **51**, 121–138.

Bažant ZP, Kazemi MT, Hasegawa T, and Mazars J (1991), "Size effect in Brazilian split-cylinder tests: measurement and fracture analysis." *ACI Materials Journal*, **88**, 325–332.

Bažant ZP and Kim J-K (1991), "Consequences of diffusion theory for shrinkage of concrete." *Materials and Structures* (RILEM, Paris), **24**, 323–326.

Bažant ZP and Xi Y (1991), "Statistical size effect in quasi-brittle structures: II. Nonlocal theory." *ASCE J. of Engineering Mechanics*, **117**, 2623–2640.

Bažant ZP, Xi Y, and Reid SG (1991), "Statistical size effect in quasi-brittle structures: I. Is Weibull theory applicable?" *ASCE J. of Engineering Mechanics*, **117**, 2609–2622.

Bažant ZP and Xu K (1991), "Size effect in fatigue fracture of concrete." *ACI Materials J.*, **88**, 390–399.

Collins MP and Mitchell D (1991), "Prestressed Concrete Structures." *Prentice Hall*, Englewood Cliffs.

Costin DM (1991), "Damage Mechanics in the post-failure regime," *Mech. of Materials*, **4**, 149–160.

Hinch K (1991), *Perturbation Methods*. Cambridge University Press, Cambridge.

Hu XZ and Wittmann FH (1991), "An analytical method to determine the bridging stress transferred within the fracture process zone: I. General theory." *Cement and Concrete Research*, **21**, 1118–1128.

Karihaloo BL and Nallathambi P (1991), "Notched beam test: Mode I Fracture Toughness." in *Fracture Mechanics Test Methods for Concrete*, eds. Shah SP and Carpinteri A, Chapman and Hall, London, 1–86.

Kemeny JM and Cook NGW (1991), "Micromechanics of deformation in rock." in *Toughening Mechanisms in Quasibrittle Materials*, ed. by Shah SP et al., Kluwer, Netherlands, 155–188.

Long QY, Suquin L, and Lung CW (1991), "Studies of fractal dimension of a fracture surface formed by slow stable crack propagation." *Journal of Physics*, **24**.

Mihashi H, Nomura N, Izumi M, and Wittmann FH (1991), "Size dependence of fracture energy of concrete." in *Fracture Processes in Concrete, Rocks and Ceramics,* van Mier, Rots and Bakker (Eds.), 441–450.

Ouyang C and Shah SP (1991), "Geometry-Dependent R-Curve for Quasi-Brittle Materials." *J. of Amer. Ceramic Soc.,* **74**, 2831-2836.

Palmer AC and Sanderson JO (1991), "Fractal crushing of ice and brittle solids." *Proceedings of the Royal Society London,* **433**, 469–477.

Petersson PE (1991), *Crack growth and development of fracture zones in plain concrete and similar materials* (Report TVBM–1006), Division of Building Materials, Lund Institute of Technology, Lund, Sweden.

Reineck K-H (1991), "Model for structural concrete members without transverse reinforcement." *Proc., IABSE Colloquium on Structural Concrete,* Stuttgart. IABSE Rep. Vol. **62**, 643–648.

Swenson, D.V. and Ingraffea, A.R. (1991). "The collapse of the Schoharie Creek bridge: A case study in concrete fracture mechanics." *Int. J. Fracture,* **51**, 73–92.

1992

ACI Committee 446 on Fracture Mechanics (1992) (Bažant, Z.P. princ. author & chair). "Fracture mechanics of concrete: concepts, models and detemination of material properties." *Fracture Mechanics of Concrete Structures (Proc. FraMCoS1—Int. Conf. on Fracture Mechanics of Concrete Structures,* Breckenridge, Colorado, June), ed. by Bažant ZP, Elsevier Applied Science, London, 1–140.

ACI Committee 446 (1992). "State-of-art-report on fracture mechanics of concrete: concepts, model and determination of material properties." in *Fracture Mechanics of Concrete Structures,* ed. by Bažant ZP, Elsevier Applied Science, London, New York, 4-144.

Bao G, Ho S, Suo Z, and Fan B (1992), "The role of material orthotropy in fracture specimens for composites." *Int. J. Solid Structures,* **29**, 1105-1116.

Bažant ZP (1992a), "Large-scale fracture of sea ice plates." *Proc. 11th IAHR Ice Symposium, Banff, Alberta,* June (ed. by T.M. Hrudey, Dept. of Civil Engineering, University of Alberta, Edmonton), vol 2, pp. 991–1005.

Bažant ZP (1992b), "Large-scale thermal bending fracture of sea ice plates." *J. of Geophysical Research,* **97** (C11), 17,739–17,751.

Bažant ZP, Editor (1992), *Fracture Mechanics of Concrete Structures*, Proc., First Intern. Conf. (FraMCoS-1), held in Breckenridge, Colorado, June 1-5, Elsevier, London (1040 p.).

Bažant ZP and Gettu R (1992), "Rate effects and load relaxation: Static fracture of concrete." *ACI Materials Journal*, **89**, 456–468.

Borodich F (1992), "Fracture energy of fractal crack, propagation in concrete and rock" (in Russian), *Doklady Akademii Nauk*, **325**, 1138–1141.

Carter BC (1992), "Size and stress gradient effects on fracture around cavities." *Rock Mech. and Rock Engng.* (Springer), **25**, 167–186.

Carter BC, Lajtai EZ, and Yuan Y (1992), "Tensile fracture from circular cavities loaded in compression." *Int. J. of Fracture*, **57**, 221-236.

Elices M, Guinea GV, and Planas J (1992), "Measurement of the fracture energy using three-point bend tests: Part 3—Influence of cutting the $P - \delta$ tail." *Materials and Structures*, **25**, 327–334.

Eligehausen, R., Bouška, P., Červenka, V., and Pukl, R. (1992), "Size effect of the concrete cone failure load of anchor bolts." *Fracture Mechanics of Concrete structures* (Proc., First Int. Conf. FraMCoS-1, held in Breckenridge, Colorado), ed. by Bažant ZP, Elsevier, London, 517–525.

Eligehausen R and Ožbolt J (1992), "Size effect in concrete structures." *Applications of Fracture Mechanics to Reinforced Concrete.* (Proc., Torino), Carpinteri A ed., Elsevier, Netherlands, 17–44.

Guinea GV, Planas J and Elices M (1992), "Measurement of the fracture energy using three-point bend tests: Part 1—Influence of experimental procedures", *Materials and Structures,*, **25**, 212–218.

He S, Plesha ME, Rowlands RE, and Bažant ZP (1992), "Fracture energy tests of dam concrete with rate and size effects." *Dam Engineering* **3**, 139–159.

Hu XZ and Wittmann FH (1992), "An analytical method to determine the bridging stress transferred within the fracture process zone: I. Application to mortar." *Cement and Concrete Research*, **21**, 559–570.

Issa MA, Hammad AM, and Chudnovsky A (1992), "Fracture surface characterization of concrete." *Proc., 9th ASCE Conference on Engineering Mechanics*, ASCE, N.Y.

Jackson KE, Kellas S, and Morton J (1992), "Scale effects in the response and failure of fiber reinforced composite laminates loaded in tension and in flexure." *J. of Composite Materials*, **26**, 2674–2705.

Levy M and Salvadori M (1992), *Why buildings fall down?* W.W. Norton, New York.

Måløy K, Hansen A, Hinrichsen E, and Roux S (1992), "Experimental measurement of the roughness of brittle cracks." *Physical Review Letters*, **68**, 213–215.

Mosolov A.B and Borodich FM (1992), "Fractal fracture of brittle bodies under compression." (in Russian), *Doklady Akademii Nauk*, **324**, No. 3, 546–549.

Ožbolt J and Bažant ZP (1992), "Microplane model for cyclic triaxial behavior of concrete." *J. of Engineering Mechanics*, ASCE **118**, 1365–1386.

Planas J and Elices M (1992), "Asymptotic analysis of a cohesive crack: 1. Theoretical background." *Int. J. of Fracture*, **55**, 153–177.

Planas J, Elices M, and Guinea GV (1992), "Measurement of the fracture energy using three-point bend tests: Part 2—Influence of bulk energy dissipation." *Materials and Structures,*, **25** pp. 305–312.

Rice, J.R. (1992). "Dislocation nucleation from a crack tip: An analysis based on the Peirls concept." *J. of the Mech. and Phys. of Solids*, **40**, 239–271.

Rice JR and Levy N (1992), "The part-through surface crack in an elastic plate." *J. Appl. Mech. ASME* **39**, 185–194.

Schlangen E and van Mier JGM (1992), "Experimental and numerical analysis of micromechanisms of fracture of cement-based composites." *Cement and Concrete Composites* **14**, 105–118.

Sluys LJ, *Wave propagation, localization and dispersion in softening solids*, PhD Dissertation, Delft University of Technology, Delft, The Netherlands (1992).

Suo Z, Bao G, and Fan B (1992), "Delamination R-curve phenomena due to damage." *J. of the Mech. and Phys. of Solids*, **40**, 1–16.

Tvergaard V and Hutchinson JW (1992), "The relation between crack growth resistance and fracture process parameters in elastic-plastic solids." *J. of the Mech. and Phys. of Solids*, **40**, 1377–1397.

Wisnom MR (1992), "The relationship between tensile and flexural strength of unidirectional composite." *J. of Composite Materials*, **26**, 1173–1180.

1993

Batto RA and Schulson EM (1993), "On the ductile-to-brittle transition in ice under compression." *Acta metall. mater.*, **41**, 2219–2225.

Bažant ZP (1993), "Scaling Laws in Mechanics of Failure." *J. of Engrg. Mech., ASCE,* ,**119** , 1828–1844.

Bažant ZP, Bai S-P, and Gettu R (1993), "Fracture of rock: Effect of loading rate." *Engineering Fracture Mechanics*, **45**, 393–398.

Bažant ZP and Jirásek M (1993), "R-curve modeling of rate and size effects in quasibrittle fracture." *Int. Journal of Fracture*, **62**, 355–373.

Bažant ZP and Schell WF (1993), "Fatigue fracture of high-strength concrete and size effect." *ACI Materials Journal*, **90**, 472–478.

Bažant ZP, Lin F-B, and Lippmann H (1993), "Fracture energy release and size effect in borehole breakout." *Int. Journal for Numerical and Analytical Methods in Geomechanics*, **17**, 1–14.

Daniel IM, Hsiao H-M, Wooh SC, and Vittoser J (1993), "Processing and compressive behavior of thick composites." *Mechanics of Thick Composites*, AMD, **162**, ASME, edited by Y.D.S. Rajapakse, June, 107–126.

Dempsey JP, Bažant ZP, Rajapakse YDS, Shyam SS, Editors (1993). "Ice Mechanics 1993" (*Proc. of a Symposium as part of ASCE-ASME-SES Joint Mechanics* Meeting held in Charlottesville, VA.), AMD Vol. **163**, Am. Soc. of Mech. Engrgs., New York.

Fleck NA and Hutchinson JW (1993), "A phenomenological theory for strain gradient effects in plasticity." *J. of the Mechanics and Physics of Solids*, **41**, 1825–1857.

Hsu TTC (1993), *Unified theory of reinforced concrete*, CRC Press.

Kim JK, Park YD, and Eo SH (1993), "Size effect in concrete specimens with dissimilar initial cracks." *Size effect in concrete structures* (Proc., Japan Concrete Institute Inter. Workshop, held in Sendai), ed. by Mihashi H, Okamura H and Bažant ZP, 181–192 (also ACI Materials Journal).

Lange DA, Jennings HM, and Shah SP (1993), "Relationship between fracture surface roughness and fracture behavior of cement paste and mortar." *Journal of American Ceramic Society*, **76**, 589–597.

Planas J and Elices M (1993), "Asymptotic analysis of a cohesive crack: 2. Influence of the softening curve." *Int. J. of Fracture*, *64*, 221–237.

Planas J, Elices M, and Guinea GV (1993), "Cohesive cracks vs. nonlocal models: Closing the gap." *Int. J. of Fracture*, **63**, 173–187.

Soutis C, Curtis PT, and Fleck NA (1993), " Compressive failure of notched fibre composites." *Proc. R. Soc. Lond. A*, **440**, 241–256.

van Mier JGM and Schlangen E (1993), "An experimental and numerical study of mode I (tensile) and mode II (shear) fracture in concrete." *J. of the Mechanical Behavior of Materials*, **4**, 179–190.

Xie H (1993), *Fractals in Rock Mechanics*, Balkema, Rotterdam.

Yuan YY, Lajtai EZ, and Ayari ML (1993), "Fracture nucleation from a compression parallel finite width elliptical flaw." *Int. J. of Rock Mech. and Mining Sci.*, **30**, 873–876.

1994

Bažant ZP (1994a), Discussion of "Fracture mechanics and size effect of concrete in tension." by Tang T, Shah SP, and Ouyang C., *J. of Structural Engineering ASCE*, **120**, 2555–2558.

Bažant ZP (1994b), "Nonlocal damage theory based on micromechanics of crack interactions." *J. of Engrg. Mech., ASCE*, **120**,, 593–617; Addendum and Errata, **120**, 1401–02.

Bažant ZP and Beissel S (1994), "Smeared-tip superposition method for cohesive fracture with rate effect and creep." *Intern. J. of Fracture*, **65**, 277–290.

Bažant ZP, Bittnar Z, Jirásek M, and Mazars J, Editors (1994). Fracture and Damage in Quasibrittle Structures: Experiment, Theory and Computer Modeling (*Proc., Europe-U.S. Workshop held at Czech Techn. Univ., Prague*, Sept. 21–23, 1994, sponsored by U.S.–NSF and European Union), E & FN Spon, London–New York.

Bažant ZP and Desmorat R (1994), "Size effect in fiber of bar pullout with interface softening slip." *J. of Engrg. Mech. ASCE*, **120**, 1945–1962.

Bažant ZP, Huet C, and Müller HS (1994), "Comment on recent analysis of concrete creep linearity and applicability of principle of superposition." *Materials and Structures* (RILEM, Paris), **27**, 359–361.

Bažant ZP and Jirásek M (1994), "Damage nonlocality due to microcrack interactions: statistical determination of crack influence function." Fracture and Damage in Quasibrittle Structures: Experiment, Theory and Computer Modeling (*Proc., Europe-U.S. Workshop held at Czech Techn. Univ., Prague*, Sept. 21-23, 1994, sponsored by U.S.-NSF and European Union), ed. by Bažant ZP, Bittnar Z, Jirásek M, and Mazars J, E & FN Spon, London–New York, 3–17.

Bažant ZP and Jirásek M (1994), "Nonlocal model based on crack interactions: A localization study." *J. of Engrg. Materials & Technology ASME*, **116** (July), 256–259.

Bažant ZP and Kwon YW (1994), "Failure of slender and stocky reinforced concrete columns: Tests of size effect." *Materials and Structures*, **27**, 79-90.

Bažant ZP and Li Y-N (1994a), "Cohesive crack model for geomaterials: stability analysis and rate effect." *Applied Mechanics Reviews*, **47** (6, Part 2, June), S91-S96.

Bažant ZP and Li Y-N (1994b), "Penetration fracture of sea ice plate: Simplified analysis and size effect." *J. of Engrg. Mech. ASCE*, **120**, 1304-1321.

Bažant ZP, Ožbolt J, and Eligehausen R (1994), "Fracture size effect: review of evidence for concrete structures." *J. of Struct. Engrg. ASCE*, **120**, 2377-2398.

Bažant ZP and Vítek JL (1994), "Stud connectors in composite beams: simplified failure modeling and size effect." Fracture and Damage in Quasibrittle Structures: Experiment, Theory and Computer Modeling (*Proc., Europe-U.S. Workshop held at Czech Techn. Univ., Prague*, Sept. 21-23, 1994, sponsored by U.S.-NSF and European Union), ed. by Bažant, ZP, Bittnar Z, Jirásek M, and Mazars J, E & FN Spon, London-New York, 333-341.

Bažant ZP and Xiang Y (1994), "Compression failure of quasibrittle materials and size effect." in *AMD-Vol 185, Damage Mechanics in Composites* (ASME Winter Annual Meeting, Chicago, Nov. 1994), ed. by D.H. Allen and J.W. Ju, 143-148.

Budianski B and Fleck NA (1994), "Compressive kinking of fiber composites: A topical review." *Applied Mechanics Reviews ASME*, **47**.

Carpinteri A (1994a), "Fractal nature of material microstructure and size effects on apparent mechanical properties." *Mechanics of Materials*, **18**, 89-101.

Carpinteri A (1994b), "Scaling laws and renormalization groups for strength and toughness of disordered materials." *International Journal of Solids and Structures*, **31**, 291-302.

Carpinteri A, B Chiaia, and G Ferro (1994), "Multifractal scaling law for the nominal strength variation of concrete structures," in *Size effect in concrete structures* (Proc., Japan Concrete Institute International Workshop, held in Sendai, Japan, 1993), ed. by Mihashi M, Okamura H, and Bažant ZP, E & FN Spon, London-New York, 193-206.

Carpinteri A and G Ferro (1994), "Size effect on tensile fracture properties: a unified explanation based on disorder and fractality of concrete microstruture." *Materials and Structures*, **27**, 563-571.

Červenka V and Pukl R (1994) "SBETA analysis of size effect in concrete structures." In *Size Effect in Concrete Structure*, in *Size effect in concrete structures* (Proc., Japan Concrete Institute International Workshop, held in Sendai, Japan, 1993), ed. by Mihashi M, Okamura H, and Bažant ZP, E & FN Spon, London–New York, 323–333.

Guinea GV, Planas J and Elices M (1994), *Materials and Structures*, **27**, 99–105 (also summaries in Proc., IUTAM Symp., Brisbane 1993 and Torino, 1994).

Jirásek M and Bažant ZP (1994), "Localization analysis of nonlocal model based on crack interactions." *J. of Engrg. Mech., ASCE*, **120**, 1521–1542.

Li Y-N and Bažant ZP (1994a), "Eigenvalue analysis of size effect for cohesive crack model." *Int. J. of Fracture*, **66**, 213–226.

Li Y-N and Bažant ZP (1994b), "Penetration fracture of sea ice plate: 2D analysis and size effect." *J. of Engrg. Mech. ASCE* **120** (7), 1481–1498.

Mihashi H, Okamura H and Bažant ZP, Editors (1994). *Size effect in concrete structures* (Proc., Japan Concrete Institute Intern. Workshop held in Sendai, Japan, Oct.31–Nov.2, 1994). E & FN Spon, London–New York.

Okamura H and Maekawa K (1994), "Experimental study of size effect in concrete structures," in *Size effect in concrete structures* (Proc., Japan Concrete Institute International Workshop, held in Sendai, Japan, 1993), ed. by Mihashi M, Okamura H, and Bažant ZP, E & FN Spon, London–New York.

Petroski H (1994), "Design paradigms: case histories of error and judgment in engineering." *Cambridge University Press*, Cambridge.

Planas J, Guinea GV, and Elices M (1994), "Determination of the fracture parameters of Bazant and Jenq-Shah based on simple tests." Report to *ACI-SEM Joint Task Group on Fracture Testing of Concrete*, Universidad Politecnica de Madrid (June).

Saouma VC and Barton CC (1994), "Fractals, fracture and size effect in concrete." *ASCE Journal of Engineering Mechanics*, **120**, 835–854.

Shioya Y and Akiyama H (1994), "Application to design of size effect in reinforced concrete structures." in *Size effect in concrete structures* (Proc., Japan Concrete Institute International Workshop, held in Sendai, Japan, 1993), ed. by Mihashi M, Okamura H, and Bažant ZP, E & FN Spon, London–New York, 409–416.

Walraven J, Lehwalter (1994), "Size effects in short beams loaded in shear." *ACI Structural Journal*, **91**, 585–593.

Xie H, Sanderson DJ and Peacock DCP (1994), "A fractal model and energy dissipation for en echelon fractures." *Engineering Fracture Mechanics*, **48**, 665–662.

1995

Barr BIG Editor, (1995), *Proc. NSF Workshop on Standards for Measurement of Mode I Fracture Properties of Concrete*, University of Wales, Cardiff.

Bažant ZP (1995a), "Scaling theories for quasibrittle fracture: Recent advances and new directions." in *Fracture Mechanics of Concrete Structures* (Proc., 2nd Int. Conf. on Fracture Mech. of Concrete and Concrete Structures (FraMCoS-2), held at ETH, Zürich), ed. by Wittmann FH, Aedificatio Publishers, Freiburg, Germany, 515–534.

Bažant ZP (1995b), "Scaling of quasibrittle fracture and the fractal question." *ASME J. of Materials and Technology*, **117**, 361–367 (Materials Division Special 75th Anniversary Issue).

Bažant ZP (1995c), "Creep and Damage in Concrete." *Materials Science of Concrete IV*, Skalny J and Mindess S eds., Am. Ceramic. Soc., Westerville, OH, 355–389.

Bažant ZP (1995d), "Size effect aspects of measurement of fracture characteristics of quasibrittle material." in *Fracture Mechanics of Concrete Structures*, Vol.3 (Proc., 2nd Int. Conf. on Fracture Mech. of Concrete Structures (FraMCoS-2), held at ETH, Zürich), ed. by F.H. Wittmann, Aedificatio Publishers, Freiburg, Germany, 1749–1772.

Bažant ZP, Gu, W-H, and Faber, KT (1995), "Softening reversal and other effects of a change in loading rate on fracture of concrete." *ACI Materials Journal*, **92**, 3–9.

Bažant ZP, Kim J-J, and Li Y-N (1995), "Part-through bending cracks in sea ice plates: Mathematical modeling." in AMD-Vol. 207, *Ice Mechanics* (ASME Summer Meeting, Los Angeles, CA, June 1995), ed. by J.P. Dempsey and Y. Rajapakse, 97–105.

Bažant, ZP and Li, Y-N (1995a), "Stability of cohesive crack model: Part I— Energy principles." *Trans. ASME, J. of Applied Mechanics*, **62**, 959–964.

Bažant, ZP and Li, Y-N (1995b), "Stability of cohesive crack model: Part II— Eigenvalue analysis of size effect on strength and ductility of structures." *Trans. ASME, J. of Applied Mechanics*, **62**, 965–969.

Bažant ZP and Li Y-N (1995c), "Penetration Fracture of Sea Ice Plate." *Int. J. Solids Structures*, **32**, 303-313.

Bažant ZP, Li Z, and Thoma M (1995), "Identification of stress-slip law for bar or fiber pullout by size effect tests." *J. of Engrg. Mech. ASCE*, **121**, 620-625.

Carpinteri A and Chiaia B (1995), in *Fracture Mechanics of Concrete Structures* (Proceedings of FraMCoS-2, held at E.T.H., Zürich), ed. by Wittmann FH, Aedificatio Publishers, Freiburg, 581-596.

Carpinteri, A Chiaia B, and Ferro G (1995a), "Size effects on nominal tensile strength of concrete structures: multifractality of material ligaments and dimensional transition from order to disorder." *Materials and Structures*, **28**, 311-317.

Carpinteri A, Ferro G, and Intervenizzi S (1995b), in *Fracture Mechanics of Concrete Structures* (Proceedings of FraMCoS-2, held at E.T.H., Zürich), ed. by Wittmann FH, Aedificatio Publishers, Freiburg, Germany, 557-570.

Dempsey JP, Adamson RM, and Mulmule SV (1995), "Large-scale in-situ fracture of ice." *Proc., 2nd Int. Conf. on Fracture Mech. of Concrete Structures (FraMCoS-2)*, held at ETH, Zürich), ed. by Wittmann FH, Aedificatio Publishers, Freiburg, 575-684.

Feng N-Q, X-H Ji, Q-F Zhuang, and J-T. Ding (1995), "A fractal study of the size effect of concrete fracture energy." *Proc., 2nd Int. Conf. on Fracture Mech. of Concrete Structures (FraMCoS-2)*, held at ETH, Zürich), ed. by Wittmann FH, Aedificatio Publishers, Freiburg, 597-606.

Fleck NA and Shu JY (1995), "Microbuckle initiation in fibre composites: a finite element study" *Journal of the Mechanics and Physics of Solids*, **43**, 1887-1918.

Jirásek M and Bažant ZP (1995a), "Macroscopic fracture characteristics of random particle systems." *Intern. J. of Fracture*, **69**, 201-228.

Jirásek M and Bažant ZP (1995b), "Particle model for quasibrittle fracture and application to sea ice." *J. of Engng. Mech. ASCE*, **121**, 1016-1025.

Kyriakides S, Ascerulatne R, Perry EJ and Liechti KM (1995). "On the compressive failure of fiber reinforced composites." *Int. J. of Solids and Structures*, **32**, 689-738.

Li Y-N, Hong AN, and Bažant ZP (1995), "Initiation of parallel cracks from surface of elastic half-plane." *Int. J. of Fracture*, **69**, 357-369.

Moran PM, Liu XH, and Shih CF (1995), "Kink band formation and band broadening in fiber composites under compressive loading." *Acta Metall. Mater.*, **43**, 2943-2958.

Mulmule SV, Dempsey JP, and Adamson RM (1995), "Large-scale in-situ ice fracture experiments—Part II: Modeling efforts." *Ice Mechanics—1995* (ASME Joint Appl. Mechanics and Materials Summer Conf., held at University of California, Los Angeles, June), AMD-MD '95, Am. Soc. of Mech. Engrs., New York.

Rocco, C G (1995), *Influencia del Tamano y Mecanismos de Rotura del Ensayo de Compresión Diametral.* Doctoral Thesis. Dep. Ciencia de Materiales, Universidad Politecnica de Madrid, ETS de Ingenieros de Caminos, Ciudad Universitaria, 28040 Madrid, Spain (in Spanish: "Size-Dependence and Fracture Mechanisms in the Diagonal Compression Splitting Test"); and a private communication in 1997 by Jaime Planas on unpublished tests of size effect on modulus of rupture of concrete at Technical University, Madrid.

Rokugo, Uchida Y, Katoh H, and Koyanagi, W (1995), "Fracture mechanics approach to evaluation of flexural strength of concrete." *ACI Materials Journal (Selected translation from Japan Concrete Institute)*, **92**, 561–566.

Schulson EM and Nickolayev OY (1995), "Failure of columnar saline ice under biaxial compression: failure envelopes and the brittle-to-ductile transition." *J. of Geophysical Research*, **100** (B11), 22,383–22,400.

Tandon, S., Faber, K.T., Bažant, Z.P., and Li, Y.-N. (1995), "Cohesive crack modeling of influence of sudden changes in loading rate on concrete fracture." *Engineering Fracture Mechanics*, **52**, 987–997.

Walraven J (1995), "Size effects: their nature and their recognition in building codes." *Studi e Ricerche* (Politecnico di Milano), **16**, 113-134.

Wittmann FH, Editor (1995). *Fracture Mechanics of Concrete Structures Proc., 2nd Int. Conf. on Fracture Mech. of Concrete Structures (FraMCoS-2)*, held at ETH, Zürich), Aedificatio Publishers, Freiburg,

Xie H and Sanderson DJ (1995), "Fractal effect of crack propagation on dynamic stress intesity factors and crack velocities." *Int. J. of Fracture*, **74**, 29–42.

1996

Bažant ZP (1996a), "Size effect aspects of measurement of fracture characteristics of quasibrittle material." *Advanced Cement-Based Materials*, **4**, 128–137.

Bažant ZP (1996b), "Is no-tension design of concrete or rock structures always safe?—Fracture analysis." *ASCE J. of Structural Engrg.*, **122**, 2–10.

Bažant ZP (1996c), "Can scaling of structural failure be explained by fractal nature of cohesive fracture?" Appendix to a paper by Bažant and Li in *Size-Scale Effects in the Failure Mechanisms of Materials and Structures* (Proc., *IUTAM Symposium*, held at Politecnico di Torino, Italy, Oct. 1994), ed. by Carpinteri A, E & FN Spon, London, 284–289.

Bažant ZP, Daniel IM, and Li Z (1996a), "Size effect and fracture characteristics of composite laminates." *ASME J. of Engrg. Materials and Technology*, **118**, 317–324.

Bažant ZP and Li Y-N (1996a), "Scaling of cohesive fracture (with ramification to fractal cracks)." in *Size-scale effects in the failure mechanisms of materials and structures* (Proc., IUTAM Symp., held at Politecnico di Torino, 1994), ed. by Carpinteri A, E & FN Spon, London 274–289.

Bažant ZP and Li Z (1996b), "Zero-brittleness size-effect method for one-size fracture test of concrete." *ASCE J. of Engrg. Mechanics*, **122**, 458–468.

Bažant ZP and Kaplan MF (1996), *Concrete at High Temperatures: Material Properties and Mathematical Models* (monograph and reference volume). Longman (Addison-Wesley), London.

Bažant ZP and Xiang Y (1996), "Size effect in compression fracture: splitting crack band propagation." *ASCE J. of Engrg. Mechanics*, **122**.

Bažant ZP, Xiang Y, Adley MD, Prat PC, and Akers SA (1996c), "Microplane model for concrete. II. Data delocalization and verification." *ASCE J. of Engrg. Mechanics*, **122**, 255-262.

Bažant Z.P, Xiang Y, and Prat PC (1996b), "Microplane model for concrete. I. Stress-strain boundaries and finite strain." *ASCE J. of Engrg. Mechanics*, **122**, 245-254.

Collins MP, Mitchell D, Adebar P, and Vecchio FJ (1996), "General shear design method." *ACI Structural Journal*, **93**, 36–45.

Hsu TTC and Zhang L-X (1996), "Tension stiffening in reinforced concrete membrane elements." *ACI Structural Journal*, **93**, 108–115.

Jirásek J (1996), "Comparison of nonlocal models." orally presented at the *33rd Annual Meeting of Society of Engrg. Science*, Tempe, Arizona.

Lehner F and Kachanov M (1996), "On modeling of "winged" cracks forming under compression." *Intern. J. of Fracture*, **77**, R69–R75.

Nixon WA (1996), "Wing crack models of the brittle compressive failure of ice." *Cold Regions Science and Technology*, **24**, 41–45.

Ožbolt J and Bažant ZP (1996), "Numerical smeared fracture analysis: Nonlocal microcrack interaction approach." *Int. J. for Numerical Methods in Engrg.*, **39**, 635–661.

Peerlings RHJ, de Borst R, Brekelmans WAM, and de Vree, JHP (1996), "Gradient-enhanced damage for quasi-brittle materials." *Int. J. of Numerical Methods in Engineering*, **39**, 3391–3403.

Ruggieri C and Dodds RH (1996), "Transferability model for brittle fracture including constraint and ductile tearing effects—a probabilistic approach." *Int. J. of Fracture*, **79**, 309–340.

Sutcliffe MPF, Fleck NA, and Xin XJ (1996), "Prediction of compressive toughness for fibre composites." *Proc. Roy. Soc. London*, Series A452, 2443–2465.

1997

Bažant ZP (1997a), "Scaling of quasibrittle fracture: Asymptotic analysis." *International Journal of Fracture*, **83**, 1–40.

Bažant ZP (1997b), "Scaling of quasibrittle fracture: The fractal hypothesis, its critique and Weibull connection." *International Journal of Fracture*, **83**, 41–65.

Bažant ZP (1997c), "Fracturing truss model: size effect in shear failure of reinforced concrete." *J. of Engng. Mech. ASCE*, **123**.

Bažant ZP and Chen EP (1997), "Scaling of structural failure." *Applied Mechanics Reviews ASME*, **50**, 593–627.

Bažant ZP and Kim J-J (1997), "Penetration of sea ice plate with part-through bending crack and size effect." *J. of Engng. Mech. ASCE*.

Bažant ZP and Li Y-N (1997), "Cohesive crack with rate-dependent opening and viscoelasticity: I. mathematical model and scaling." *Int. J. of Fracture*, **86**, 247–265.

Bažant ZP and Xiang, Y (1997), "Size effect in compression fracture: splitting crack band propagation." *J. of Engng. Mech. ASCE*, **123**, 162–172.

Budianski B, Fleck NA and Amazigo JC (1997). "On compression kink band propagation." Report MECH 305, Harvard University, Cambridge.

Christensen RM and De Teresa SJ (1997), "The kink band mechanism for compressive failure of fiber composite materials." *J. of Applied Mech. ASME*, **64**, 1–6.

Fleck NA and Hutchinson JW (1997), "Strain gradient plasticity."" In *Advances in Applied Mechanics*, vol. **33**, Hutchinson JW and Wu TY eds., Academic Press, New York, 295–361.

Hutchinson JW (1997), "Linking scales in mechanics." In *Advances in Fracture Research*, Karihaloo BL, Mai Y-W, Ripley MI, and Ritchie RO eds., Pergamon Press, New York, 1–14.

Li Y-N and Bažant ZP (1997), "Cohesive crack with rate-dependent opening and viscoelasticity: II. numerical algorithm, behavior and size effect." *Int. J. of Fracture*, **86**, 267–288.

1998

Bažant ZP (1998a), "Size effect in tensile and compression fracture of concrete structures: computational modeling and design." *Fracture Mechanics of Concrete Structures* (Proc., 3rd Int. Conf., FraMCoS-3, held in Gifu, Japan), Mihashi H and Rokugo K eds., Aedificatio Publishers, Freiburg, Germany, 1905–1922.

Bažant ZP (1998b), "Compression fracture—mechanics of damage localization and size effect." *Material Instabilities in Solids* (Proc., IUTAM Symp., Delft, June 1997). Eds. de Borst R and van der Giessen E, John Wiley, Chichester, 355–367.

Bažant ZP and Kim J-J (1998), "Size effect in penetration of sea ice plate with part-through cracks. I. Theory." and "II. Results." *J. of Engrg. Mechanics ASCE*, **124**, 1310–1315 and 1316–1324.

Bažant ZP and Planas J (1998), *Fracture and Size Effect in Concrete and Other Quasibrittle Materials*. CRC Press, Boca Raton and London.

Koide H, Akita H, and Tomon M (1998), "Size effect on flexural resistance due to bending span of concrete beams." *Fracture Mechanics of Concrete Structures* (Proc., 3rd Int. Conf., FraMCoS-3, held in Gifu, Japan), Mihashi H and Rokugo K eds., Aedificatio Publishers, Freiburg, Germany, 2121–2130 (and private communication of additional data by Koide).

Kuhlmann U and Breuninger U (1998), "Zur Tragfähigkeit von horizontaler liegenden Kopfbolzdübeln." *Stahlbau*, **67**, 547–550.

Mihashi H and Rokugo K eds. (1998), *Fracture Mechanics of Concrete Structures* (Proc., 3rd Int. Conf., FraMCoS-3, held in Gifu, Japan), publ. by Aedificatio Publishers, Freiburg.

Pattison K (1998), "Why did the dam burst?" *Invention and Technology*, **14**, 23–31.

1999

ACI Standard 318 (1999), *Building Code Requirements for Structural Concrete and Commentary*, Am. Concrete Institute, Farmington Hills, Michigan.

Bažant ZP (1999a), "Size effect." *International Journal of Solids and Structures*, **37**, 69–80; special issue of inivited review articles on *Solid Mechanics* edited by Dvorak GJ for U.S. Nat. Comm. on Theor. and Appl. Mech., publ. as a book by Elsevier Science, Ltd.

Bažant ZP (1999b), "Size Effect on Structural Strength: A Review." *Archives of Applied Mechanics* (Ingenieur Archiv, Berlin), **69**, 703–725.

Bažant ZP (1999c), "Size effect in conrete structures: nuisance or necessity?" (plenary keynote lecture), in *Structural Concrete: The Bridge Between People, Proc.*, fib *Symp. 1999* (held in Prague), Fédération Internationale du Béton, publ. by Viacon Agency, Prague, 43–51.

Bažant ZP and Becq-Giraudon E (1999), "Effects of size and slenderness on ductility of fracturing structures." *J. of Engrg. Mechanics ASCE*, **125**, 331–339.

Bažant ZP, Kim J-J, Daniel IM, Becq-Giraudon E, and Li G. (1999), "Size effect on compression strength of fiber composites failing by kink band propagation." *Int. J. of Fracture*, **95**, 103–141.

Bažant, Z.P., and Rajapakse, Y.D.S., Editors (1999). *Fracture Scaling* (Proc., ONR Workshop on Fracture Scaling, University of Maryland, College Park, June 10–12, 1999; special issue reprinted from *Int. J. of Fracture*, **95**, 1999), Kluwer Academic Publishers, Dordrecht.

Bažant ZP and Vítek JL (1999), "Compound size effect in composite beams with softening connectors. I. Energy Approach." and "Differential equations and behavior." *J. of Engrg. Mech. ASCE*, **125**, 1308–1314 and 1315–1322.

Gao H, Huang Y, Nix WD, and Hutchinson JW (1999a), "Mechanism-based strain gradient plasticity—I. Theory." *J. of the Mechanics and Physics of Solids*, **47**, 1239–1263.

Gao H, Huang Y, and Nix WD (1999b), "Modeling plasticity at the micrometer scale." *Naturwissenschaften*, **86**, 507–515.

Koide H, Akita H, and Tomon M (1999), "Probability model of flexural resistance on different lengths of concrete beams." *Proc., ICASP-8* (8th Int. Conf. on Application of Safety and Probability), Sydney.

Şener S, Bažant ZP, and Becq-Giraudon E (1999), "Size effect on failure of bond splices of steel bars in concrete beams." *J. of Structural Engrg.*, **125**, 653-661.

de Borst R and Gutiérrez MA (1999), "A unified framework for concrete damage and fracture models including size effects." *Inter. J. of Fracture*, **95**, 261-277.

Dempsey JP, Adamson RM, and Mulmule SV (1999), "Scale effect on the in-situ tensile strength and failure of first-year sea ice at Resolute, NWR." *Int. J. of Fracture*, **95**, 347-366.

2000

Bažant ZP (2000), "Asymptotic Scaling of Cohesive Crack Model and K-Version of Smeared-Tip Method." *Report*, Civil Engrg. Dept., Northwestern University.

Bažant ZP, Caner FC, Carol I, Adley MD, and Akers SA (2000), "Microplane model M4 for concrete: I. Formulation with work-conjugate deviatoric stress." *J. of Engrg. Mechanics ASCE*, **126**, 944-953.

Bažant ZP and Novák D. (2000a), "Probabilistic nonlocal theory for quasibrittle fracture initiation and size effect. Parts I and II" *J. of Engrg. Mech. ASCE*, **126**, 166-174 and 175-185.

Bažant ZP and Novák D (2000b), "Energetic-statistical size effect in quasibrittle failure at crack initiation." *ACI Materials Journal*, **97**, 381-392.

Gao H and Huang Y (2000), "Taylor-based nonlocal theory of plasticity." *J. of the Mech. and Physics of Solids*.

Huang Y, Gao H, Nix WD, and Hutchinson JW (2000), "Mechanism-based strain gradient plasticity—II. Analysis." *J. of the Mechanics and Physics of Solids*, **47**, 99-128.

Phoenix SL and Beyerlein IJ (2000), "Statistical strength theory for fibrous comosite materials." *Comprehensive Composite Materials*, A. Kelly and C. Zweben, eds., Vo. **1**, chapter 1.20, Pergamon-Elsevier Science (in press).

Zi G, and Bažant ZP (2000), "Smeared-Tip Method for Cohesive Crack Scaling and and Its Numercal Calibration." *Preliminary Report*, Civil Engrg, Dept., Northwestern Unversity.

Relevant Author's Works after the First Printing (up to 2004)

Bažant, ZP (2001a), "Probabilistic modeling of quasibrittle fracture and size effect" (principal plenary lecture), (Proc., 8th Int. Conf. on Structural Safety and Reliability (ICOSSAR), held at Newport Beach, Cal., 2001), RB Corotis, GI Schueller and M Shinozuka, eds., Swets & Zeitinger (Balkema), 1–23.

Bažant, ZP (2001b), "Scaling of failure of beams, frames and plates with softening hinges." *Meccanica* (Kluwer Acad. Publ.) 36, 67–77, 2001 (special issue honoring G Maier).

Bažant, ZP (2001c), "Size effects in quasibrittle fracture: Apercu of recent results." *Fracture Mechanics of Concrete Structures* (Proc., FraMCoS-4 Int. Conf., Paris), R de Borst, J Mazars, G Pijaudier-Cabot and JGM van Mier, eds., Swets & Zeitlinger (A.A. Balkema Publishers), Lisse, 651–658.

Bažant, ZP (2001d), "Scaling laws for sea ice fracture." *Proc., IUTAM Symp. on Scaling Laws in Ice Mechanics and Ice Dynamics* (held in Fairbanks, June 2000), ed. by JP Dempsey and HH Shen, Kluwer Academic Publ., Dordrecht, 195–206.

Bažant, ZP, Caner, FC, and Červenka, J (2001), "Vertex effect and confinement of fracturing concrete via microplane model M4." *Fracture Mechanics of Concrete Structures* (Proc., FraMCoS-4 Int. Conf., Paris), R de Borst, J Mazars, G Pijaudier-Cabot and JGM van Mier, eds., Swets & Zeitlinger (A.A. Balkema Publishers), Lisse, 497–504.

Bažant, ZP, Červenka, J, and Wierer, M (2001), "Equivalent localization element for crack band model and as alternative to elements with embedded discontinuities." *Fracture Mechanics of Concrete Structures* (Proc., FraMCoS-4 Int. Conf., Paris), R de Borst, J Mazars, G Pijaudier-Cabot and JGM van Mier, eds., Swets & Zeitlinger (A.A. Balkema Publishers), Lisse, 765–772.

Bažant, ZP and Novák, D (2001a), "Proposal for standard test of modulus of rupture of concrete with its size dependence." *ACI Materials Journal* 98 (1), 79–87.

Bažant, ZP and Novák, D (2001b), "Nonlocal Weibull theory and size effect in failures at fracture initiation." *Fracture Mechanics of Concrete Structures* (Proc., FraMCoS-4 Int. Conf., Paris), R de Borst, J Mazars, G Pijaudier-Cabot and JGM van Mier, eds., Swets & Zeitlinger (A.A. Balkema Publishers), Lisse, 659–664.

Bažant, ZP and Novák, D (2001c), "Nonlocal model for size effect in quasibrittle failure based on extreme value statistics." (Proc., 8th Int. Conf. on Structural Safety and Reliability (ICOSSAR), Newport Beach, Cal., 2001), RB Corotis, ed., Swets & Zeitinger (Balkema), 1–8.

Brocca, M and Bažant, ZP (2001), "Size effect in concrete columns: Finite element analysis with microplane model." *J. of Structural Engineering ASCE* 127 (12), 1382–1390.

Carol, I, Jirásek, M, and Bažant, ZP (2001), "A thermodynamically consistent approach to microplane theory. Part I: Free energy and consistent microplane stresses." *Int. J. of Solids and Structures* 38 (17), 2921–2931.

Planas, J, Bažant, ZP, and Jirásek, M (2001), "Reinterpretation of Kariha-loo's size effect analysis for notched quasibrittle structures." *Intern. J. of Fracture* 111, 17–28.

Bažant, ZP, Zhou, Y, Novák, D, and Daniel, IM (2001), "Size effect in fracture of sandwich structure components: foam and laminate." *Proc., ASME Intern. Mechanical Engrg. Congress* (held in New York), Vol. AMD-TOC (paper 25413), Am. Soc. of Mech. Engrs., New York, pp. 1–12.

Bažant, ZP (2002a), "Reminiscences on four decades of struggle and progress in softening damage and size effect". *Concrete Journal (Tokyo)* 40 (2), 16–28 (JCI Anniversary); Expanded version republished in *Mechanics (Am. Academy of Mech.)* 32 (5–6), 2003, 1–10.

Bažant, ZP (2002b), "Size effect theory and its application to fracture of fiber composites and sandwich plates." *Continuum Damage Mechanics of Materials and Structures*, O Allix and F Hild, eds., Elsevier, Amsterdam, pp. 353–381.

Bažant, ZP (2002c), "Size effect theory and its application to fracture of fiber composites and sandwich plates." *Continuum Damage Mechanics of Materials and Structures*, O Allix and F Hild, eds., Elsevier, Amsterdam, pp. 353–381.

Bažant, ZP (2002d), "Scaling of sea ice fracture—Part I: Vertical penetration." *J. of Applied Mechanics ASME* 69 (Jan.), 11–18.

Bažant, ZP (2002e), "Scaling of sea ice fracture—Part II: Horizontal load from moving ice." *J. of Applied Mechanics ASME* 69 (Jan.), 19–24.

Bažant, ZP (2002f), "Concrete fracture models: testing and practice." *Engineering Fracture Mechanics* 69 (2), 165–206 (special issue on Fracture of Concrete and Rock, M Elices, ed.).

Bažant, ZP (2002g), "Scaling of dislocation-based strain-gradient plasticity." *J. of the Mechanics and Physics of Solids* 50 (3), 435–443 (based on Theor. & Appl. Mech. Report No. 2000-12/C699s, Northwestern University 2000; see also Errata, Vol. 52, 2004, p. 975).

Bažant, ZP and Becq-Giraudon, E (2002), "Statistical prediction of fracture parameters of concrete and implications for choice of testing standard." *Cement and Concrete Research* 32 (4), 529–556.

Bažant, ZP and Frangopol, DM (2002), "Size effect hidden in excessive dead load factor." *J. of Structural Engrg. ASCE* 128 (1), 80–86.

Bažant, ZP and Guo, Z (2002a), "Size effect and asymptotic matching approximations in strain-gradient theories of micro-scale plasticity." *Int. J. of Solids and Structures* 39, 5633–5657; see also Errata, Vol. 40 (2003), p. 6215).

Bažant, ZP and Guo, Z (2002b), "Size effect on strength of floating sea ice under vertical line load." *J. of Engrg. Mechanics* 128 (3), 254–263.

Bažant, ZP and Jirásek, M (2002), "Nonlocal integral formulations of plasticity and damage: Survey of progress." *ASCE J. of Engrg. Mechanics* 128 (11), 1119–1149 (ASCE 150th anniversary article).

Bažant, ZP, Yu, Q, and Zi, G (2002), "Choice of standard fracture test for concrete and its statistical evaluation." *Int. J. of Fracture* 118 (4), Dec., 303–337.

Caner, FC and Bažant, ZP (2002), "Lateral confinement needed to suppress softening of concrete in compression." *J. of Engrg. Mechanics ASCE* 128 (12), 1304–1313.

Jirásek, M and Bažant, ZP (2002), *Inelastic Analysis of Structures*. J. Wiley & Sons, London and New York (735 + xviii pp.).

Lin, F-B, Yan, GY, Bažant, ZP, and Ding, F (2002), "Nonlocal strain-softening model of quasibrittle materials using boundary element method." *Engrg. Analysis with Boundary Elements* 26, 417–424.

Bažant, ZP (2003a), "Asymptotic matching analysis of scaling of structural failure due to softening hinges. I. Theory." *J. of Engrg. Mech. ASCE* 129 (6), 641–650.

Bažant, ZP (2003b), "Asymptotic matching analysis of scaling of structural failure due to softening hinges. II. Implications." *J. of Engrg. Mech. ASCE* 129 (6), 651–654.

Bažant, ZP and Novák, D (2003), "Stochastic models for deformation and failure of quasibrittle structures: recent advances and new directions."

Computational Modelling of Concrete Structures (Proc., EURO-C Conf., St. Johann im Pongau, Austria), N Bićanić, R de Borst, H Mang and G Meschke, eds., A.A. Balkema Publ., Lisse, Netherlands, 583–598.

Bažant, ZP, Zhou, Y, Zi, G, and Daniel, IM (2003), "Size effect and asymptotic matching analysis of fracture of closed-cell polymeric foam." *Int. J. of Solids and Structures* 40, 7197–7217.

Bažant, ZP and Zi, G (2003), "Asymptotic stress intensity factor profiles for smeared-tip method for cohesive fracture." *Int. J. of Fracture* 119 (2), 145–159.

Bažant, ZP, Zi, G and McClung, D (2003), "Size effect law and fracture mechanics of the triggering of dry snow slab avalanches." *J. of Geophysical Research* 108 (B2), 2119–2129.

Cusatis, G, Bažant, ZP, and Cedolin, L (2003a), "Confinement-shear lattice model for concrete damage in tension and compression: I. Theory." *J. of Engrg. Mechanics ASCE* 129 (12), 1439–1448.

Cusatis, G, Bažant, ZP, and Cedolin, L (2003b), "Confinement-shear lattice model for concrete damage in tension and compression: I. Computation and validation." *J. of Engrg. Mechanics ASCE* 129 (12), 1449–1458.

Novák, D, Bažant, ZP, Vořechovský, M (2003), "Computational modeling of statistical size effect in quasibrittle structures." *Applications of Statistics and Probabiltity in Civil Engineering* (Proc., 9th Int. Conf., ICASP-9, held in San Francisco), A Der Kiureghian, S Madanat and JM Pestana, eds., Millpress, Rotterdam, 621–628.

Sládek, J, Sládek, V, and Bažant, ZP (2003), "Non-local boundary integral formulation for softening damage." *Int. J. for Numerical Methods in Engrg.* 57, 103–116.

Zi, G and Bažant, ZP (2003), "Eigenvalue method for computing size effect of cohesive cracks with residual stress, with application to kink bands in composites." *Int. J. of Engrg. Science* 41 (13–14), 1519–1534 (special issue honoring K Willam, ed. by ZP Bažant, I Carol and P Steinmann).

Bažant, ZP, Caner, FC, Cedolin, L, Cusatis, G, and Di Luzio, G (2004), "Fracturing material models based on micromechanical concepts: Recent advances." *Fracture Mechanics of Concrete Structures* (Proc., FraMCoS-5, 5th Int. Conf. on Fracture Mech. of Concrete and Concr. Structures, Vail, Colo.), Vol. 1, VC Li, KY Leung, Willam, KJ, and Billington, SL, eds., IA-FraMCoS, 83–90.

Bažant, ZP, Pang, SD, Vořechovský, M, Novák, D and Pukl, R (2004), "Statistical size effect in quasibrittle materials: Computation and extreme value theory." *Fracture Mechanics of Concrete Structures* (Proc., FraMCoS-5, 5th Int. Conf. on Fracture Mech. of Concrete and Concr. Structures, Vail, Colo.), Vol. 1, VC Li, KY Leung, Willam, KJ, and Billington, SL, eds., IA-FraMCoS, 189–196.

Bažant, ZP and Yu, Q (2004), "Size effect in concrete specimens and structures: New problems and progress." *Fracture Mechanics of Concrete Structures* (Proc., FraMCoS-5, 5th Int. Conf. on Fracture Mech. of Concrete and Concr. Structures, Vail, Colo.), Vol. 1, VC Li, KY Leung, Willam, KJ, and Billington, SL, eds., IA-FraMCoS, 153–162.

Bažant, ZP, Zhou, Y, Novák, D, and Daniel, IM (2004), "Size effect on flexural strength of fiber-composite laminate." *J. of Engrg. Materials and Technology ASME* 126 (Jan.), 29–37.

Bažant, ZP (2004a), "Scaling theory for quasibrittle structural failure." *Proc., National Academy of Sciences* 101, in press.

Bažant, ZP (2004b), "Probability distribution of energetic-statistical size effect in quasibrittle fracture." *Probabilistic Engineering Mechanics*, in press.

Bažant, ZP and Caner, FC (2004), "Microplane model M5 with kinematic and static constraints for concrete fracture and anelasticity." *J. of Engrg. Mech. ASCE*, in press.

Bažant, ZP, Guo, Z, Espinosa, H, Zhu, Y and Peng, B (2004), "Epitaxially Influenced Boundary Layer Model for Size Effect in Thin Metallic Films." Theor. & Appl. Mech. Report, Northwestern University; *J. of Appl. Physics*, submitted to.

Bažant, ZP and Yu, Q (2004), "Designing against size effect on shear strength of reinforced concrete beams without stirrups." *J. of Structural Engineering ASME*, in press.

Bažant, ZP and Yavari, A (2004), "Is the cause of size effect on structural strength fractal or energetic-statistical?" *Engrg. Fracture Mechanics* 43, in press.

Additional References for Addendum (up to 2004)

Acharya, A and Bassani, JL (2000), "Lattice incompatibility and a gradient of crystal plasticity." *J. of the Mechanics and Physics of Solids* 48, 1565–1595.

Carpinteri, A (1987), "Stress-singularity and generalized fracture toughness at the vertex of re-entrant corners." *Engineering Fracture Mechanics* 26, 143–155.

Daniels, HE (1945), "The statistical theory of the strength of bundles & threads." *Proc. Royal Soc.* A183, London, 405–435.

Duan, K and Hu, XZ (2003), "Scaling of quasi-brittle fracture: Boundary effect.", submitted to *Engineering Fracture Mechanics*.

Duan, K, Hu, XZ, and Wittmann, FH (2002), "Explanation of size effect in concrete fracture using non-uniform energy distribution." *Materials and Structures* 35, 326–331.

Duan, K, Hu, XZ, and Wittmann, FH (2003), "Boundary effect on concrete fracture and non-constant fracture energy distribution." *Engrg. Fracture Mechanics* 70, 2257–2268.

Dunn, ML, Suwito, W, and Cunningham, SJ (1997), "Stress Intensities at notch singularities." *Engineering Fracture Mechanics* 57(4), 417–430.

Dunn, ML, Suwito, W, and Cunningham, SJ (1997), "Fracture initiation at sharp notches: correlation using critical stress intensities." *Int. J. Solids Structures* 34(29), 3873–3883.

Elices, M, Guinea, GV, and Planas, J (1992), "Measurement of the fracture energy using three-point bend tests: Part 3-influence of cutting the P-d tail." *Materials and Structures* 25, 327–334.

Fleck, NA and Hutchinson, JW (1993), "A phenomenological theory for strain gradient effects in plasticity." *J. of the Mechanics and Physics of Solids* 41, 1825–1857.

Fleck, NA and Hutchinson, JW (1997), "Strain gradient plasticity." *Advances in Applied Mechanics*, vol. 33, JW Hutchinson and TY Wu, eds., Academic Press, New York, 295–361.

Fleck, NA, Olurin, OB, Chen, C, and Ashby, MF (2001), The Effect of Hole Size upon the Strength of Metallic and Polymeric Foams. *Journal of the Mechanics and Physics of Solids* 49, 2015–2030.

Gibson, LJ and Ashby, MF (1997), *Cellular Solids: Structure and Properties*, 2nd ed. Cambridge University Press, Cambridge, UK.

Guinea, GV, Planas, J, and Elices, M (1994a), "Correlation between the softening and the size effect curves." *Size Effect in Concrete* (edited by Mihashi, H, Okamura, H, and Bažant, ZP), E&FN Spon, London, 233–244.

Guinea, GV, Planas, J, and Elices, M (1994b), "A general bilinear fitting for the softening curve of concrete." *Materials and Structures* 27, 99–105.

Guinea, GV, Elices, M, and Planas, J (1997), "On the initial shape of the softening function of cohesive materials." *Int. J. of Fracture* 87, 139–149.

Hu, XZ (1997), "Toughness measurements from crack close to free edge. "*International Journal of Fracture* 86(4), L63–L68.

Hu, XZ (1998), "Size effects in toughness induced by crack close to free edge." *Fracture Mechanics of Concrete Structures* (ed. by Mihashi, H, and Rohugo, K), Proc. FraMCoS-3, Japan, 2011–2020.

Hu, XZ and Wittmann, FH (1991), "An analytical method to determine the bridging stress transferred within the fracture process zone: I. General theory." *Cement and Concrete Research* 21, 1118–1128.

Hu, XZ and Wittmann, FH (1992a), "An analytical method to determine the bridging stress transferred within the fracture process zone: II. Application to mortar." *Cement and Concrete Research* 22, 559–570.

Hu, XZ and Wittmann, FH (1992b), "Fracture energy and fracture process zone." *Materials and Structures* 25, 319–326.

Hu, XZ and Wittmann FH (2000), "Size effect on toughness induced by crack close to free surface." *Engineering Fracture Mechanics* 65, 209–221.

Jackson, KE (1992), "Scaling Effects in the Flexural Response and Failure of Composite Beams," *AIAA Journal*, Vol. 30(8), pp, 2099–2105.

Johnson, DP, Morton, J, Kellas, S, and Jackson, KE (2000), "Size Effect in Scaled Fiber Composites Under Four-Point Flexural Loading," *AIAA Journal*, Vol. 38(6), 1047–1054.

Jirásek, M (2003), Keynote lecture presented at EURO-C, St. Johann in Pongau, Austria.

Karihaloo, BL (1999), "Size effect in shallow and deep notched quasi-brittle structures." *International Journal of Fracture* 95, 379–390.

Karihaloo, BL, Abdalla, HM, and Imjai, T (2003), "A simple method for determining the true specific fracture energy of concrete." *Magazine of Concrete Research* 55(5), 471–481.

Karihaloo, BL, Abdalla, HM, and Xiao, QZ (2003), "Size effect in concrete beams." *Engineering Fracture Mechanics* 70, 979–993.

Lavoie, JA, Soutis C, and Morton, J (2000), "Apparent Strength Scaling in Continuous Fiber Composite Laminates." *Composite Science and Technology*, Vol. 60, pp. 283–299.

Mahesh, S, Phoenix, SL and Beyerlein, IJ (2002), "Strength distributions and size effects for 2D and 3D composites with Weibull fibers in an elastic matrix." *Int. J. of Fracture* **115**, 41–85.

Mindlin, RD (1965), "Second gradient of strain and surface tension in linear elasticity." *Int. J. Solids Struct.* 1, 417–438.

Phoenix, SL and Beyerlein, IJ (2000), "Distribution & size scalings for strength in a one-dimensional random lattice with load redistribution to nearest & next nearest neighbors." *Phys. Review E* **62**, 1622–1645.

Reineck, K, Kuchma, DA, Kim, KS, and Marx, S (2003), "Shear database for reinforced concrete members without shear reinforcement." *ACI Structural Journal* 100, No. 2, 240-249; with discussions by Bažant and Yu, Vol. 101, No. 1 (Jan.-Feb. 2004).

Rice, RJ (1968), "Mathematical analysis in the mechanics of fracture." *Fracture-An Advanced Treatise*, Vol. 2 (ed. by Liebowitz, H), Academic Press, New York, 191–308.

RILEM Committee QFS (ZP Bažant, chairman) (2004), "Quasibrittle fracture scaling and size effect." *Materials and Structures* (Paris) 37, in press.

Tang, T, Bažant, ZP, Yang, S, and Zollinger, D (1996), "Variable-notch one-size test method for fracture energy and process zone length." *Engrg. Fracture Mechanics* 55(3), 383–404.

Toupin, RA (1962), "Elastic materials with couple stresses." *Arch. Ration. Mech. Anal.* 11, 385-414.

Williams, ML (1952), "Stress singularities resulting from various boundary conditions in angular corners of plates in extension." *Journal of Applied Mechanics* 74, 526–528.

Wisnom, MR (1991), "The Effect of Specimen Size on the Bending Strength of Unidirectional Carbon Fiber-Epoxy," *Composite Structures*, Vol. 18, pp. 47–63.

Wisnom, MR and Atkinson, JA (1997), "Reduction in Tensile and Flexural Strength of Unidirectional Glass Fiber-Epoxy with Increasing Specimen Size," *Composite Structures*, Vol. 38, pp. 405–412.

Wittmann, FH, Mihashi, H, and Nomura, N (1990), "Size effect on fracture energy of concrete." *Engrg. Fracture Mech.* 35, 107–115

Zenkert, D and Bäcklund J (1989), PVC Sandwich Core Materials: Mode I Fracture Toughness. *Composite Science and Technology* 34, 225–242.

Zenkert, D (1989), PVC Sandwich Core Materials: Fracture Behaviour under Mode II and Mixed Mode Loading. *Materials Science and Engineering* 108, 233–240.

Subject Index

activation energy, 151, 152
asymptotic analysis, 21–49
    power scaling law
        large sizes, 14
        small sizes, 14
    small sizes, 16
    smeared-tip method, 177–217
    transitional size, 17
boundary layer
    concrete, cohesive softening law tail, 255
    effect, 10
    of cracking, 42, 43, 45
    Prandtl's theory, 1, 25, 193
    stress redistribution, 38, 43, 44
boundary value problem, xi, 18, 78, 181
bridging stress
    and crack opening, 165, 180
    and stress intensity factor, 205
    deformations in the FPZ, 181
    distribution, 166, 180
    in kink band, 137, 145
    residual, 133
brittleness
    concept, 26
    influence of loading rate, 153, 154
    of floating sea ice, 35
brittleness number
    analysis of size effect tests, 29, 30, 47
    Bažant's, 25, 26, 47

characteristic length
    absence, 11, 18
characteristic material length
    absence, 7, 11–13, 17, 20, 56
    and fracture process zone, 21, 22, 156
    discrete elements, 174
    effect of viscosity, 156
    finite elements, 172
    floating sea ice, 78, 84
    in nominal strength, 23
    in stress distribution, 28
    inelastic hinges, 90
    Irwin's, 93, 188
    localization limiter, 174
    microplane model, 167, 172
    small structure sizes, 14
characteristic structure size, 11, 56, 89, 94, 99, 111, 135, 137, 182
characteristic time, 156
closed cell polymeric foam fracture, 254
CMOD, see crack mouth opening displacement
cohesive crack model, 167, 177, 178
    and Rice's J-integral, 187
    asymptotic analysis, 24, 26
    basic hypothesis, 177
    broad size range, 210
    conversion to an eigenvalue problem, 165
    crack opening, 165
    energetic aspect, 178

fitting formula, 193
for fiber composites, 148
fracture energy, 32, 180,
    210, 211
in historical context, 7
kink band failure, 147
limitations, 179, 181, 193
nonstandard, 185, 187,
    215, 216
plastic analysis, 202
small size asymptotics, 204
softening curve, 145
standard, 181, 182, 198, 199,
    203, 204, 215
standard and nonstandard
    equivalence, 185
stress profile resolution, 165
stress-displacement curve, 185
stress-displacement law,
    179, 180
time dependent formulation,
    152, 153
compliance
    calculation from K_I
        expressions, 88
    Green's functions, 179, 204
    method, for equivalent crack
        length, 21, 40
compression failure
    basic mechanisms, 121
    buckling phenomena, 122
    by splitting crack band, 91,
        122, 124, 133
    crushing phenomena, 92, 101,
        116, 130, 131
    hinge softening, 90
    internal buckling
        phenomena, 122
    of concrete beams, 90, 130, 131
    of ice plate, 87
    size effect on nominal
        strength, 124

concrete structures
    design codes, 110–117
    no-tension design, 117–119
    plain, *see* unreinforced
        concrete
    reinforced, *see* reinforced
        concrete
crack band model
    basic idea, 170, 178
    comparison with cohesive
        crack model, 177, 178
    comparison with test data, 45
    in historical context, 7–9
    mesh dependence, 125, 172
crack growth
    fatigue, *see* fatigue crack
        growth stable, 38, 47, 53,
        55, 57, 59, 70, 71, 111
crack influence function, 171
crack initiation failures, 245
crack mouth opening
    displacement, 153
crack opening profile
    and stress intensity factor, 184
    calculation form K_I
        expressions, 87
    calculation from K_I
        expressions, 182, 183
creep, 10, 153
    viscoelasticity, 153, 156, 209
critical crack tip opening
    displacement, 32, 199
critical stress intensity factor, *see*
    fracture toughness
CS theory, 237
cycling loading, *see* fatigue crack
    growth
design codes, fractal theory, 264
dimensionless variables, 230
discrete element method, 37, 170,
    174–176
dry snow slab avalanches,
    triggering, 267

eigenvalue analysis, 165, 166
energy balance, 16, 122, 124
energy release
　analysis of steel-concrete
　　composite beams, 102,
　　104, 106, 107
　　complementary potential
　　　energy, 106
　　complementary shear strain
　　　energy, 106
　　slip of studs, 107, 109
　and deterministic size effect, 2,
　　15, 55
　asymptotic analysis, 21, 29, 58
　axial splitting cracks,
　　122–124, 131
　complementary energy
　　analysis, 22, 23
　crack band, 124
　for large structure sizes, 15
　for small structure sizes, 16
　in approximate analysis, 7
　in breakout of boreholes in
　　rock, 132
　in continuum damage model,
　　170, 172, 173
　in kink bands, 138, 140 142
　inelastic hinge, 90–92, 95
　potential energy analysis, 22
　strain energy density, 29
　stress redistributions, 55, 57
energy release function, 27, 40, 135,
　　179, 182, 196, 198, 206
energy release rate, 17, 22, 23
　and material orthotropy, 135
　at zero crack length, 40
　critical, 19
　dimensionless expression,
　　23, 24
　for a system of loads, 48
　for floating ice plates
　　vertical penetration, 82, 84
　in nonlocal LEFM, 206

equivalent LEFM analysis, 21, 27,
　　29, 40, 69, 86, 107, 133,
　　145, 148, 199, 203, 206, 216
Espinosa, H, 265
failure load (pdf), 264
fatigue crack growth, 7, 38,
　　154, 189
fibre-composite laminates, flexural
　　strength, 252
finite angle notches, size effect, 249
Fleck and Hutchinson's
　　formulation, 228
floating sea ice
　force applied to fixed
　　structures, 87–89
　line load, 99–101
　thermal bending fracture,
　　77–78
　vertical penetration, 78–86
FPZ, see fracture process zone
fractal
　curve, 69, 70, 72
　dimension, 71–74
　fracture energy, 71
　lacunar, 72–76
fractal and multifractal scaling law,
　　42, 49, 73, 75, 111, 134
fracture energy, 29, 32, 76, 84,
　　138, 211
　according to fractal theories,
　　70–72
　and broad-range size effect
　　law, 210, 217
　and characteristic material
　　length, 22
　definition, 15
　for a splitting crack band, 91,
　　124, 131
　for concrete, 32, 211
　from the load-deflection curve
　　of a notched beam, 211
　from the load-deflection curve
　　of a notched beam, 32, 211

in slip of studs, 107
in the cohesive crack model,
    32, 180, 187, 206, 210
in the crack band model, 170
of a kink band, 141, 143, 145
relation to the J-integral, 68
fracture mechanics
  in historical context, 5, 7
  linear elastic, see linear elastic
    fracture mechanics
  of fractal cracks, 71
fracture process zone
  and characteristic material
    length, see characteristic
    material length, fracture
    process zone
  and fractal theories, 72
  and inhomogeneity size, 15,
    38, 86
  and stress profile, 165, 167,
    180, 181
  and transition of behavior, 27
  and Weibull theory, 60, 66
  effective length of the
    R-curve, 21
  equivalent crack length, 21, 40
  flux of energy, 28
  in a composite beam, 103
  in fractal theories, 72
  in historical context, 6
  J-integral path, 67, 139, 187
  Poisson effect, 10
  size, 60, 66, 77, 86, 92,
    107, 153, 156, 167, 177,
    183, 188
fracture testing, 260
fracture toughness, 17, 88, 118,
    156, 183
$\mathcal{G}$, see energy release rate
G_b, see fracture energy
G_F, see fracture energy
G_f, see fracture energy

geometry dependence, see shape
    dependence
Hall-Petch formula, 266
incremental plasticity, scaling
    strain-gradient, 236
J-Integral, 27
  and the cohesive crack model,
    178, 180, 187
  and Weibull size effect, 67
  application to kink bands,
    137–140, 142, 143
  calculation of a flux of
    energy, 28
  critical value, 138, 143
  dimensionless expression, 29
  expression, 142, 143
  path, 28, 140
    independence, 67
K_I, see stress intensity factor
K_c, see fracture toughness
large notch or crack, 241
LEFM, see linear elastic fracture
    mechanics
linear elastic fracture mechanics
  equivalent analysis, see
    equivalent LEFM analysis
  generalization to
    orthotropy, 135
  nonlocal, 206, 217
  power law exponent $-1/2$, 7,
    12, 13, 24, 59, 60, 78,
    109, 112
  stress intensity factor, 17
localization, see strain softening
localization limiter, see strain
    softening, localization
    limiter
material characteristic length, 229
mesh
  bond slip, 125
  dependence, 170, 172, 173, 175
  discrete elements, 175
  inclined crack band, 125, 172

reinforced concrete columns,
  125, 127
remeshing, 170, 172
MFSL, *see* fractal and multifractal
  scaling law
MSG theory, 237
micromechanical model, 72, 76,
  121, 170, 172, 230
microplane model, 125,
  167–169, 172
multifractal scaling law, *see* fractal
  and multifractal
  scaling law
nominal strength, 23, 109
  and elastic buckling, 19
  asymptotic, 16, 21, 23, 26, 40,
    66, 124, 198, 202
  data, 31, 34–38, 95, 125,
    126, 153
  dimensionless, 89
  expression, 40, 41, 43, 68, 91,
    99, 124, 186
  for transitional size range,
    13, 14
  in compression failure, 124
  in historical context, 4, 5, 7
  of a composite beam, 103
  of sea ice, 20, 81
  of stud, 109
  pullout, 109
  residual, 26
  shear loading, 131
  time dependence, 153
  Weibull-type scaling law, 13, 68
nonlocal damage mechanics,
  see nonlocal model
nonlocal model
  averging rule, 171
  and finite element size, 172
  characteristic length, 172
  cohesive law, 181
  finite element analysis, 15, 172,
    176, 177

of cohesive crack, 26
for damage, 57, 171, 172, 174,
  176, 177, 193, 205, 217
in historical context, 5, 9
generalization for Weibull
  statistical size effect, *see*
  Weibull, statistical size
  effect, nonlocal
  generalization
in historical context, 172
nonlocal LEFM, *see* linear
  elastic fracture mechanics,
  nonlocal
physical justification, 171
theories for strain localization,
  58, 62
with crack interactions, 37
notch length, fracture energy, 258
R-curve, 21–24, 48, 169, 170, 199,
  203, 206
random particle model, *see* discrete
  element method
rate dependence, *see* time
  dependence
reinforced concrete
  beams
    compression failure, 90, 91
    diagonal shear failure, 35, 37,
      55, 57, 111, 113, 130, 131
    torsional failure, 113
  columns, 112, 113, 124–126, 128
  cylinders
    double-punch loading, 36
    fiber-reinforced concrete, 133
  size effect laws for, 110, 112
  slabs
    punching shear failure, 113
relaxation, 153, 209, 212
scaling and size effect, 232
shape dependence, 14, 24, 27, 40,
  56, 62, 67, 69, 91, 110,
  133, 148, 183–186, 188,

198, 199, 204, 207,
215–217
shrinkage, 10
SIF, *see* stress intensity factor
small size asymptotic
    load-deflection response,
235
snapback, 16, 17, 96, 98, 157, 158,
161, 162, 167
splitting crack band, *see*
    compression failure, by
    splitting crack band
steel-concrete composite beams,
101–110
strain-gradient plasticity, scaling
of, 228
    Fleck and Hutchinson
    formulation, 228
strain softening
    and distributed cracking, 43
    and energy dissipation, 91, 173
    and loading rate, 10, 156,
157, 173
    and wave speed
    propagation, 173
    effective numerical
    methods, 172
    efficient numerical
    methods, 169
    elasto-plastic stability, 89, 90
    in cohesive crack model,
181, 217
    in elasto-plastic stability, 20
    in historical context, 7
    in microplane model, 168
    loading rate, 173
    localization limiter, 170,
173, 174
    nonlocal theories, 58, 60
stress intensity factor, 87, 194, 204,
207, 208, 215
    at fracture initiation, 196
    case of many loads, 133

critical, *see* fracture toughness
density profiles, 179, 183, 184,
187, 207
dimensionless, 182
ductility analysis, 162
eccentric loading, 118
fatigue loading, 154
function, 179, 189, 198
of cohesive crack, 166, 178,
183, 184, 204, 207
sea-ice radial bending crack, 81
shape dependence, 198
sharp craft without FPZ, 135
stress relief zone, 15
stud, *see* steel-concrete composite
beams
surface energy, 72, 151
thermal bending fracture, 77, 78
thin metallic films, boundary layer
effect, 265
three-point bend notched beams
    experimental results, 8, 21, 26,
30, 61, 162
time dependence, 151
    creep, 10, 153
    due bond ruptures, 154
    due to bond ruptures, 151, 152
    due to diffusion phenomena, 10
    due to viscosity, 10, 151, 154,
156, 157
    dur to bond ruptures, 154
    fatigue, *see* fatigue
    crack growth
    of cohesive crack, 152, 153
    of fracture growth, 151
    of length scale, 10, 156
TNT theory, 237
toughness, *see* fracture toughness
universal size effect law, 248
unreinforced concrete
    beams, 113
        torsional failure, 113
    beams or plates, 90

dams, 57, 113, 117
retaining walls, 113, 117
viscoelasticity, 151, 152, 156, 209, 211, 212
Weibull
    distribution, 53, 55, 56, 60, 73
    energetic-statistical approach, 66–68
    nominal strength, 68
    size effect law parameters, 57, 59, 62, 66, 74
    statistical size effect, 4, 6, 7, 13
        and fractality, 72–74, 76
        asymptotic analysis, 66
        basic hypothesis, 53
        for concrete, 35, 57, 111
        for quasibrittle structures, 53, 56
        limitations, 13, 38, 56, 59, 62
        nonlocal generalization, 58–60, 62
    stress, 54, 67
    weakest link model, 6, 55, 67, 198
Weibull modulus, 53, 57, 66, 67, 74
Weibull theory, 247
weight function, 169, 171, 181, 206
yield plateau, 138

Reference Citation Index

Acharya and Bassini (2000), 236, 237
ACI 318 (1989), 112
Argon (1972), 134
Ashby and Hallam (1986), 121
Assur (1963), 78

Bao, Ho, Suo and Fan (1992), 135
Barenblatt (1959), 7, 177
Barenblatt (1962), 7, 177
Barenblatt (1979), 12, 17, 25, 193
Barenblatt (1987), 17, 20, 193
Barenblatt (1979, 2003), 245
Barenblatt (1996, 2003), 241
Barr (1995), 9
Bassini (2001), 245
Bažant (1967), 122
Bažant (1976), 7, 161, 170, 177
Bažant (1982), 9, 170
Bažant (1983), 18, 19, 25, 145, 191, 216
Bažant (1984), 5, 15, 18, 25, 145
Bažant (1984a), 7, 10, 86, 191, 216
Bažant (1984b), 9, 171, 172
Bažant (1984c), 168
Bažant (1985), 258
Bažant (1985a), 193
Bažant (1985b), 193, 209
Bažant (1986), 29
Bažant (1987), 25
Bažant (1987a), 144
Bažant (1990a), 179, 184
Bažant (1992a), 79

Bažant (1992b), 77, 78
Bažant (1993), 12, 13, 78, 122, 124, 152
Bažant (1994b), 171
Bažant (1995), 42
Bažant (1995c), 152
Bažant (1995d), 32, 38, 46, 69
Bažant (1996), 130, 255, 256
Bažant (1996a), 23, 32, 46, 131
Bažant (1996b), 113, 117, 118
Bažant (1997a), 21, 57, 133, 191, 206
Bažant (1997b), 9, 42, 71
Bažant (1997c), 42
Bažant (1998), 42
Bažant (1998a), 133, 197, 216
Bažant (1998b), 133
Bažant (1999a), 11
Bažant (1999b), 11, 212
Bažant (2000), 114, 115, 179, 191, 237
Bažant (2001, 2002), 235, 259
Bažant (2001a), 51, 212
Bažant (2001b), 219, 223
Bažant (2001c), 55
Bažant (2001d), 212
Bažant (2001e), 230
Bažant (2002), 245, 258, 259, 260
Bažant (2004a), 241
Bažant and Xi (1991), 62
Bažant and Şener (1997), 113
Bažant and Becq-Giraudon (1999), 162

Bažant and Becq-Giraudon (2001), 32
Bažant and Becq-Giraudon (2002), 258, 260
Bažant and Beissel (1994), 184
Bažant and Cao (1986a), 130
Bažant and Cao (1986b), 130
Bažant and Cao (1987), 113, 130
Bažant and Cao (2002a), 238
Bažant and Cedolin (1979), 7, 177
Bažant and Cedolin (1980), 7
Bažant and Cedolin (1991), 7, 17, 20, 90, 98, 158, 161, 173
Bažant and Chen (1997), 1, 12, 191, 248
Bažant and Estenssoro (1979), 10
Bažant and Frangopol (2000), 114, 117
Bažant and Gettu (1990), 32
Bažant and Gettu (1992), 151, 153, 173
Bažant and Jirásek (2001), 92
Bažant and Jirásek (1994a), 171
Bažant and Jirásek (1994b), 171
Bažant and Jirásek (2002), 258
Bažant and Kazemi (1990), 259
Bažant and Kazemi (1991), 216
Bažant and Kazemi (1991a), 35, 130
Bažant and Kazemi (1991b), 32, 206
Bažant and Kim (1991), 10
Bažant and Kim (1997), 79, 80
Bažant and Kim (1998), 85, 86
Bažant and Kwon (1994), 112, 113, 125
Bažant and Li (1995), 145, 259
Bažant and Li (1995), 43, 80, 165
Bažant and Li (1995a), 166, 178, 197
Bažant and Li (1995b), 36, 166, 178
Bažant and Li (1996), 47, 259

Bažant and Li (1996), 47, 248, 259, 260
Bažant and Li (1997), 151–153, 197, 210
Bažant and Lin (1988a), 9, 36
Bažant and Lin (1988b), 9, 172
Bažant and Novák (2000a), 57, 62
Bažant and Novák (2000b), 13, 56, 62, 66, 73, 197, 198
Bažant and Novák (2000a), 60, 125
Bažant and Novák (2000b), 125
Bažant and Oh (1983), 9, 36, 170, 177
Bažant and Pang (2004), 264
Bažant and Pfeiffer (1987), 25
Bažant and Pijaudier-Cabot (1988), 9, 172
Bažant and Planas (1998), 3, 24, 29, 30, 32, 43, 47, 57, 62, 93, 145, 177, 178, 184, 186, 191, 193, 197, 199, 202, 206, 210
Bažant and Raftshol (1983), 10
Bažant and Rajapakse (1999), 9, 11
Bažant and Schell (1993), 156
Bažant and Sun (1987), 130
Bažant and Vítek (1999), 101, 102, 107
Bažant and Wu (1993), 151
Bažant and Xi (1991), 58, 60
Bažant and Xiang (1996), 124
Bažant and Xiang (1997), 122, 124, 144
Bažant and Xu (1991), 156
Bažant and Yavari (2004), 265
Bažant and Yu (2002), 244, 258
Bažant and Yu (2003), 263
Bažant and Yu (2004), 245, 249, 255
Bažant, Yu and Zi (2002), 261, 262
Bažant et al. (1984), 172
Bažant et al. (1999), 137, 142, 145, 147, 148

Bažant et al. (1999a), 144
Bažant et al. (1999b), 144
Bažant, Belytschko and Chang
  (1984), 171
Bažant, Belytschko and Chang
  (1984), 9
Bažant, Caner, Carol, Adley and
  Akers (2000), 168
Bažant, Daniel and Li (1996), 135
Bažant, ed. (1992), 9
Bažant, Guo, and Faber (1995), 173
Bažant, Lin and Lippmann
  (1993), 132
Bažant, Tabbara et al. (1990), 36
Bažant, Tabbara, Kazemi and
  Pijaudier-Cabot
  (1990), 174
Bažant, Xi and Reid (1991), 74
Bažant, Şener and Prat (1988), 113
Bažant, Şener and Prat (1988), 130
Bažant and Planas (1998), 26
Bažant, Pijaudier-Cabot and Pan
  (1987a), 161
Bažant, Pijaudier-Cabot and Pan
  (1987b), 161
Bažant, Zhou and Daniel (2003),
  252, 253
Bažant, Zhou, Novak and Daniel
  (2004), 247, 254
Bažant, Zi and McClung
  (2003), 262
Bender and Orszag (1978), 25
Beremin (1983), 54, 67
Biot (1965), 122
Borodich (1992), 9, 69, 71
Bouchaud, Lapasset et Planes
  (1990), 9, 69
Bridgman (1922), 17
Brocca and Bažant (2000), 125
Brocca et al. (2001), 254
Brown (1987), 9, 69
Buckingham (1914), 17
Buckingham (1915), 17

Budianski (1983), 134
Budianski and Fleck (1994), 134
Budianski, Fleck and Amazigo
  (1997), 134

Cahn (1989), 9
Cahn (1989), 69
Carpinteri (1984), 18
Carpinteri (1986), 6, 9, 18
Carpinteri (1987), 249
Carpinteri (1989), 6
Carpinteri (1994a), 9, 10, 69
Carpinteri (1994b), 9, 69
Carpinteri and Chiaia (1995), 9, 10,
  42, 69, 73
Carpinteri and Chiaia (1995a), 9
Carpinteri and Ferro (1994), 9, 10,
  42, 69
Carpinteri, Chiaia and Ferro
  (1994), 9, 10, 42, 69
Carpinteri, Chiaia and Ferro
  (1995), 9, 10, 42, 69
Carpinteri, Ferro and Intervenizzi
  (1995), 9, 10, 42, 69
Carter (1992), 121, 132
Carter, Lajtai and Yuan
  (1992), 132
Červenka and Pukl (1994), 9, 170
Chelidze and Gueguen (1990),
  9, 69
Chen and Runt (1989), 9, 69
Christensen and DeTeresa
  (1997), 134
Collins (1978), 130
Collins and Mitchell (1980), 130
Collins, Mitchell, Adebar and
  Vecchio (1990), 130
Costin (1991), 121
Cotterell (1972), 121
Cottrell (1963), 32
Cundall (1971), 174
Cundall and Strack (1979), 174
Cusatis et al. (2003a,b), 260

Daniels (1945), 264
da Vinci (1500s), 3
de Borst and Gutiérrez (1999), 174
Dempsey, Adamson and Mulmule (1995), 33
Dempsey, Adamson and Mulmule (1999), 33, 35
Duan and Hu (2003), 248, 259
Duan, Hu and Wittmann (2002, 2003), 255
Dugdale (1960), 7, 177
Dunn et al. (1997a,b), 249

Elices et al. (1992), 258
Eligehausen and Ožbolt (1992), 109
Evans and Marathe (1968), 173

Fairhurst and Cornet (1981), 121
Feng, Ji, Zhuang and Ding (1995), 9, 70
Fisher and Tippett (1928), 5, 6, 56
Fleck and Hutchinson (1993), 219
Fleck and Hutchinson (1997), 219
Fleck and Hutchinson (1997), 137, 219, 228
Fleck and Hutchinson (1963, 1997), 237
Fleck and Olurin (2001), 254
Fleck and Shu (1995), 137
Freudenthal (1968), 5, 6
Fréchet (1927), 5, 56

Galileo (1638), 4
Gao and Huang (2000), 219
Gao, Huang and Nix (1999), 219, 237
Gao, Huang, Nix and Hutchinson (1999), 219
Gibson and Ashby (1997), 254
Giles (1962), 17
Griffith (1921), 5

Guinea, Planas and Elices (1994a,b), 258, 259
Guinea et al. (1997), 260

Hadamard (1903), 173
Haimson and Herrick (1989), 132
Hawkes and Mellor (1970), 121
Hill (1962), 173
Hillerborg (1985a), 32
Hillerborg (1985b), 32
Hillerborg, Modéer and Petersson (1976), 7, 32, 165, 178
Hinch (1991), 25
Horii and Nemat-Nasser (1982), 121
Horii and Nemat-Nasser (1986), 121
Hornbogen (1989), 9, 69
Hsu (1988), 130
Hu (1997, 1998), 255
Hu and Wittmann (1991, 1992a), 255
Huang, Gao, Nix and Hutchinson (2000), 219
Hutchinson (1997), 219, 228

Iguro, Shiyoa, Nojiri and Akiyama (1985), 130
Ingraffea (1977), 121, 170, 172
Ingraffea and Heuzé (1980), 121
Irwin (1958), 22, 135
Issa, Hammad and Chudnovsky (1992), 9, 69

Jackson (1992), 252
Jenq and Shah (1985), 32
Jirásek and Bažant (1995a), 36, 174
Jirásek and Bažant (1995b), 36, 174, 175
Jirásek (1996), 171
Jirásek (2003), 258
Johnson et al. (2000), 252

Kachanov (1985), 171
Kachanov (1987), 171
Kani (1967), 130
Kaplan (1961), 6
Karihaloo, Abdalla and Imjai
    (2003), 255
Kemeny and Cook (1987), 132
Kemeny and Cook (1991), 121, 132
Kendall (1978), 122
Kesler, Naus and Lott (1971), 7
Kfouri and Rice (1977), 7, 178
Kittl and Diaz (1988), 6
Kittl and Diaz (1989), 6
Kittl and Diaz (1990), 6
Knauss (1973), 7, 178, 179
Knauss (1974), 7, 178, 179
Kuhlmann and Breuninger
    (1998), 102
Kupfer (1964), 130
Kyriakides et al. (1995), 134

Lange, Jennings and Shah (1993),
    9, 69
Lehner and Kachanov (1996), 121
Leicester (1969), 6, 7
Leonhardt (1977), 130
Leonhardt and Walther (1962), 130
Li and Bažant (1997), 151, 152
Long, Suquin and Lung (1991),
    9, 69

Mahesh et al. (2002), 264
Maier and Zavelani (1970), 98
Måløy, Hansen, Hinrichsen and
    Roux (1992), 9, 69
Mandel (1964), 173
Mandelbrot (1984), 9
Mandelbrot et al. (1984), 69
Mariotte (1686), 4
Marti (1980), 130
Marti (1989), 36
Mecholsky and Mackin (1988),
    9, 69

Mihashi (1983), 6
Mihashi and Izumi (1977), 6
Mihashi and Rokugo (1998), 9
Mihashi and Zaitsev (1981), 6
Mindlin (1965), 219
Mörach (1922), 130
Mosolov and Borodich (1992), 9,
    69, 71
Mulmule et al. (1995), 33
Murakami (1987), 87

Nakayama (1965), 32
Needleman (1990), 178
Nesetova and Lajtai (1973),
    121, 132
Nielsen and Braestrup (1975), 130
Nixon (1996), 121

Ožbolt and Bažant (1996), 36, 172

Palmer and Rice (1973), 138, 140,
    143, 144
Paris and Erdogan (1967), 154
Peerlings, de Borst, Brekelmans
    and de Vree (1996), 174
Peirce (1926), 5
Peng and Tian (1990), 9, 69
Petersson (1981), 9, 166, 178
Pheonix and Beyerlein (2000), 264
Pijaudier-Cabot and Bažant
    (1987), 9, 172
Planas and Elices (1986), 179, 184
Planas and Elices (1988), 9
Planas and Elices (1989a), 9
Planas and Elices (1989b), 9
Planas and Elices (1992), 178, 179,
    184, 199, 216
Planas and Elices (1993), 9, 10,
    178, 179, 184, 199, 216
Planas, Bažant and Jirásek
    (2001), 11
Porter (1933), 17
Prandtl (1904), 1, 193

Reineck et al. (2003), 263
Reinhardt (1985), 178, 179
Rice (1968a), 140, 178, 187, 256
Rice (1968b), 143, 144
Rice (1992), 178
Rice and Levy (1972), 80
RILEM (1990), 32
Ritter (1899), 130
Rocco (1995), 67
Rokugo, Uchida, Katoh and
    Koyanagi (1995), 67, 197
Rosen (1965), 134
Rots (1988), 9
Ruggieri and Dodds (1996), 54
Rüsch and Hilsdorf (1963), 173

Sammis and Ashby (1986), 121
Sanderson (1988), 121
Saouma and Barton (1994), 9, 70
Saouma, Barton and Gamal-el-Din
    (1990), 9, 39
Schlaich, Schafer and Jannewein
    (1987), 130
Schlangen and van Mier (1992), 175
Schulson (1990), 121
Schulson and Nickolayev
    (1995), 121
Sedov (1959), 20, 193
Şener, Bažant and Becq-Giraudon
    (1999), 113
Shioya and Akiyama (1994), 130
Shioya, Iguro, Nojiri, Akiayama
    and Okada (1989), 130
Shipsa et al. (2000), 254
Slepyan (1990), 78, 79
Sluys (1992), 10
Smith (1974), 178, 179
Soutis, Curtis and Fleck (1993), 147
Steif (1984), 121
Streeter and Wylie (1975), 17
Suo, Bao, and Fan (1992), 178

Tandon, Faber, Bažant and Li
    (1995), 173
Tang et al. (1996), 259
Tattersall and Tappin (1966), 32
Taylor (1938), 168
thermal bending fracture, 77
Thürlimann (1976), 130
Timoshenko (1953), 5
Tippett (1925), 5
Toupin (1962), 219
Tvergaard and Hutchinson
    (1992), 178

van Mier and Schlangen (1993), 175
von Mises (1936), 5, 56

Walraven (1978), 130
Walraven (1995), 130
Walraven and Lehwalter
    (1994), 130
Walsh (1972), 7
Walsh (1976), 7
Weibull, 4, 66
Weibull (1939), 6, 10, 53, 62
Weibull (1949), 6, 53
Weibull (1951), 53
Weibull (1956), 6, 53
Wells (1961), 32
Williams (1952), 249
Willis (1967), 178, 179
Wisnom (1991), 252
Wisnom and Atkinson (1997), 252
Wittmann (1990), 255
Wittmann and Zaitsev (1981), 121
Wittmann, ed. (1994), 9
Wnuk (1974), 7, 178, 179

Xie (1987), 9, 69
Xie (1989), 9, 69
Xie (1993), 9, 69
Xie and Sanderson (1995), 9, 69
Xie, Sanderson and Peacock (1994),
    9, 69

Young (1807), 5
Yuan, Lajtai and Ayari (1992), 132
Yuan, Lajtai and Ayari (1993), 121

Zaitsev (1985), 121
Zaitsev and Wittmann (1974), 6
Zech and Wittmann (1977), 6, 13, 56, 67
Zenkert and Bäcklund (1989), 254
Zubelewicz and Bažant (1987), 175